Mobile Technologies: Principles, Design and Models

Mobile Technologies: Principles, Design and Models

Editor: Adam Houle

NY RESEARCH
PRESS

New York

Published by NY Research Press
118-35 Queens Blvd., Suite 400,
Forest Hills, NY 11375, USA
www.nyresearchpress.com

Mobile Technologies: Principles, Design and Models
Edited by Adam Houle

© 2018 NY Research Press

International Standard Book Number: 978-1-63238-578-9 (Hardback)

Cataloging-in-Publication Data

Mobile technologies : principles, design and models / edited by Adam Houle.
p. cm.
Includes bibliographical references and index.
ISBN 978-1-63238-578-9
1. Mobile communication systems. 2. Wireless communication systems.
3. Cell phone systems. 4. Telecommunication. I. Houle, Adam.
TK5103.2 .M63 2018
621.384 5--dc23

Contents

Preface... IX

Chapter 1   **Factors that may Contribute to the Establishment of Mobile Learning in Institutions – Results from a Survey**..1
O. Zawacki-Richter , T. Brown and R. Delport

Chapter 2   **Mobility through Location-based Services at University**... 6
S. Martín, E. Sancristobal, R. Gil, M. Castro and J. Peire

Chapter 3   **An Online Environment for Academics and Professionals to Locate Collaborators and Refine Ideas**... 13
David Guralnick and Christine Levy

Chapter 4   **A Self-Regulated Learning Approach: A Mobile Context-aware and Adaptive Learning Schedule (mCALS) Tool**... 16
J.Y-K. Yau and M. S. Joy

Chapter 5   **Learning Mathematics in an Authentic Mobile Environment: The Perceptions of Students**...22
N. Baya'a and W. Daher

Chapter 6   **A Framework for Caching Relevant Data Items for Checking Integrity Constraints of Mobile Database**... 31
Zarina Dzolkhifli, Hamidah Ibrahim, Lilly Suriani Affendey and Praveen Madiraju

Chapter 7   **Investigating Mobile Devices Integration in Higher Education in Cyprus: Faculty Perspectives**...37
N. Eteokleous and D. Ktoridou

Chapter 8   **The MOBO City: A Mobile Game Package for Technical Language Learning**.........................48
F. Fotouhi-Ghazvini, R. Earnshaw, D. Robison and P. Excell

Chapter 9   **The Barcelona Mobile Cluster: Actors, Contents and Trends**....................................54
C. A. Scolari, H. Navarro Güere, I. García, H. Pardo Kuklinski and J. Soriano

Chapter 10   **M-Learning: The usage of WAP Technology in E-Learning**...62
J. Al-Sadi and B. Abu-Shawar

Chapter 11   **The Role of Podcasts in Students' Learning**..69
I. Nataatmadja and L. E. Dyson

Chapter 12   **Simplistic is the Ingredient for Mobile Learning**.. 74
Issham Ismail, Hanysah Baharum and Rozhan M. Idrus

Chapter 13 **Using USB Keys to Promote Mobile Learning**...79
M. Rosselle, D. Leclet and B. Talon

Chapter 14 **Using Mobility to Enhance Routing Process in MIS System**...84
K. Oudidi, A. Habbani and M. Elkoutbi

Chapter 15 **Acceptance on Mobile Learning via SMS: A Rasch Model Analysis**...93
Issham Ismail, Siti Sarah Mohd Johari and Rozhan Md. Idrus

Chapter 16 **Contextual Mobile Learning A Step Further to Mastering
Professional Appliances**...100
B. T. David, R. Chalon, O. Champalle, G. Masserey and C. Yin

Chapter 17 **Transformable Menu Component for Mobile Device Applications: Working with
both Adaptive and Adaptable user Interfaces**..105
V. Glavinic, S. Ljubic and M. Kukec

Chapter 18 **Managing Social Activity and Participation in Large Classes with Mobile Phone
Technology**...111
A. Thatcher and G. Mooney

Chapter 19 **Mobile Learning in Context – Context-aware Hypermedia in the Wild**.......................122
Frank Allan Hansen and Niels Olof Bouvin

Chapter 20 **A Collaborative Webbased Framework with Optimized Mobile
Synchronisation: Upgrading to Medicine 2.0**...138
P. L. Kubben

Chapter 21 **Developing a Mobile Application via Bluetooth Wireless Technology for
Enhancing Student-Instructor Communication**...141
Dr. Sahar Idwan

Chapter 22 **Computer-based Wireless Advertising Communication System (CBWACS)**......................145
Yahya S. H. Khraisat and Anwar Al-Mofleh

Chapter 23 **Mobile Education: Towards Affective Bi-modal Interaction for Adaptivity**.......................153
E. Alepis, M. Virvou and K. Kabassi

Chapter 24 **Let's Meet at the Mobile – learning Dialogs with a Video Conferencing
Software for Mobile Devices**..159
Hans L. Cycon, Thomas C. Schmidt, Gabriel Hege, Matthias Wählisch and
Mark Palkow

Chapter 25 **Simulation and Proposed Handover Alert Algorithm for Mobile Communication
Networks**...164
Muzhir Al-Ani and Wael Al-Sawalmeh

Chapter 26 **A Review of the Navigation HCI Research during the 2000's**....................................170
Teija Vainio

Chapter 27 **Extending LIP to Provide an Adaptive Mobile Learning**...177
Mona Laroussi and Alain Derycke

Chapter 28   **A Framework for Educational Collaborative Activities based on Mobile Devices** 187
Cruz-Flores, René and López-Morteo and Gabriel

Chapter 29   **Towards for analyzing alternatives of Interaction Design based on Verbal Decision Analysis of user Experience** 197
Marília Mendes, Ana Lisse Carvalho, Elizabeth Furtado and
Placido Rogerio Pinheiro

Chapter 30   **Adoption of Mobile Learning among Distance Education Students in Universiti Sains Malaysia** 204
Issham Ismail, Rozhan M. Idrus, Azidah Abu Ziden and Munirah Rosli

Chapter 31   **Mobile Learning via SMS among Distance Learners: Does Learning Transfer Occur?** 209
Aznarahayu Ramli, Issham Ismail and Rozhan Md. Idrus

Chapter 32   **An Adaptation of E-learning Standards to M-learning** 215
Daoudi Najima and Ajhoun Rachida

Permissions

List of Contributors

Index

Preface

The main aim of this book is to educate learners and enhance their research focus by presenting diverse topics covering this vast field. This is an advanced book which compiles significant studies by distinguished experts in the area of analysis. This book addresses successive solutions to the challenges arising in the area of application, along with it; the book provides scope for future developments.

Mobile technology is a rapidly expanding field of science and technology. Wireless networking, code division multiple access (CDMA), GPS navigation are some of the swiftly growing branches of mobile technologies. This book presents the complex subject of mobile technology in the most comprehensible and easy to understand language. It aims to present researches that have transformed this discipline and aided its advancement. This book is a resource guide for experts as well as students.

It was a great honour to edit this book, though there were challenges, as it involved a lot of communication and networking between me and the editorial team. However, the end result was this all-inclusive book covering diverse themes in the field.

Finally, it is important to acknowledge the efforts of the contributors for their excellent chapters, through which a wide variety of issues have been addressed. I would also like to thank my colleagues for their valuable feedback during the making of this book.

Editor

Factors That May Contribute to the Establishment of Mobile Learning in Institutions – Results From a Survey

O. Zawacki-Richter [1], T. Brown [2], R. Delport [3]

[1] University of Oldenburg, Oldenburg, Germany
[2] Midrand Graduate Institute, Midrand, South Africa
[3] University of Pretoria, Pretoria, South Africa

Abstract—This paper reports on a survey investigating the role that differences in expectations and perceptions of mobile learning and mobile devices play in establishing mobile learning at an educational institution, Responses from institutions with no institutional plans for mobile learning and others that do plan or currently have developed mobile learning programmes were compared. Various factors that may contribute to the establishment of mobile learning in educational institutions were therefore also investigated. These factors include, amongst other, expectations concerning the impact of mobile technologies on teaching and learning, and perceptions concerning mobile learning applications and mobile learning activities.

Index Terms—distance education, education innovation, mobile learning, m Learning, mobile devices, Introduction

I. INTRODUCTION

Ellen Wagner, Director, Global Education Solutions, Macromedia, proclaimed "2005 is the year that mobile learning comes of age. Mobile learning brings a true 'anytime, anywhere' dimension to e-learning. Mobile learning will feature smart phones and personal communicators, while continuing to link learners with resources via laptop, notebook and tablet computers in a variety of physical settings"[1]. Whether this has indeed become current reality needs to be investigated.

An international survey was recently conducted which sought to explore current expectations of mobile learning for distance education. It was distributed within various professional distance education networks and was also sent to faculty and alumni of the Master of Distance Education programme at the University of Maryland University College (UMUC) in the U.S. The main findings have been reported [2].

II. METHOD

This paper reports on further analysis of the survey results, investigating the role that differences in expectations and perceptions of mobile learning and mobile devices play in establishing mobile learning at an educational institution. Responses from institutions with no institutional plans for mobile learning and others that do plan or currently have developed mobile learning programmes were compared. Various factors that may contribute to the establishment of mobile learning in educational institutions were therefore also investigated.

These factors include, amongst other, expectations concerning the impact of mobile technologies on teaching and learning, and perceptions concerning mobile learning applications and mobile learning activities.

The following objectives were set for this study:

- To evaluate expectations of respondents from institutions with no institutional plans for developing course materials for use on mobile devices and respondents from other institutions that plan or currently have developed mobile learning programmes concerning the impact of mobile technologies on teaching and learning,

- To evaluate perceptions concerning mobile learning applications and mobile learning activities, and

- To identify envisaged constraints (weaknesses of mobile devices) that might hinder the distribution of mobile learning.

III. RESULTS AND DISCUSSIONS

A. Demographics

Eighty-eight responses were received from 27 countries. Table I provides information on the countries of origin and the number of respondents, while Table II presents information on the number of responses from all partaking higher education institution types. The highest percentage of respondents (59%) was from institutions that offer both face-to-face (contact-based) and distance learning programmes (mixed-mode institutions).

TABLE I
NUMBERS OF RESPONDENTS FROM DIFFERENT COUNTRIES

Country	Responses	Country	Responses
Albania	1	Israel	1
Australia	2	Lativa	1
Austria	1	Malta	1
Barbados	1	Mexico	1
Canada	9	Netherlands	3
Colombia	2	Norway	1
Cyprus	1	Portugal	1
Finland	1	Romania	1
France	1	South Africa	15
Georgia	1	Sweden	1
Germany	15	Switzerland	1
Great Britain	8	Turkey	2
Hungary	1	USA	12
Ireland	3	Total	88

TABLE II
PLANS FOR THE DEVELOPMENT OF MOBILE LEARNING COURSE
MATERIALS FOR INDIVIDUAL INSTITUTION TYPES

Institution type	Number	Percentage
A traditional distance teaching institution (single-mode)	9	10.3
A purely online teaching institution or virtual university	3	3.4
An institution offering both, face-to-face (contact-based) and distance learning programmes (mixed-mode/hybrid)	52	59.8
A traditional face-to-face or contact-based teaching institution (single-mode)	8	9.2
A corporate university or training institution	4	4.6
Other [a]	11	12.6
Totals	87	100

[a] The institutions that were referred to as 'other' included a community college, an e-learning service provider, a telecom vendor and a research centre.

B. Expectations concerning the impact of mobile technologies on teaching and learning

Responses were grouped according to non-existence of, or development in some or other form of course materials for use on mobile devices. The item: 'No, there are no institutional plans for developing course materials for use on mobile devices' was classified as 'Non-existent' while the following bulleted list of items were classified as 'Existent'.

- Yes, there are institutional plans for developing course materials for use on mobile devices, but there has been little done.

- Yes, our institution is now developing course materials for use on mobile devices. These are developed specifically for mobile devices.

- Yes, our institution is now developing course materials for use on mobile devices in a standard format for output on a variety of mobile and stationary devices.

Table III gives an indication of the number of institutions that were represented within the institution types that reported non-existence, or alternatively development in some or other form of course materials for use on mobile device.

No significant association was observed, following application of the chi-square test, between institution type and non-existence of, or development in some or other form of course materials for use on mobile devices. Therefore further analyses were performed without taking institution type into account.

To assess whether expectations concerning the possible impact of technology in general on teaching and learning differed between the 'non-existence' and 'existence' groups, a further comparison was performed. Items ranged from no impact on teaching and learning strategies and methodologies to radical changes being envisaged.

TABLE III
PLANS FOR THE DEVELOPMENT OF MOBILE LEARNING COURSE
MATERIALS FOR INDIVIDUAL INSTITUTION TYPES

Institution type	Number	Non-existent	Existent
A traditional distance teaching institution (single-mode)	9	4 (44%)	5 (56%)
A purely online teaching institution or virtual university	3	1 (33%)	2 (67%)
An institution offering both, face-to-face (contact-based) and distance learning programmes (mixed-mode/hybrid)	52	25 (67%)	17 (33%)
A traditional face-to-face or contact-based teaching institution (single-mode)	8	6 (75%)	2 (25%)
A corporate university or training institution	4	1 (25%)	3 (75%)
Other	11	4 (36%)	7 (64%)
Totals	87	41	46

Not one respondent thought that technological changes should not have an impact on our teaching and learning strategies and methodologies, while approximately 50% of the respondents of the 'non-existent' group and 15% of the 'existent' group (as defined) thought that technology changes should have an impact, although this is not currently the case (Table IV).

The 51 respondents that were of the opinion that teaching and learning strategies and methodologies adapt continuously due to new affordances that technology affords, were almost equally distributed between institutions with non-existent or existent programmes for mobile learning; the respective percentages were 59% and 64% for the institution groups (Table IV). Only 7% and 19% respondents from the 'non-existent' and the 'existent' groups respectively anticipated radical changes being introduced by technology (Table IV). The

TABLE IV
PERCEPTIONS CONCERNING IMPACT OF TECHNOLOGY ON TEACHING
AND LEARNING OF 'NON-EXISTENT' AND 'EXISTENT' GROUPS

Impact on teaching and learning	(n)	Non-existent	Existent
Technology changes should not have an impact on our teaching and learning strategies and methodologies.	0	0 (0%)	0 (0%)
Technology changes should have an impact on our teaching and learning strategies and methodologies, but this is currently not the case.	21	14 (34%)	7 (17%)
Teaching and learning strategies and methodologies adapt continuously due to new affordances that technology provides.	51	24 (59%)	27 (64%)
Technology changes bring about radical changes to our teaching and learning strategies and methodologies.	11	3 (7%)	8 (19%)
Totals	83	41	42

association between group and expectation concerning the impact of mobile technologies on teaching and learning was not observed to be statistically significant.

A comparison was performed to establish what the impact of the attributes and the opportunities that mobile technologies could afford were anticipated to be. The most significant finding was that most respondents (64 of 83) were of the opinion that mobile technology would be very helpful in enhancing teaching and learning independent of time and space, 31 of which were from the 'non-existent' and 33 from the 'existent' group (Table V).

Once again, no significant association was observed between group and expectation concerning the impact of mobile technologies on teaching and learning.

The 'Other' opinions that were voiced by respondents from the 'non-existent' group were:

- Mobile devices will make learning even more flexible and spontaneous than "traditional" e-learning.
- Mobile technologies could allow education to be brought more effectively into different environments where technology is used appropriately/effectively.

Assessment of opinion on whether mobile learning would facilitate new strategies and methodologies for learner support as well as content development and delivery in distance education, showed no difference in opinion distribution between 'non-existent' and 'existent' group. Respondents from both groups were equally distributed between opinion groups, as is evident from Table VI.

No differences in expectations and perceptions of respondents from institutions with no institutional plans for mobile learning and respondents from institutions that do plan or currently have developed mobile learning programmes were thus observed.

TABLE V
PERCEPTIONS CONCERNING IMPACT OF MOBILE TECHNOLOGIES ON TEACHING AND LEARNING

The expected impact of the attributes and opportunities that mobile technologies afford	(n)	Non-existent	Existent
Have no impact on teaching and learning.	1	1	0
Be widely applied mainly for administrative services and/or assessment purposes.	6	4	2
Be very helpful in enhancing teaching and learning independent of time and space.	64	31	33
Completely change the way we teach and learn.	10	3	7
Other	2	2	0
Totals	83	41	42

TABLE VI
RESPONSES WITH RESPECT TO NEW STRATEGIES AND METHODOLOGIES BEING FACILITATED BY MOBILE LEARNING

Mobile learning will facilitate new strategies and methodologies for learner support and content development and delivery in distance education.	(n)	Non-existent	Existent
• Yes, mobile learning affords new opportunities for learner support and content development and delivery.	60	29	31
• No, mobile learning will not lead to anything entirely new. It's just another medium or channel for learner support and content delivery among others	23	12	11
Totals	83	41	42

C. Perceptions concerning mobile learning applications and mobile learning activities

Respondents were requested to rate

- the importance of learning 'tools' for students on mobile phones or smartphones;
- the importance of learning activities which are appropriate for mobile devices;
- the importance of applications (software) on mobile devices; and
- the usefulness of mobile learning 'tools' for learning and teaching

Ratings are reported as expressed by respondents, irrespective of 'non-existent' and 'existent' groupings. The percentage of respondents that rated the listed items as <3 on a scale of one to five, where one is the highest rating, are reported in Table VII.

To summarise the most significant findings reported in Table VII, items that were perceived as being important (rating of 1 or 2 on a scale of 1 to 5, where 1 is the highest rating by 50% or more of the respondents) were:

- 'Being connected anywhere, anytime' was perceived as being both the most import and the most useful learning 'tool' for students on mobile phones or smartphones;
- 'Accessing class notes, schedules, documents, websites, etc via wireless connections' was also regarded as a useful mobile learning tool, which links up closely with the preceding comment; and
- 'Collaborative learning' and 'Field work' were regarded as the most important learning activity for mobile devices.

TABLE VII
RATINGS OF MOBILE LEARNING APPLICATIONS AND MOBILE LEARNING ACTIVITIES

Rating of importance of learning 'tools' for students on mobile phones or smartphones	Percentage
Text messaging (SMS) for communication and interaction. (#: 86)	47
Voice calls for communication and interaction. (#: 87)	40
Text messaging to e-mail and vice versa. (#: 86)	47
Sharing texts, notes and documents. (#: 86)	31
Being connected anywhere, anytime. (#: 86)	69
Rating of usefulness of the mobile learning 'tool' that were perceived as being most useful	
Sharing texts, notes and documents via Bluetooth or wireless connections. (#: 82)	42
Accessing class notes, schedules, documents, websites, etc via wireless connections. (#: 82)	50
Using the scheduling and diary applications for organising their learning environments. (#: 81)	44
Using mobile Office or the like applications for their normal learning activities. (#: 82)	31
Being connected anywhere, anytime. (#: 82)	62
Rating of importance of learning activities which are appropriate for mobile devices	
Coursework (accessing and reading learning materials) (#: 85)	21
Assessment (quizzes, tests, questions-and-answers, etc) (#: 85)	34
Collaborative learning (interaction with tutor, discussion with other students, group work) (#: 85)	54
Field work (location-based learning: gathering and sharing on the site information) (#: 84)	58
Information retrieval (search in databases and encyclopaedias) (#: 85)	45
Rating of importance of applications (software) on mobile devices	
Mobile Office (Word, Excel, PowerPoint, etc). (#: 85)	48
Diary and scheduling. (#: 77)	49
Audio and video applications. (#: 84)	43
Imaging. (#: 75)	33
Additional accessories (notes, calculator, etc.). (#: 78)	31
Browser for internet connection/online data services. (#: 85)	61

Immediate accessibility of information from any location, particularly information that can be accessed via a browser, seems to be the affordance respondents appreciate most about mobile learning. New possibilities in terms of active and authentic learning appear to be anticipated with the use of mobile devices. Successful implementation of mobile learning at an institution will probably imply that learning opportunities be created that challenge students to source information with a certain degree of immediacy.

D. Identification of weaknesses of mobile devices that might hinder the distribution of mobile learning

Respondents were requested to either agree or disagree with defined statements concerning mobile devices. Table VIII lists the statements and also the percentage of respondents that agreed with a rating of >3 on a scale of 1 to 5 where 1 = strongly disagree and 5 = strongly agree.

Although 62% of respondents agreed that screens are currently too small to present complex learning material, the general expectation is that sufficient memory for small images, audio and video clips, as well as sufficient data transmission capacity, will be available in future. However, limited battery life of mobile devices was regarded as a problem for extensive use by 59% of respondents. Approximately half of the respondents felt that screen size is not as important as mobile devices should rather be used for communication and interaction purposes rather than for content distribution. Half of the respondents were also not convinced that cost of networks will not play an important role in the future.

Whether these factors will affect implementation of mobile learning in distance education needs to be investigated further.

E. Conclusion

The following conclusions are drawn from this study:

- Perceptions concerning the impact of mobile technologies on teaching and learning appear not to have an influence on an institution's planning for, or development of course materials for use on mobile devices.
- Opinions concerning mobile learning applications and mobile learning activities clearly express the significance of learning supported by mobile devices.
- No prohibiting technical constraints of mobile devices, as defined in the questionnaire, were identified in this survey.
- Having institutional plans for mobile learning in place does appear to have an impact on the involvement in projects and the implementation of mobile learning at institutions

REFERENCES

[1] Predictions for 2005. By Lisa Neal, Editor-In-Chief, eLearn Magazine. http://www.elearnmag.org/subpage.cfm?section=opinion&article=30-1 (01/07/2005)

[2] Zawacki-Richter, O.; Brown, T.H.; Delport, R. (2006). Mobile learning - a new paradigm shift in distance education? Paper delivered at mLearn 2006, the 5th international world conference on mobile learning, Banff, Canada, 22 - 25 October 2006..

AUTHORS

Olaf Zawacki-Richter, Dr. is with the University of Oldenburg, Germany (e-mail: olaf.zawacki.richter@uni-oldenburg.de).

Tom Brown, Dr. is from the Midrand Graduate Institute, Midrand, South Africa (e-mail: tom.brown@mgi.ac.za).

Rhena Delport; Prof. Dr. is from the University of Pretoria, South Africa (e-mail: rhena.delport@up.ac.za).

TABLE VIII
RATING ON STATEMENTS CONCERNING MAJOR WEAKNESSES OF MOBILE DEVICES THAT MIGHT HINDER THE DISTRIBUTION OF MOBILE LEARNING

Statement	Percentage
Displays and screens are too small to present complex learning material. (#: 85)	62
Screen size should not be important as mobile devices should be used for communication and interaction purposes rather than for content distribution. (#: 84)	48
Costs of mobile network services will continue to decrease and should not play an important role. (#: 85)	55
Technological advancements make it possible to have sufficient memory for small images, audio and video clips. (#: 85)	78
Device capabilities and mobile network infrastructures are improving to provide sufficient data transmission capacity (e.g. 3G and HSDPA). (#: 83)	71
Limited battery life of mobile devices is a problem for extensive use. (#: 85)	59

#: Number of responses

Mobility through Location-based Services at University

S. Martín, E. Sancristobal, R. Gil, M. Castro and J. Peire

UNED (Spanish University for Distance Education, Madrid, Spain

Abstract—Location tracking systems are becoming more relevant in many new environments, due to the fact they the core of context aware applications. This new concept can improve the way universities provide services and a wide number of companies do business. Inside university area, users location (both students, teachers and staff) gives rise to a new kind of services based on their profile and on the area in which the user is in each moment, allowing a personalization of the offered contents. The present paper shows how location-based applications can be developed for mobile devices through a middleware that allows different location methods, such as Wi-Fi and RFID. Finally some location-based applications are given showing possible examples in different environments.

Index Terms—Location-based services, Mobility, M-learning, RFID.

I. INTRODUCTION

Nowadays mobile technologies are being applied on educational environments successfully. One of the main reasons of this success is the improvement of the technical features of the devices. New generations of mobile devices have wider screens with better resolution, built-in digital cameras, and connectivity not only with GPRS but also with Wi-Fi or UMTS. In some of them it is even possible to find GPS receivers, RFID readers or smartcards integrated.

All these new technologies inside a small and portable device are giving rise to a new generation of applications in all kind of environments. These kinds of applications are called M-Learning inside university environment. Here mobile devices are supporting collaborative and mobile work and enabling the students to learn anywhere and any moment, especially through games or courses designed for these small devices.

Location-based systems can now also find a place on this mobile and university environment thanks to the integration of GPS receivers. The motivation for the use of this kind of services at universities is because it allows knowing more information about the context. For example, a student in a cafeteria will have different needs than in a laboratory or in a secretary, or a teacher in a classroom will need different information than in an office. Knowing where the user is in every moment it is possible to offer personalized information through the mobile device depending not only on his/her profile but also on his/her location.

The present paper uses a real application developed by UNED (Spanish University for Distance Education) to show the evolution of location-based systems at university, making special emphasis on indoor location technologies, such as Wi-Fi and RFID.

II. OBJECTIVE

The result of this research is an application manager able to run different applications inside the mobile device of a user depending on his/her profile and the place where he/she is in.

In this way, a user entering into an area equipped with devices such as laptops, mobile phones or PDAs, will receive the pertinent information obtained from the information services of the organization. For example, a user in a conference hall will receive automatically the documentation related to it (slides of the presentation, additional documents, CV of the speaker, etc). In the same way, a student will receive the information about a practice entering or approaching to the corresponding laboratory.

III. ARCHITECTURE

In the design of the tool it had a great importance the construction of an open and modular architecture that allowed the integration on diverse environments depending on market requirements. The figure 1 represents the logical architecture of the system, showing the three main elements that constitute it: the location system, the adaptation middleware and the application manager. The location system is in charge of finding where the user is. Usually, there is a program installed in the mobile device that interacts with the location system obtaining the coordinates of the position.

The second layer in this architecture is an adaptation middleware that will ensure independence of the location system. Thanks to this middleware it will be possible to use several methods of location, for instance GPS, Wi-Fi or Radio Frequency Identification (RFID). It provides an abstract interface that must be implemented by every location system. In consequence, the application manager does not have to worry about what kind of location system is being used; it only has to interact with the adaptation middleware (Figure 1).

Finally the application manager will decide the application to show to the user depending on the user's location, the moment and the user's profile. This module will send a command to the mobile device to show the suitable application.

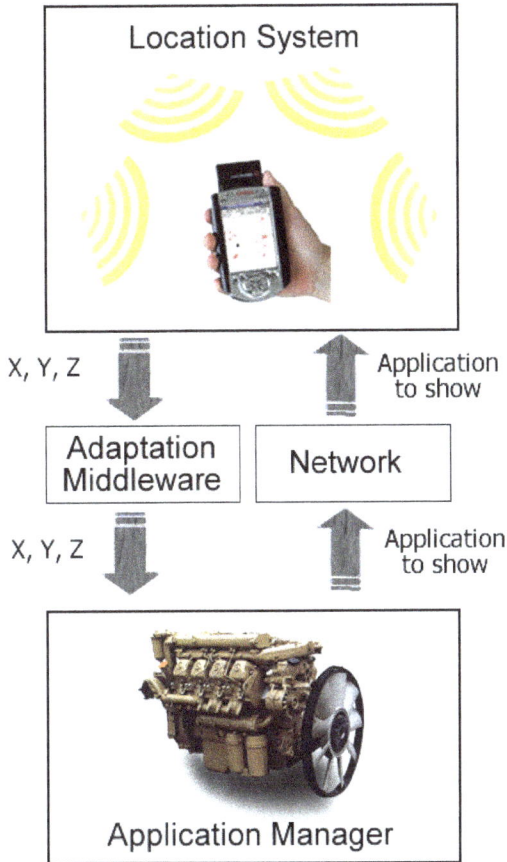

Figure 1. Logical architecture of the system

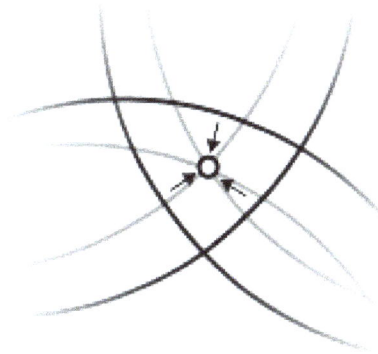

Figure 2. How GPS works

IV. EVOLUTION THROUGH LOCATION TECHNOLOGIES IN MOBILE DEVICES

For a long time, the most famous positioning method has been GPS (Global Positioning System), but other systems have emerged in the last years, fundamentally based on wireless technologies. Some examples are the Wi-Fi networks, Bluetooth or the Infrared sensors. In addition, other Radio based technology such as RFID can also be used to locate any kind of things: people, goods, animals, etc.

A. GPS: Used worldwide

It is a fact that the main and more widely used location method in mobile devices is the Global Positioning System (GPS). This technology offers a quite reasonable accuracy in outdoor environments.

Each GPS satellite transmits data that indicates its location and the current time. The signals, moving at the speed of light, arrive at a GPS receiver at slightly different times because some satellites are farther away than others. The distance to the GPS satellites can be determined by estimating the amount of time it takes for their signals to reach the receiver. When the receiver estimates the distance to at least four GPS satellites, it can calculate its position in three dimensions (Figure 2).

Although four satellites are required for normal operation, fewer may be needed in some special cases. If one variable is

already known (for example, a sea-going ship knows its altitude is 0), a receiver can determine its position using only three satellites. Also, in practice, receivers use additional clues (Doppler shift of satellite signals, last known position, dead reckoning, inertial navigation, and so on) to give degraded answers when fewer than four satellites are visible.

The position calculated by a GPS receiver requires the current time, the position of the satellite and the measured delay of the received signal. The position accuracy is primarily dependent on the satellite position and signal delay.

To measure the delay, the receiver compares the bit sequence received from the satellite with an internally generated version. By comparing the rising and trailing edges of the bit transitions, modern electronics can measure signal offset to within about 1% of a bit time, or approximately 10 nanoseconds for the Coarse/Acquisition code. Since GPS signals propagate at the speed of light, this represents an error of about 3 meters.

However, the use of this technology inside buildings is not possible because the receiver needs to have direct contact with the satellites. For that reason, technologies such as Wi-Fi or RFID are appearing in location systems for indoor environments.

B. Wi-Fi location for mobile devices

Wi-Fi based location uses a small program installed in the mobile device that will recollect the powers received from the access points every 250 ms, and it will send them to a server where they will be processed to obtain the user's coordinates. The inconveniency of this method is that it requires a previous calibration of the system.

The location manager is based on a positioning engine able to locate wireless clients, such as mobile devices, PDAs, laptops or others devices that fulfil the standard 802.11b with a reasonable precision (currently the most advanced system have an error margin of one meter, enough to the application that we use).

In first place, the system requires the installation of at least three wireless access points in the covered area, which will emit in different channels. Once the access points are working properly, it is necessary to calibrate the application measuring the power and noise in several points of the map (Figure 3).

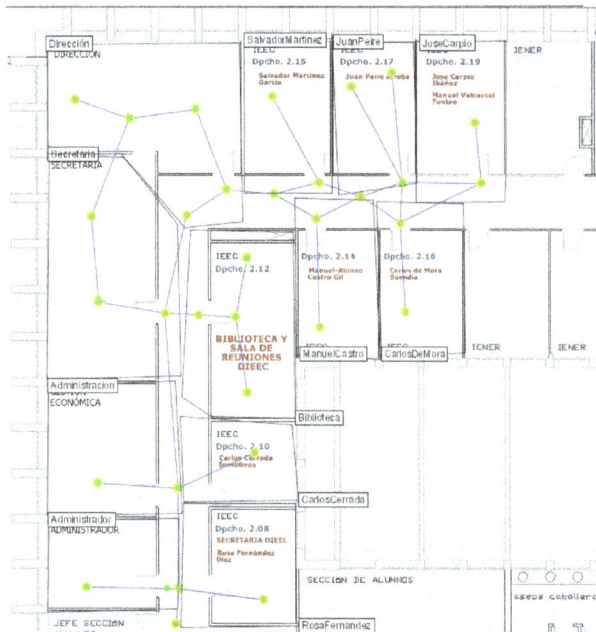

Figure 3.   Measuring the power in the covered map.

These measurements will give as a result the optimal position of the access points in the covered area. It will also be used to setup and calibrate the neuronal networks.

For this calibration, a positioning pattern will be used with the more significant geographic points of every room, which will register the received powers of each access point.

Another aspect to consider is the definition of areas in the map, which allows deciding the application showed to the user depending of his/her position. The definition of the areas also guarantees the integrity of the system controlling the access to these areas and allowing the compilation of statistics of movements of the users.

Ideally, the precision of location would be perfect, however, the received power by the access points is not constant and can be affected by electrical fields (for example a microwave), physics (walls in rooms, offices and zones of step that contain isolation materials) or thermal effects (people affluence).

C.   RFID location for mobile devices

On the other hand, Radio-frequency identification (RFID) is an automatic identification method, relying on storing and remotely retrieving data using devices called RFID tags or transponders. This identification is usually used for identification, as the substitute of bar codes, tracking any kind of products or even animals or people. Some tags can be read from several meters away and beyond the line of sight of the reader. In addition this identification method can also be used to locate people in an area. But it only locates people when the RFID reader reads the tag that the user carries.

Nowadays it is possible to find RFID readers integrated in many mobile phones. This system is being used together with smartcards mainly to allow payments. But it

also opens a wide mosaic of applications in the location field. Inside RFID-based location there are two kinds of systems, depending on if the user carries the tag and is identified by several readers in the building (Passive location) or if what the user carries is the reader, and he/she reads the tags integrated in the environment (Active location). Both will be described in the following points.

1)   Passive RFID location

In this case, there is no need of any action by the user to be identified. The tag that the user carries will be read and the user identified only crossing a door. If the system has several RFID readers it is possible to track the movements of the user by a map, knowing in which room the user is at every moment and in which rooms he/she has been in the past.

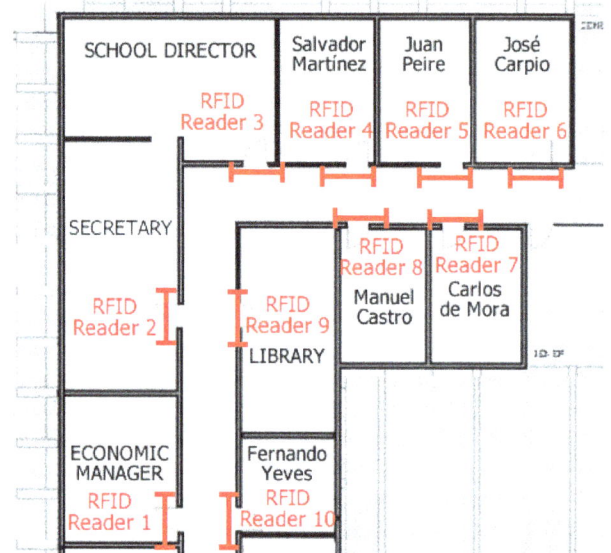

Figure 4.   RFID readers at the entrance of every room

As it can be seen in the figure 4, every time the user enters in a room he/she will be identified and located through the RFID readers integrated at the entrance of every room.

Once the system knows where and who is the user it is possible to offer him personalized contents thanks to the application manager that will decide the most suitable application to be shown.

2)   Active RFID location

Active location requires the user carry out some action. In this case there will be several RFID tags spread along the map, for example at the entrance of rooms, on informative panels, etcetera.

When the user approaches his/her mobile phone to some RFID tags it will read the information stored into the tag, what will be used to know where the user is this moment (Figure 5).

Once the tag is read, the mobile phone will be able to connect to a server and obtain personalized information depending on its profile and its location inside the map.

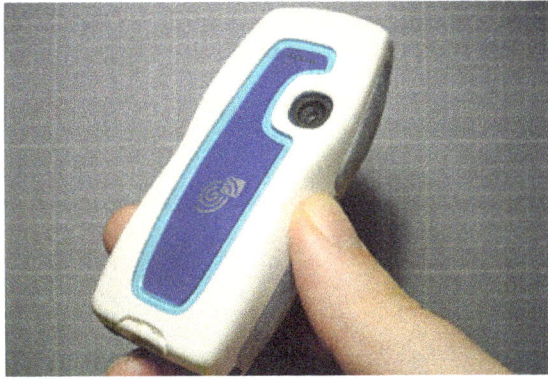

Figure 5. An RFID reader in a mobile phone.

V. APPLICATIONS ON UNIVERSITY ENVIRONMENT

Inside the location field with mobile devices, UNED has developed a system to offer personalized information depending on where and who the user is (Figure 6).

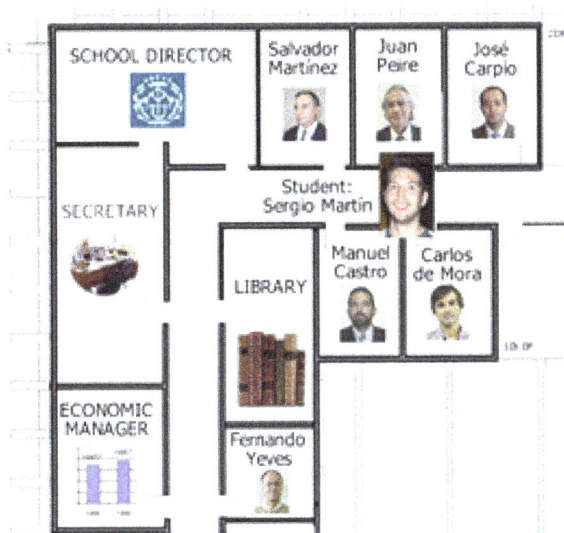

Figure 6. A student approaching the office of a teacher.

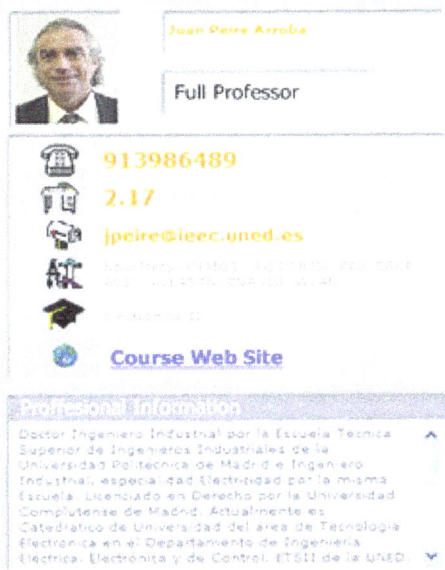

Figure 7. A sample screen of the application

Basically, the system obtains the information about the location of the user thanks to the geographic location system that offers the coordinates (X, Y, Z) of all the devices connected to the system. From them, it is possible to determine in which area or room of the building are these terminals. With this data, the tool determines what information must be offered to each terminal and is possible to track the user's movements.

In figure 7 can be seen a screen of the application, showing the associated information to the area in which the user is in this moment, in this case, it is information related to a teacher.

Figure 8. Information in a PDA screen.

In this case, due to the fact the system is a prototype the information associated is very simple, only the information related to the teacher: subjects, timetables, curriculum vitae, e-mail, and etcetera. This view is for a mobile device with a wide screen, such as Ultra Mobile PC (UMPC) or a laptop.

The application gives also support to more reduced devices such as mobile phones or PDAs in order to provide access to anyone anywhere. In figures 7 and 8, it is possible to see a screen of the application in a PDA, showing the information related to the room where the user is this moment. In this case it is the information of a teacher because the user is close to this teacher's room.

VI. OTHER APPLICATIONS

One of the main advantages of location-based tools is the great capacity of integration in diverse areas:

- Security (emergency, attendance in highway, forest monitoring, etc.).
- Services search (vehicles, people).
- Tourist information points (museums, art galleries, etc.)

- Routing of calls to the closest centre (shipments of food, services of technical attendance, etc.).
- Customized information services (yellow pages, tourist information, located publicity, etc.).
- Hospitals can improve patient care by keeping constant track of doctors, nurses, support staff and other medical resources;
- Retail stores, stadiums, and other service-oriented facilities can adjust staffing levels and product inventory to best accommodate consumer patterns;
- Corporations can track assets more effectively.

As an example of this versatility and great capacity of integration in other environments, in the development of the prototype were prepared two more environments in addition to the university: an Egyptian museum and a hospital.

In the Egyptian museum environment were developed several information pages about the art pieces in every room, such as "Box with Carved Scenes of King Tutankhamun and His Queen" shown in figure 9, that were shown in the user's mobile device every time he/she accessed to the room.

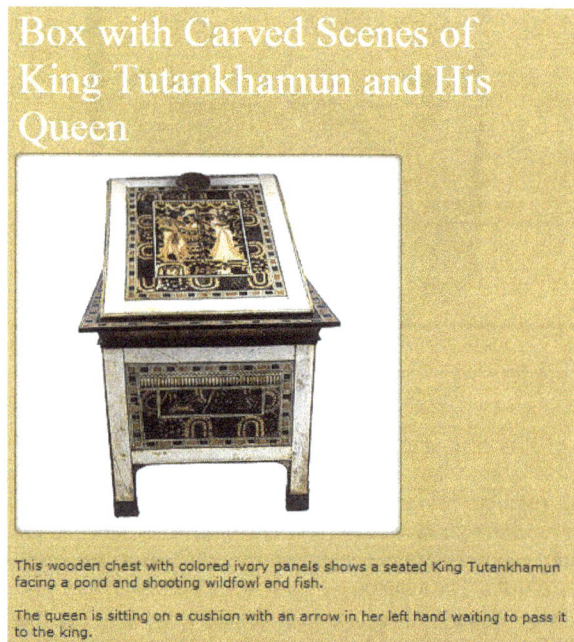

Figure 9.   One of the applications in the Egyptian museum environment.

The other environment developed to show the versatility of the system was a Hospital, where doctors and nurses were located inside the building to improve the efficiency.

In addition, when doctors were visiting their patients they obtained their medical history in the mobile device just approaching to them. In figure 10 a sample of a patient's medical history shown to a doctor can be seen.

VII. PRIVACY ISSUES

While all of these applications and services promise enormous consumer benefit, privacy concerns abound,

and must be addressed if new services and applications are going to be accepted by consumers [16].

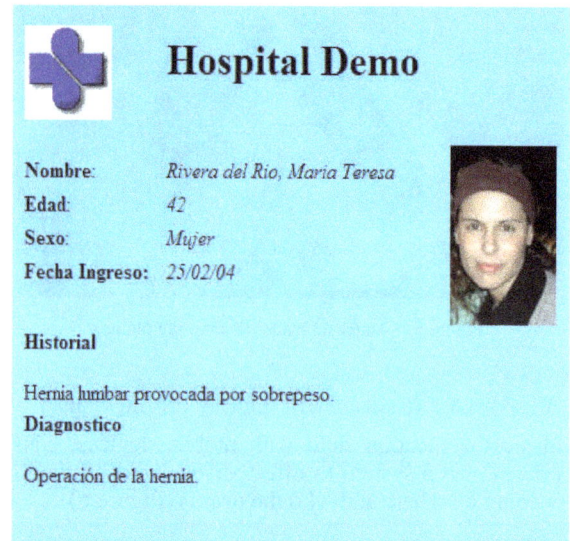

Figure 10. One of the applications in the Hospital environment.

Location-sensing technology raises interesting problems, such as the expectations users have for privacy in particular places or while engaged in specific activities [17]. For example, some people will not mind to be monitored at work but they will prefer a higher level of privacy in his private life, or sometimes they will not want to be monitored when they are in a bar or in the bathroom.

In addition, to the location aspect of place, the social context of place also has some affect on their willingness to share location information. Some studies reveal that people is far more willing to share location information when they are alone. The influence of being with friends is statistically stronger when people are at home or in the library. These studies suggest that when people are at home or in the library are more interested in enabling social contact- that is, having others find them-and possibly less concerned about privacy, even when they were at home [17].

In the case of RFID technology, a primary security concern is the illicit tracking of RFID tags. Tags which are world-readable pose a risk to personal location privacy [18]. This case is deeply different from Wi-Fi or GPS location privacy problems, because it is possible not only to know who and where the user is, but also other information such as what kind of clothes he wears, due to the fact not only humans can wear RFID tags, but also any kind of products. This means that if a user enters a store with a pack of gum in his/her pocket, the reader can identify that pack of gum, the time and date he/she bought it, where he/she bought it, and how frequently he/she comes into the store. If the user used a credit card or a frequent shopper card to purchase it, the manufacturer and store could also tie that information to his/her name, address, and e-mail. He/she could then receive targeted advertisements by gum companies as he/she walks down the aisle, or receive mailings through e-mail or regular mail about other products [19]. For that reason there are several kinds of mechanism to avoid reading and writing in the RFID tags, such as Faraday Cages (Figure 11).

These techniques block the electromagnetic signals sent by the reader to power the RFID tag. A manifestation of the increasing fear that is appearing in this sense is that it is possible to find these Faraday Cages embedded into wallets and even trousers (Figure 12) to prevent not allowed readings from malicious users.

Figure 11. The Faraday's cage acts as an RFID Shield intercepting the electromagnetic signal which normally powers the RFID tag.

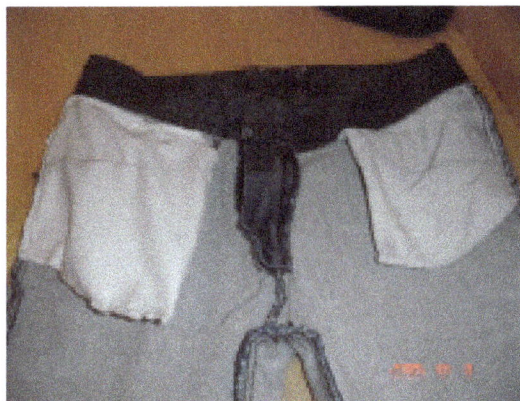

Figure 12. A Faraday's cage embedded in a pair of trousers and a shield protector for passports.

In summary, location-aware systems must develop privacy policies that will clearly inform the users about how their personal information will be used, who is going to manipulate it and the purpose. All this information should be specified in a contract signed by the user and the company in charge of the personal information. In any case, this information can be read without the explicit permission of users and must be only used for the specified purposes.

VIII. CONCLUSIONS

Location technologies embedded into mobile devices are nowadays a reality. Users can easily purchase PDAs or mobile phones with full connectivity: 3G, GPS, Infrared, Wi-Fi and Bluetooth; and in a near future they will also offer other technologies such as Wi-Max or ZigBee. The use of these technologies will allow location-aware systems become more popular and spread in everyday life, reaching other areas of the society, such as the sanitary environment, museums, monitoring of buildings, tourist information points, conferences, customer service departments, and etcetera.

On the other hand, there are some privacy issues that must be addressed in order to ensure users trust these systems, what means location-aware technologies will have a very promising future.

ACKNOWLEDGMENT

The authors would like to acknowledge the Spanish Science and Education Ministry and the Spanish National Plan I+D+I 2004-2007 the support for this paper as the project TSI2005-08225-C07-03 "MOSAICLearning: Mobile and electronic learning, of open code, based on standards, secure, contextual, personalized and collaborative".

REFERENCES

[1] Martín, S., Castro, M., Gil, R., Colmenar A., Peire J., "Ubiquitous and biometric applications on distance education. An alternative to the traditional examination". Alcalá de Henares (Spain): I International Conference on Ubiquitous Computing: Applications, Technology and Social Issues, 39-42. 2006.

[2] Martín, S., Castro, M., Peire J., "Nuevas Aplicaciones de la Computación Ubicua en la Enseñanza Personalizada". Alcalá de Henares (Spain): II Congreso Iberoamericano sobre Computación Ubicua, 89-96. 2006.

[3] Guardo, E., López, E., Rueda, J.M. "Smart organizations: Benefits of Mobile Internet to business processes". eBusiness and eWork. 2006.

[4] Castro, M., Gil, R., Martin, S., Peire, J., "New Project on Secure Education Services for On-Line Learning". San Juan (Puerto Rico): 9th International Conference on Engineering Education. 2006.

[5] Martín, S., Castro, M., Peire J., "Experiencias e introducción de dispositivos móviles en la Enseñanza a Distancia". Granada (Spain): Published by Thomson. Simposio sobre computación ubicua e inteligencia ambiental. 2005.

[6] Arjona M., Gonzalez, M., Reglero, J., Peire, J., "New Pedagogical Tools for Mobile Learning Groups" IST MOBILE, Sitges (Spain), 2001.

[7] Seong, K. and Ng, M. "Synchronization of RFID readers for dense RFID reader environments". Proceedings of the International Symposium on Applications and the Internet Workshops (SAINTW'06). 2006.

[8] Carbunar, B., Ramanathan M., "Redundant reader elimination in RFID systems". Santa Clara, California: Second Annual IEEE Communications Society Conference on SECON. 2006.

[9] Leong K., Ng, M., "Positioning Analysis of Multiple Antennas in a Dense RFID Reader Environment". Proceedings of the International Symposium on Applications and the Internet Workshops (SAINTW'06). 2006.

[10] Niebert N., "Ambient Networks: An Architecture for Communication Networks Beyond 3G", IEEE Wireless Commun., vol. 11, no. 2, pp. 14–21. 2004.

[11] Ahlgren. B., "Ambient Networks: Bridging Heterogeneous Network Domains", Proc. 16th IEEE Symp. Pers. Indoor and Mobile Radio Commun., Berlin, Germany. 2005.

[12] Niebert, N., "Ambient Networks: a Framework for Future Wireless Internetworking", Proc. IEEE VTC 2005 Spring, Stockholm, Sweden. 2005.

[13] Saha, D., and Mukherjee A., "Pervasive Computing: A Paradigm for the 21st Century", IEEE Computer Society, pp. 25–31. 2003.

[14] Henricksen, K., "Middleware for Distributed Context- Aware Systems", Proc. DOA 2005, LNCS 3760, 2005, pp. 846–63.

[15] Roman, M., "GAIA: A Middleware Infrastructure to Enable Active Spaces", IEEE Pervasive Computing, vol. 1, no. 4, pp. 74–83. January 2002.

[16] Centre for Technology and Democracy, "Data privacy. Wireless Location". http://www.cdt.org/privacy/issues/location/. Queried on-line on: 20th of December, 2007.

[17] Anthony, D., Henderson, T., Kotz, D., "Privacy in Location-Aware Computing Environments". IEEE Pervasive Computing, Vol. 6, No. 4, pp. 64-72, October-December 2007.

[18] Martin, S, Gil, R., Sancristobal, E., Castro, M., Peire, J., "Increasing throughput and personalizing the examination process in universities using RFID". 1st Annual RFID Eurasia Conference & Exhibitions, Istambul, Turkey, 5-7 September 2007.

[19] McIver, R., "RFID Privacy Issues. How RFID Will Impact Consumer Privacy". RFID Gazette, March 2005. Queried online on: 20th of December, 2007.

AUTHORS

S. Martín is with the Electrical and Computer Engineering Department, UNED (Spanish University for Distance Education), Madrid, Spain (e-mail: smartin@ieec.uned.es).

E. Sancristobal is with the Electrical and Computer Engineering Department, UNED (Spanish University for Distance Education), Madrid, Spain (e-mail: elio@ieec.uned.es).

R. Gil is with the Electrical and Computer Engineering Department, UNED (Spanish University for Distance Education), Madrid, Spain (e-mail: rgil@ieec.uned.es).

M. Castro is with the Electrical and Computer Engineering Department, UNED (Spanish University for Distance Education), Madrid, Spain (e-mail: mcastro@ieec.uned.es).

J. Peire is with the Electrical and Computer Engineering Department, UNED (Spanish University for Distance Education), Madrid, Spain (e-mail: jpeire@ieec.uned.es).

An Online Environment for Academics and Professionals to Locate Collaborators and Refine Ideas

David Guralnick and Christine Levy

Kaleidoscope Learning, New York, New York, USA

Abstract—Academic and industry conferences have been used for years as a key method for sharing knowledge and ideas among academics and professionals in specific areas of study. Conferences provide a rare opportunity for people to form relationships with colleagues around the world, and not only to exchange ideas within the context of formal presentations, but to get to know one another informally through other conference activities such as dinners and receptions.

While conferences do indeed have tremendous value and contribute substantially to the growth of research in their fields, we have identified some ways that we can use technology to improve the impact of conferences on research and results, to make better use of the time between conferences, and to allow more involvement from people who cannot attend conferences.

In this paper, we describe a community-based Web site for academics and professionals, and to be rolled out first for an international e-learning association.

Index Terms—e-learning, collaboration, conferences, web technology, social networking, professional networking

I. INTRODUCTION

Academic and industry conferences have been used for years as a key method for sharing knowledge and ideas among academics and professionals in specific areas of study. Conferences provide a rare opportunity for people to form relationships with colleagues from around the world, and not only to exchange ideas within the context of formal presentations, but to get to know one another informally through other conference activities such as dinners and receptions. Relationships formed at conferences may lead to future collaboration well beyond the conference itself, and to further the field of study. Thus the role of conferences is, ideally, to improve the relevant field not solely during the conference duration, but after the conference and into the future.

While conferences do indeed have tremendous value and contribute substantially to the growth of research in their fields, in today's world we have identified some ways that we can use technology to improve the impact of conferences on research and results, to make better use of the time between conferences, and to allow more involvement from people who cannot attend conferences.

In the remainder of this paper, we describe the detailed reasons for this project, describe the features of our

community-based Web site for academics and professionals, and discuss future extensions.

II. MOTIVATION, PART I: IMPROVING IN-CONFERENCE AND BETWEEN-CONFERENCE NETWORKING AND COLLABORATION

Conferences certainly have their value when it comes to academic and professional collaboration; however, they also have a few limitations:

- Many people who would contribute greatly cannot attend some or all conferences—funding is often limited, and travel to conferences can be expensive; also, schedules do not always allow people to attend all of the conferences they wish to attend.
- For those who do attend, there may be missed opportunities in terms of meeting people with common interests. Attendees can, and do, read the conference program to select which talks they would like to attend, but as we discuss below, there are likely more people that attendees may not know they have something in common with—for example, if a speaker giving a talk about simulations also does research on mobile learning, that may not be clear to an attendee who was interested in mobile learning, and a connection opportunity would be missed.
- Even in the best of cases, conference time is always limited and rushed; there's simply not enough time, as a practical matter, for colleagues to meet and discuss everything they would like, both with previous colleagues and new ones.
- People have the best of intentions regarding continuing collaborative work after a conference and into the future; but in practice, often "real life" takes over and people do not end up following up with each other post-conference, again resulting in a missed opportunity.

In today's world, "professional" social networking sites such as LinkedIn (www.linkedin.com) abound and are gaining popularity. Further, instant-communications sites such as Twitter (www.twitter.com) are also gaining popularity, as a way to allow regular, inexpensive communication by people around the world. These sites and other similar technologies have helped demonstrate people's interest in further communication, but are general in nature.

In order to improve the value of conferences and between-conference time, toward a stronger academic and professional community structure, we have begun the design of a Web network that combines social networking with specific information such as people's academic research interests, helps people find others with similar interests, and provides information and support via a library of papers, videos, and demonstrations of products.

III. MOTIVATION, PART II: IMPROVING THE E-LEARNING FIELD

E-learning as a field has, particularly over the past several years, become very broad and also very fragmented, with numerous subareas. These subareas often could share more knowledge than they do in practice: for example, academics who study the use of e-learning in university classrooms rarely interact with practitioners who design e-learning products for job performance improvement of corporate employees. The e-learning industry has a variety of conferences: some are broad in scope and some more specific; some are more focused on high-level ideas; and many focused on technology specifically. Overall, e-learning commands huge interest internationally, and is still a rapidly-growing and exploding field.

However, the field of e-learning has met with fairly limited success relative to its expectations and possibilities. People across subareas and countries remain excited about e-learning and its potential but much of that potential remains unrealized. Conferences generate great discussions, but often the discussion threads die out after the conferences, as people return to their real jobs with little continuity. People do learn from each other to some degree, but more in fits and starts than in a continuous fashion.

These factors, plus the fact that e-learning as a field may want to practice what it preaches in terms of innovative uses of the Web and mobile devices for collaboration, make the field of e-learning the ideal candidate for the first site to try to improve collaboration within a field of study.

The features and methodology we describe in this paper can theoretically apply to any area; for the reasons above, we are rolling out the first version in our own field, e-learning, via the International E-learning Association (IELA, www.ielassoc.org.). In the following sections, we define the IELA site design components.

IV. SITE DESCRIPTION AND KEY FEATURES

The new site, to be first implemented for the IELA, includes a variety of components which all work together in service of the high-level goals of furthering collaboration and communication among academics and professionals in the e-learning field. Since the IELA is a membership organization, members can opt to provide personal information on their interests and ways they may be contacted by other members. This personal information is crucial in order for members to make connections, though of course the option remains for members who prefer to be private to remain that way.

Key features of this site include:

- A library of papers, videos and audios (some of which are from IELA-sponsored conferences), and papers and articles, which are indexed so they can easily be found, and have reviews and comments associated with them.

- A list and descriptions of upcoming events in the field, including conferences, seminars, presentations, and other related events, and including both in-person and online events. Associated with each events, where applicable, is a list of IELA members (with links for more information on the people and their interests and contact information) showing who's attending the event.

- Live and recorded Webcasts of news, events, and discussions in the field. Recorded events are archived in the library described in the first bullet point above, and are augmented with comments and appropriate indexing terms.

- A networking section to connect with people with similar interests; this section lets people fill out a profile with their interests, then search for others with similar interests and explore their work (for whatever work and links are included in the IELA site), or (if the other person has allowed it) contact the person directly. People can be located using keywords, by name, by location, or by organization.

- A way for members to locate other members, or other members with specific interests, from a specific organization, etc., who will be attending a particular upcoming event.

- A single place to get in touch with people in multiple ways—such as phone, email, instant messaging and text messaging—for those who want to be contacted.

- Discussion forums to provide an additional way for members to interact.

- Comments and reviews are linked directly to the items they relate to, and all reviews and discussions are set up to be easily found in a search.

With a variety of features that are linked together and all designed to serve the goals of communication, collaboration, and furthering ideas and knowledge in the field, the site described above aims to use technology to improve connections between people, by employing design principles that come from research in the fields of e-learning and cognitive psychology (e.g., Schank, 1977; Schank, 1980; Guralnick, 1994; Guralnick, 2000).

V. SITE DESIGN AND THE ROLE OF TECHNOLOGY

One of the crucial points about the design of this site is that it is based on specific design goals, which in turn came from an analysis of the types of tasks that users may need to perform. As a result of this rigorous design process, the features above are closely linked together rather than simply compiled into a site. This does require more effort, on both the design and implementation sides, than simply linking to and using existing sites such as Twitter. We suggest here that this level of customization and additional work is absolutely crucial to the success of the site, that it is not sufficient to simply take existing technologies and compile them into one place. Instead, we have followed a goal-based methodology for the site design and interface, as defined in Guralnick (2003); Guralnick (2007); Norman (1988); and Cooper, Reimann, and Cronin (2007).

VI. IMPLICATIONS, FUTURE ADDITIONS, AND OTHER USES

This site is indeed a first release, and will benefit tremendously from the feedback from its member community of users. Our philosophy here, from a scientific standpoint, is to generate and test; that is, we are creating this site based on an analysis of the goals of the users, and are producing this site and rolling it out to those users, after a prototype phase which will undergo usability testing, and after substantial technical testing. It will be instructive to study (and to report the results of the study in the future) the site in use and the users' view of it. Data will be tracked in the aggregate form to help us understand usage (but not at the level of each person, in order to protect users' privacy).

At a future point, changes and extensions will certainly be designed and implemented, and the site's design and methodology may well be adapted to other fields as well.

VII. CONCLUSION

Technology has the ability to support and aid in collaboration in many ways, from creating social networks to providing a knowledge base to supporting direct communication between people. Oftentimes, however, each technical piece is separate, which fails to provide the seamless usage, and resultant efficiency and ease-of-use, than an integrated system can provide. Academics and professionals, who tend to network at conferences and form connections there, are a group that could tremendously benefit from an inter-conference networking and collaboration suite, one that ties together conferences, journals, and people, with the goals of improving a particular field of study. Such an online system can help people maintain contact and project work between in-person meetings and conferences, and can provide a way to include people for whom finances or time prevent their participation in some conferences. The networking and collaboration site we have described in this paper has the potential to influence the way people collaborate, and to widen the group of people whose work can impact a field of study. By beginning with the field of e-learning for the first implementation of this suite, we hope to see an impact and also to gather data for the next version and for expansion into other fields of study.

REFERENCES

[1] A. Cooper, R. Reimann, & D. Cronin. *About Face 3: The Essentials of Interaction Design.* Indianapolis, Indiana: Wiley Publishing, Inc., 2007.

[2] D. Guralnick. *An Authoring Tool for Procedural Task Training.* Evanston, Illinois: Northwestern University Press, 1994.

[3] D. Guralnick. "Integrated knowledge management tool suites: a user-centered approach to collaborative web content development and distribution," in *Journal of Interactive Instruction Development*, Winter 2001.

[4] D. Guralnick, "Designing effective e-learning user interfaces.," in Proceedings of the IADIS International Conference on the WWW and the Internet, 2007.

[5] D. Guralnick, "A goal-oriented approach to e-learning authoring tools," in Proceedings of the Worldwide Conference on Educational Multimedia and Hypermedia, 2003.

[6] D. A. Norman, *The Design of Everyday Things.* New York: Basic Books, 1988.

[7] R. C. Schank and R. P. Abelson. *Scripts, Plans, Goals, and Understanding: An Inquiry Into Human Knowledge Structures.* New York: Lawrence Erlbaum, 1977.

[8] R. C. Schank, *Dynamic Memory.* London: Cambridge University Press, 1982.

AUTHORS

David Guralnick is with Kaleidoscope Learning, 304 Park Avenue South, 11th Floor, New York, NY 10010 USA (e-mail: dguralnick@kaleidolearning.com), and is President of the International E-learning Association (www.ielassoc.org).

Christine Levy is also with Kaleidoscope Learning, 401 North Michigan Avenue, Suite 1200, Chicago, IL 60611 (email: clevy@kaleidolearning.com), and is Secretary of the International E-learning Association (www.ielassoc.org).

A Self-Regulated Learning Approach: A Mobile Context-aware and Adaptive Learning Schedule (mCALS) Tool

J.Y-K. Yau and M.S. Joy

University of Warwick, Coventry, UK

Abstract—Self-regulated students are able to create and maximize opportunities they have for studying or learning. We combine this learning approach with our Mobile Context-aware and Adaptive Learning Schedule (mCALS) tool which will create and enhance opportunities for students to study or learn in different locations. The learning schedule is used for two purposes, a) to help students organize their work and facilitate time management, and b) for capturing the users' activities which can be retrieved and translated as learning contexts later by our tool. These contexts are then used as a basis for selecting appropriate learning materials for the students. Using a learning schedule to capture and retrieve contexts is a novel approach in the context-aware mobile learning field. In this paper, we present the conceptual model and preliminary architecture of our mCALS tool, as well as our research questions and methodology for evaluating it. The learning materials we intend to use for our tool will be Java for novice programmers. This is appropriate because large amounts of time and motivation are necessary to learn an object-oriented programming language such as Java, and we are currently seeking ways to facilitate this for novice programmers.

Index Terms—Context-aware, Learning Schedule, Mobile Learning, Learning Java programming, Learning Objects, Self-Regulated Learning

I. INTRODUCTION

A self-regulated student can be characterized by his/her "*active participation in learning from the meta-cognitive, motivational, and behavioral point of view*", and the characteristics of such a student coincide with those attributes of higher-performance and higher-capacity students [1]. Montalvo and Torres [1] noted that a self-regulated student is able to

a) use cognitive strategies to organize, transform, elaborate and recover information;

b) direct their mental processes toward the achievement of personal goals through plan and control;

c) show positive emotions towards tasks and a high sense of academic self-efficacy, and have the ability to control these to adapt to the requirements of the task and of the specific learning situation;

d) plan and control the time and effort on tasks, and create and structure preferable learning environments such as identifying a suitable place for study and

obtaining help from teachers and students when they experience difficulties;

e) use strategies to maintain their concentration, effort and motivation, and avoid external and internal distractions, whilst performing tasks.

Initial investigation for the requirements of a mobile learning organizer established that there was a demand by users for institutional support of mobile learning, especially for timetabling information and providing course content [2]. Learning organizers have been used in other mobile learning systems [2, 6], however this has not been for the purpose of capturing and retrieving users' contexts at a later stage. Several mobile learning projects are underway in Europe and one definition of mobile learning which has become prevalent from these projects is that mobile learning is not only about learning using handheld computer devices, but about learning across contexts [3]. It is especially important to capture these contexts (in one way or another) as previous assumptions which apply to stationary learning applications no longer apply to applications which function on mobile devices [4].

Prekop and Burnet [5] divide contexts into *Internal* and *External* dimensions.

- The *Internal* dimensions include human factors such as *users* (emotional/physical state, personal events, beliefs, previous experiences), *social environment* (work context, business processes, communication), and *activities* (goals, tasks).

- The *External* dimensions include the *physical environment* (light, sound, movement, touch, acceleration, temperature, air pressure, proximity to other objects, time), *infrastructure* and *location*, and *technological* features (device and product design).

In this paper, our Mobile Context-aware and Adaptive Learning Schedule (mCALS) is described as a learning tool which we believe can be effective for self-regulated learners, as the learning schedule can a) help them organize their work and facilitate time management, and b) be used for capturing and retrieving contexts and allowing our tool to create and enhance opportunities for students (who are willing to learn) to learn in various locations. Self-regulated learners require both *will* and *skill* for achievement of learning/studying [1]. Our mCALS tool aims to support the *skill* part for students by determining which learning materials (in terms of learning objects) would be appropriate for them at that location with those contextual attributes which can affect their ability to learn/study. We believe that by taking these attributes into consideration, students' learning/studying process can be enhanced.

Our research for this model is motivated by the current lack of pedagogic knowledge in how different contexts in mobile environments that affect students' learning/studying; and whether by using contexts and making the mobile learning system context-aware we can increase the learning effectiveness of such mobile learning systems.

The paper is structured as follows -- in section 2, a literature review is presented, and the resulting outstanding problems and issues discussed. In section 3, the conceptual model and the preliminary architecture of our mCALS tool is illustrated and discussed, and our research questions and methodology for evaluation are then examined. Finally in section 4, we present our conclusions and suggestions for future work.

II. LITTERATURE REVIEW

A number of context-aware mobile learning tools have been proposed by different authors. A location-aware learning reminder/organizer was developed by Ryu and Parsons [6] to help students find their way round the university campus -- for example, to go to a particular location for a lecture. Contextual help is also provided when they walk past certain buildings such as the library – books which have become available are notified to the student. A mobile location-aware handheld event planner was designed and evaluated by Fithian et al. [7], which allows the user to view which colleagues are located near him/her, if he/she decides to have a spontaneous meeting. Normal events can also be scheduled and notifications sent to participants.

A system prototype has been developed by Bomsdorf [8] allowing learning materials to be selected depending on a given situation – this takes into account learner profiles such as their location, time available for learning, concentration level and frequency of disruptions. Similarly, a situation-aware framework/mechanism has been developed by Bouzeghoub et al. [9] which takes into account time, place, user knowledge, user activity, user environment and device capacity for adaptation to the user. A Java Learning Object Ontology has been developed by Lee et al. [10] as an adaptive learning tool to facilitate different learning strategies/paths for students, which can be chosen dynamically.

Existing problems/issues relating to context-awareness (whether within the mobile learning paradigm or not) include the following.

- How to capture users' contexts without the users having to provide them and interfering with what they are doing, so that it becomes implicit [11]?
- How to make mobile devices aware of the environment that they are situated in, for example, the level of noise, lighting, temperature, without attaching a number of sensors to the devices? Wearable/pervasive computing can provide a solution for detecting attributes within the users' environment such as temperature or light; however wearable computing might cause inconvenience or discomfort to the users [11].
- How to detect people's emotions and intentions? Currently this is still a very difficult task and some

sensors might be able to record these, however these may not be entirely accurate [12].
- How to maintain users' privacy and personal integrity with the use of location-aware technology [13]?

III. OUR MOBILE CONTEXT-AWARE AND ADAPTIVE LEARNING SCHEDULE (MCALS) TOOL

Figure 1 depicts the conceptual model of our Mobile Context-aware and Adaptive Learning Schedule (mCALS) tool, consisting of three components –

- Learner's Schedule/Profile,
- Adaptation Mechanism, and
- Learning Object Repository.

The goal of the system is to select appropriate learning objects for students based on their current user contexts and user preferences. The user context attributes include their location (we use the elements concentration level and frequency of interruption), and the user preferences attributes include their knowledge level for that topic – in our case Java -- and their available time. The user contexts are captured via a learner schedule (otherwise known as a learning organizer), which is a novel way of capturing contexts in such a context-aware mobile learning system. We believe that this learning schedule approach can be a successful time management technique and an effective self-regulated learning approach for motivated students.

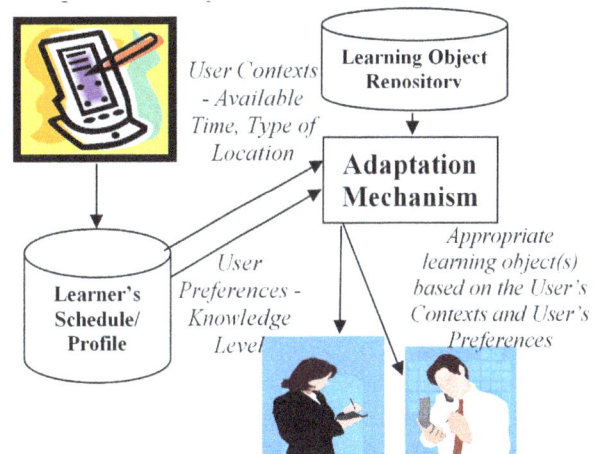

Figure 1. Conceptual Model of mCALS

A. Our Preliminary Architecture of mCALS

The system architecture of our tool, illustrated in Figure 2, is logically divided into three layers – Learner Model layer, Adaptation layer and Learning Objects layer. (Note that this is a more detailed version of the one illustrated in [14]).

1) Learner Model Layer

There are three system components within this layer –

- Learner Profile,
- Learner Schedule,
- UpdateKnowledgeLevel.

Two methods are in place to check that the Learner Schedule is up-to-date, namely Software Verification and User Verification. The Student Database stores details about the learner profile, and their schedule, and the

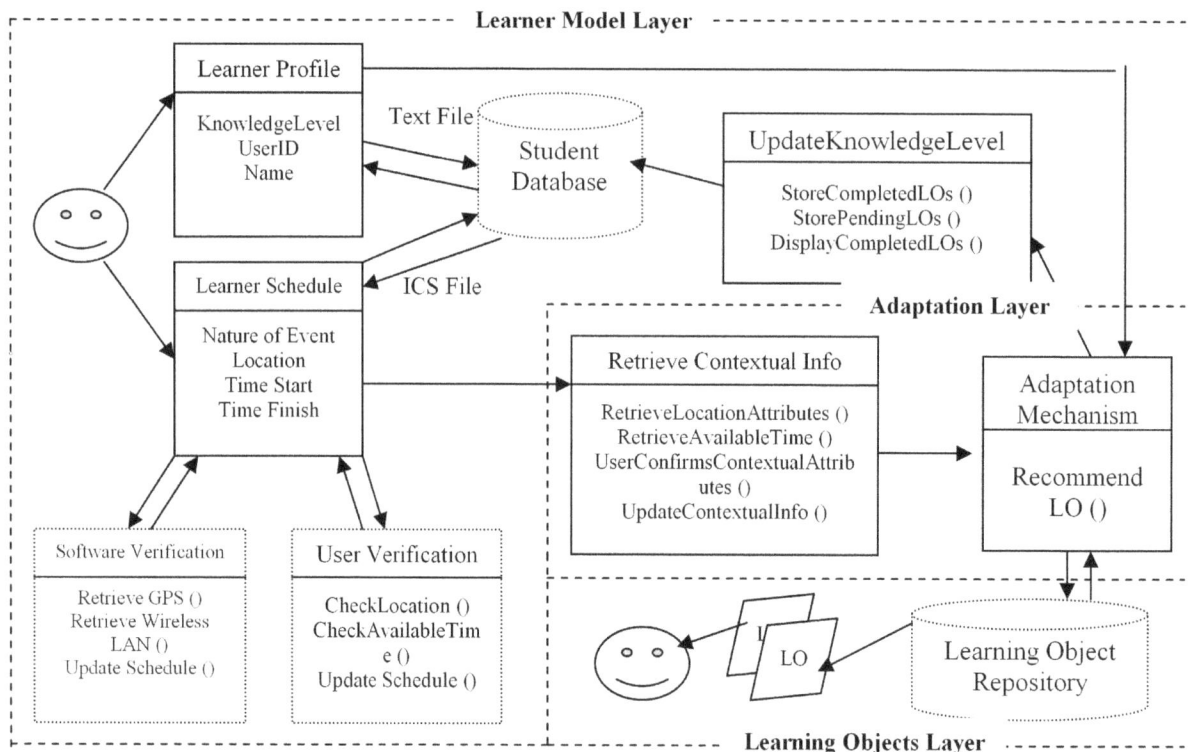

Figure 2.   System Architecture of the Mobile Context-aware and Adaptive Learning Schedule tool Framework

student's updated knowledge level of Java. For the learner profile, their knowledge level of Java – novice, intermediate or advanced (in the form of a string) is updated when they have successfully attempted some learning materials. A unique user ID is created for each student, and the students' personal details, although not essential, can also be entered here (such as name, age, gender) if required.

For the learner schedule, students are required to enter their scheduled events; the primary attributes are the nature of the event (such as lecture), location, and start and end times (in the form of a string). A graphical-based calendar will be displayed here for ease of entry.

These events are stored into the learner schedule and for the purpose of retrieving the event details with ease, the calendar is transformed into an ICS text file.

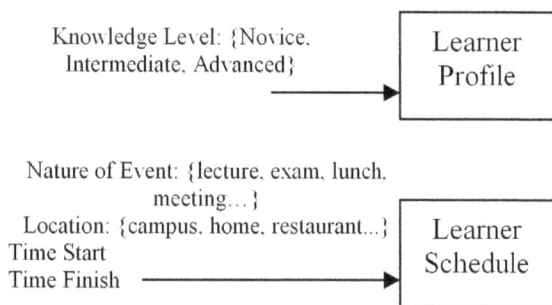

Figure 2.1: Input for the Learner Profile and Learner Schedule Components

Before the Learner Schedule information is retrieved and accessed by the *Retrieve Contextual Info* component, two verification methods – software and user verification – are in place to check whether the learner's schedule is still accurate.

For the software verification, we expect that GPS will detect the appropriate outdoor locations and WirelessLAN will detect the indoor locations. Therefore two methods – *RetrieveGPS()* and *RetrieveWirelessLAN()* retrieve this information, feed it to our schedule and update where necessary. Using this information, the learner's actual location can be identified and this can be used to confirm whether the learner is keeping to his/her schedule.

The user verification prompts the user to check a) his/her location and b) his/her available time, and this information is used to update the schedule, if necessary.

2)   Adaptation Layer

Pre-adaptation – Retrieve Contextual Information

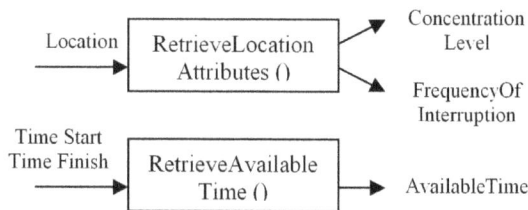

Figure 2.2: RetrieveLocationAttributes and RetrieveAvailableTime Methods

We use the location attribute to calculate two default values for the level of concentration and frequency of interruption typical for that type of location. The values of these attributes in relation to the location have been obtained by a study performed by Cui and Bull [15] where they found that different students perceived the same level of concentration and frequency of interruption in the same location, although the noise levels may have been different. These are used as default levels for the students' concentration level and frequency of interruption which are used as grounds for adaptation of the recommended learning objects (amongst other factors). Using the Time Start and Time Finish attributes, we obtain the available time that the student has at a particular point in time. See Figure 2.2.

Figure 2.3 shows the input and output of the *Retrieve Contextual Info* component, the function of which is to first retrieve the contextual information from the learner schedule, and then transfer this into actual approximate values which can be used by the Adaptation Mechanism. The attributes taken from the learner schedule include Location, Time Start and Time Finish.

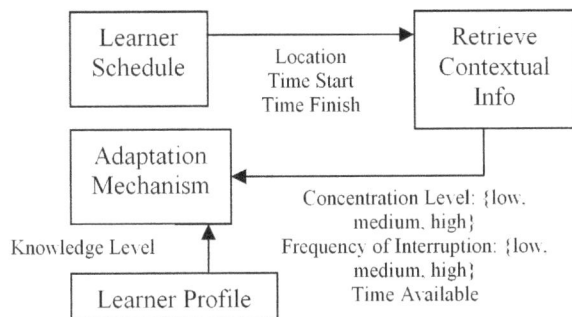

Figure 2.3: Input and Output of the Retrieve Contextual Info Component

A method *UserConfirmsContextualAttributes()* gives the user the option to view and confirm the values of these attributes, or change these values, if necessary. These attributes include Concentration level, Frequency of Interruption and Available Time. The method *UpdateContextualInfo()* is used to update this contextual information, as necessary. Figure 2.2 shows the input and output of the *RetrievedLocationAttributes()* and *RetreiveAvailableTime()* methods.

The parameters fed into the mechanism include knowledge level, concentration level, frequency of interruption, and available time. Specific adaptation rules are defined within the adaptation mechanism which then take the parameters into consideration and select the appropriate learning objects for the learner. The Learning Object Repository contains learning objects for Java, in the form of text for factual information and examples and multiple choice questions such as exercises and tests. Graf and Kinshuk [16] noted that active learners prefer testing and experimenting and the use of exercises and tests would be ideal for them; whereas for reflective learners, they prefer to read materials and therefore content containing objects and examples would be appropriate for them.

3) Learning Objects Layer

The learning objects which have been recommended to the students are stored along with the following information –

- Whether the student has completed it.
- In the case of a test or exercise, whether the student answered it correctly.

This information is transferred to the student database and when the student has attempted an appropriate amount

of material (and accurately), their knowledge level is increased.

B. Our Research Questions

Our research questions are divided into five sections – Learning Schedule, Learning Environment/Space, Contextual Attributes, Our Adaptation Mechanism and Usability of our tool. We defined Learning Schedule as a set of activities related to both learning and non-learning events which have been planned by the user.

1) Learning Schedule

We want to examine whether students use a learning schedule for time management. One group we will target for our experiment will be first year computer scientists because we are currently seeking ways to facilitate novices to learn Java programming. We want to find out whether students find that using a learning schedule is an effective way of managing their time.

- How likely is it that these students observe/follow their learning schedule?
- If students have not previously used a learning schedule, do they think that they can benefit from using one for managing their time?
- If they do not use one, is that because they foresee that they cannot observe/follow their learning schedule?
- Are there any particular personal characteristics/learning styles relating to the student which affect whether they use a learning schedule for time management, and their ability to observe and follow it? (Characteristics such as conscientiousness, motivation)
- Are students willing to provide their daily learning-related and non-related activities which are accurate and up-to-date and which they will conform to?

2) Learning Environment/Space

- Which places do first year computer science/programming students learn/study/do programming in?
- Are there specific places where they will go to perform certain learning activities, such as to a computer lab for programming?
- For each of these places, how well can they learn/study/do programming in?
- Are there restrictions in any of these places which hinder their ability to perform a certain learning task, or with the mobile devices that they are using?
- Which factors in each of these places affect their concentration level for learning/studying/doing programming and to what degree?
- How well can students learn/study/do programming if the place that they are in is prone to frequent interruptions?
- How do interruptions affect the students in performing certain tasks?

3) Contextual Attributes (Knowledge Level, Location - Concentration Level and Frequency of Interruption, Available Time)

- Is it motivating for computer science students to learn/study if they are using learning materials which are adapted and are appropriate for their knowledge level?

- Do they think that the location that they are learning/studying/doing programming in can affect how well they learn/study/do programming?
- What are the factors within the location which can affect their ability to study (such as noise level, movement) and how do these affect learning/studying?
- If students are learning in a place prone to interruptions, would it be helpful for them if the learning materials selected for them had greater granularity so that they can frequently stop as necessary?

4)   Adaptation Mechanism

- Would students find it more convenient and more helpful to their learning when learning materials selected for them are appropriate for them in the given contextual situation?
- Will they be able to carry out the activities selected for them?
- Do they think it would be more helpful towards their learning/studying/doing programming if their approximate concentration level and/or the frequency of interruption in the location that they are situated in and/or the available time that they have for studying during that period of studying time is taken into consideration?
- Are students able to study/learn/do programming more effectively (or with more ease of use) when these contextual features are taken into consideration?

5)   Usability of our tool

- Would students use such a device for learning in various places?
- Would students find it convenient and time-saving that the learning materials selected for them are appropriate for them in the given contextual situation?
- Do they like having activities selected for them? Would students prefer to input the location and time available manually themselves at the point of using it, rather than keeping a learning schedule for the system to retrieve the contexts later when necessary?
- Would the students mind the tool knowing where they are?

C.   Our Methodology for evaluating mCALS

Our evaluation methods are divided into two phases – an interview research activity and a system prototype evaluation. We have currently completed phase 1 of our research activity and are analyzing the interview data. 37 students participated in our semi-structured interview and, each interview took approximately 20 minutes to complete and four areas were explored – a) studying in different environments, b) personal information management, c) learning characteristics, and d) learning preferences. Phase 2 of our research activity will be underway once we have completed our interview data analysis and obtained a deep understanding of what students required i.e. forming the user requirements of our context-aware mobile learning system tool. A system prototype will then be implemented and evaluated via the means of simulation tests for

checking the technical workings of the system including that the adaptation mechanism is able to accurately retrieve the up-to-date context and select appropriate materials for learners based on their current contexts. We will also ask a number of students to evaluate our tool in order to obtain learner-centered feedback via questionnaires or interviews.

D.   Preliminary Interview Data Analysis

Preliminary analysis of our interview data shows that the participants were able to identify places which were most effective for them for studying and had their preferred learning environment(s), whether in the library, computer lab, or their bedroom etc. There were many differences between the students' preferred learning environments, for example, one student found that he/she was able to concentrate very well in the library because of the quietness, but another found it extremely distracting with lots of people around. Factors which the students noted could distract them from concentration included noise (whether constant or irregular), temperature, number of people around, and their proximity to them.

In terms of personal information management, most of our participants used a diary for managing their studies and noted that it was effective for time-management especially since it reminds them not to forget important meetings and deadlines. It allows them to schedule their workload, assign time for specific modules to maximize effectiveness and efficiency of their studies and some of the participants highlighted the fact that they were still able to fit in their social and leisure activities. Our findings show that participants who used a diary regularly were fairly likely to keep their scheduled events unless something unexpected or more urgent came up. On the other hand, it was also noted by a few of our participants that they used their diaries as a reminder tool to make them aware or remind them of possible events/tasks that they can attend/perform if they had sufficient time such that not everything in the diary would be strictly adhered to.

IV. CONCLUSIONS/FUTURE WORK

In this paper, we have described our Mobile Context-aware and Adaptive Learning Schedule tool as a system which could aid self-regulated learners whilst they are learning in mobile learning environments. Their learning opportunities at different locations could be created or enhanced by our tool. The originality of our research includes the capturing of different learning contexts which are created in different environments and the use of our learning schedule for storing and retrieving this information. We plan to expand the scope of our research not limited to self-regulated learners. Thereafter, we hope to generalize these requirements for a generic context-aware mobile learning system.

ACKNOWLEDGMENT

This journal paper also appears in the proceedings of the *International Conference on Interactive Mobile and Computer Aided Learning*, April 2008.

REFERENCES

[1] Montalvo, F. and Torres, M. (2004) *"Self-Regulated Learning: Current and Future Directions"*. Electronic Journal Research in Educational Psychology, 2 (1), 1-34.

[2] Corlett, D. Sharples, M., Bull, S. and Chan, T. (2005) *"Evaluation of a Mobile Learning Organiser for University Students"*. Journal of Computer Assisted Learning, Vol. 21, No. 3, pp. 162-170(9).

[3] Sharples, M. (2006) "Big Issues in Mobile Learning", Kaleidoscope Network of Excellence Report.

[4] Lavoie, M. (2006) "I, Mlearning: Identifying design recommendations for a context-aware mobile learning system", IADIS International Conference on Mobile Learning, pp. 265-269.

[5] Prekop, P. and Burnett, M. (2003) *"Activities, context and ubiquitous computing"*. Computer Communications, Vol. 26, Num. 11, pp. 1168-1176.

[6] Ryu, H. and Parsons, D. (2008) *"A learner-centred design of a location-aware learning reminder"*, Int. J. Mobile Learning and Organization, *to appear*.

[7] Fithian, R., Iachello, G., Moghazy, J., Pousman, Z. and Stasko, J. (2003) "The design and evaluation of a mobile location-aware handheld event planner", International Conference on Human Computer Interaction with mobile devices and services, pp. 145-160.

[8] Bomsdorf, B. (2005) "Adaptation of Learning Spaces: Supporting Ubiquitous Learning in Higher Distance Education", Dagstuhl Seminar Proceedings 05181 Mobile Computing and Ambient Intelligence: The Challenge of Multimedia.

[9] Bouzeghoub, A., Do, K. and Lecocq, C. (2007) "Contextual Adaptation of Learning Resources", IADIS International Conference Mobile Learning, pp. 41-48.

[10] Lee, M., Ye, D. and Wang, T. (2005) "Java Learning Object Ontology", International Conference on Advanced Learning Technologies, pp. 538-542.

[11] Schmidt, A. (2002) "Ubiquitous Computing – Computing in Context". PhD thesis.

[12] Schmidt, A. (2000) *"Implicit Human Computer Interaction Through Context"*. Personal Technologies, Vol 4(2), June 2000.

[13] Synnes, K., Nord, J. and Parnes, P. (2003) "Location Privacy in the Alipes Platform", International Conference on System Sciences, pp. 10-19.

[14] Yau, J. and Joy, M. (2007) "Architecture of a Context-aware and Adaptive Learning Schedule for Learning Java", International Conference on Advanced Learning Technologies, pp. 252-256.

[15] Cui, Y. and Bull, S. (2005) *"Context and learner modeling for the mobile foreign language learner"*. Science Direct, System 33, pp. 353-367.

[16] Graf, S. and Kinshuk (2006) "An Approach for detecting learning styles in learning management systems", International Conference on Advanced Learning Technologies, pp. 161-163.

AUTHORS

J. Y-K. Yau is with Department of Computer Science, University of Warwick (e-mail: j.y-k.yau@warwick.ac.uk)

M.S. Joy is with Department of Computer Science, University of Warwick (e-mail: m.s.joy@warwick.ac.uk).

5

Learning Mathematics in an Authentic Mobile Environment: The Perceptions of Students

N. Baya'a and W. Daher

[1] Al-Qasemi Academic College of Education, Baqa El-Garbiah, Israel

Abstract—This research reports an experiment which took place in an Arab middle school in Israel. The experiment was led by three pre-service teachers who were carrying out their final project in the field of teaching mathematics using mobile phones. The pre-service teachers worked with 32 eighth grade students to carry out authentic real life outdoor activities in the nature. We used discourse analysis and grounded theory to analyze the perceptions of the students regarding their mathematics learning using mobile phones. We found that the novelty of the experiment and the use of mobile phones in mathematics learning were the main characteristics perceived by the students as influencing their decision to join the experiment. Furthermore, the students perceived various qualities of the mathematics learning that were enabled by the use of mobile phones: (1) exploring mathematics independently (2) learning mathematics through collaboration and team work; where the collaboration is on equal terms (3) learning mathematics in a societal and humanistic environment (4) learning mathematics in authentic real life situations (5) visualizing mathematics and investigating it dynamically (6) carrying out diversified mathematical actions using new and advanced technologies (7) learning mathematics easily and efficiently. In the overall, the students were positively impressed by the potentialities and capabilities of the mobile phones used in the mathematics learning process. This indicates that mathematics education could benefit form utilizing these new technological tools.

Index Terms—Mobile learning, mathematics education, middle school students, pre-service teachers, new technologies, mobile phones, perceptions of mathematics learning.

I. INTRODUCTION

A. Mathematics mobile education

In spite of the burst of mobile phones in most of our daily lives aspects, the utilization of these devices in education is still new [1] and in its infancy [2]. This is true for the use of mobile phones in education in general and especially for its use in teaching mathematics. On the other hand, we are aware recently of the increasing commonness of mobile and wireless devices, especially mobile phones, amongst the young generation. This provides new possibilities, opportunities and challenges for the educational environment [3]. Furthermore, some recent studies examined the use of mobile phones in learning mathematics amongst university pre-service teachers ([4], [5]). This implies new era for mobile phone integration in the mathematics class in which the diversified mobile features are utilized to build mathematical knowledge.

Ref. [5] studied learning processes and experiences within a mobile phone learning environment. Doing so they aimed to examine how socio-cultural and situated learning aspects were reflected in these processes and experiences. They found that the contribution of the mobile phone environment "lies not only in making dynamic mathematical application more available, but also in supporting the execution of tasks that are closer to the students' experiences and more relevant to them, which has the potential to enhance experiential learning." In addition the participants' learning experiences contributed to personal learning processes, which may motivate the participant learning.

B. Students' perceptions of their learning

Ref. [6] says, "Understanding students' perceptions of accounting is an important first step in the effort to attract the best to the accounting profession." We think that understanding students' perception of mathematics learning using mobile phones is an important step to understand how to attract middle school students to learn mathematics in this new environment. This understanding would help us know what factors influence students' learning of mathematics using mobile phones and how to motivate them to do this learning successfully and with enjoyment.

Ref. [7] studied students' perceptions of online and distance learning and found that students perceive online learning to have a significant relative advantage to traditional learning. These advantages include saving their time, fitting better in their schedules, and enabling them to take more courses. The students did not believe that they learn more in online learning courses. In additions, they had some concerns related to being able to contribute to the forum discussions.

C. Authentic learning

Ref. [8] claims that students better understand and apply studied materials when they are engaged in real world issues and situations. Ref. [9] points out that authentic situations and scenarios provide a stimulus for students' learning and thus create greater motivation and excitement for this process. This ref. also states that representing and simulating real-world problems, provides an important context for students' thinking. Regarding the contribution of online education to authentic learning, [10] asserts that technology and online instruction can facilitate learning by providing simulation of real-life context in order to simplify and illustrate it for the learners who face solving complex authentic problems. Regarding using mobile devices in authentic learning, [11] declares that mobile devices extend the learning environment in which the students work, and integrate it in real life situations where learning can occur in authentic contexts. Ref. [12] says, "Mobile learning can guide a learner to an authentic learning context and incorporate the field objects with closely related information in the handheld device to initiate the process of knowledge acquisition." In our research, we want to study the students' perceptions of mathematics learning when their knowledge acquisition is initiated by authentic activities during which they use mobile phone features in real-life situations.

Little research has been done regarding students' mathematical learning using mobile phones, especially research that involves middle school students. Our research aims to understand how middle school students perceive such learning.

D. Research questions

- What characteristics of the mobile phone environment do middle school students perceive as important for their positive decision to join an experiment involving mathematics learning using mobile phones?
- What characteristics of the mobile phone environment do middle school students perceive during and after they learn mathematics in authentic learning context using mobile phones?
- What qualities of mathematics learning do students perceive as being enabled by the use of the mobile phones in the learning process?

II. RESEARCH METHODOLOGY

A. Research setting

The experiment took place in an Arab middle school in the city of Umelfahm in Israel, and was led by three pre-service teachers who were carrying out their final project in the field of teaching mathematics using mobile phones. This project was part of their tasks in a mathematics didactics seminar. The pre-service teachers selected 32 eighth grade students to participate in the experiment. The selection was done based on the interest of the students and the ownership of appropriate cellular phone. The learning was done by carrying out outdoor activities that involved exploring and investigating mathematics concepts and relations of real-life phenomena. The students utilized various features and qualities of the mobile phone to do such exploration and investigation.

The students used mathematical midlets from the site of the Institute for Alternatives in Education that operates within the Faculty of Education at the University of Haifa, Israel: www.math4mobile.com. These midlets support the learning of algebra and geometry. In our experiment, the students used the algebraic midlets that enabled them to see the graphs of several templates of linear functions. They could see the change in the corresponding straight line as the result of changing parameters in the algebraic form. They also had the opportunity to set points in a coordinate system and to check if a straight line could connect all of them; indicating a linear relation in the real-life phenomenon.

Carrying out the activities the students exploited the mobility, dynamics, availability and accessibility properties of the cellular phones and used various tools and technologies embedded in them, such as: taking pictures, recording video, recording audio, measuring time, transferring of information, voice and text communication, for-

warding screen content to learning mates and sending SMS or MMS messages to them.

B. Data collecting means and tool:

The pre-service teachers used the following means and tools to collect data regarding the participants' learning of mathematics using mobile phones:

Blog: the students were required to comment on and document their mathematics learning using mobile phones. They used a pre-constructed blog in order to suggest ideas regarding the use of mobile phones in the mathematics learning process and to inquire about this use.

The blog started with a welcoming text, written by one of the pre-service teachers who led the experiment. The pre-service teacher informed the students that they could ask questions in the blog, add remarks, add feedback, comment, document events and actions and write about their feelings regarding the experiment. The pre-service teacher also asked the students to write about their expectations and the activities that they would like to be engaged with.

After the first post of the teacher and before the students contributed to the blog, some visitors inquired about the experiment, requesting information about it. In addition, throughout the experiment, several visitors gave remarks and inquired about it in the blog.

Interviews with the students: The pre-service teachers interviewed each participant for thirty minutes about her/his mathematics learning using mobile phones. The interviews were semi structured.

We used the interviews that the pre-service teachers held with the students and the blog in which the students remarked on, documented and discussed their mathematics learning using mobile phones to analyze and characterize the students' perceptions of this learning.

C. Data processing and analysis

We used discourse analysis [13] and the grounded theory approach [14] to identify the students' perceptions regarding the mathematics learning using mobile phones.

Ref. [15] describes discourse analysis as an "analysis which studies practices of producing knowledge and meanings in concrete contexts and institutions", and adds: "Discourse analysis systematizes different ways of talking in order to make visible the perspectives and starting points on the basis of which knowledge and meanings are produced in a particular historical moment. It pays attention to the way in which discourses produce and transform social reality, and makes it possible to evaluate the practical consequences of different ways of approaching a particular phenomenon."

We used discourse analysis to analyze the students' writing regarding the way in which they perceived the mathematics learning using mobile phones.

In general, the grounded theory approach has three stages:

- *Open coding*: Identification of repeated behaviors that can be characterized. This is done by dividing each type of the gathered data into segments and examining these segments for similarity and difference. At this stage, the target is to identify categories of behaviors that occur among the mobile phone mathematics learning community. This stage concludes with putting the similar behaviors in the same category and characterizing each category.

- *Axial coding*: after identifying the categories and characterizing them, we examine the relations among these categories and their subcategories. Here we characterize the behaviors according to the context in which they occur and according to the conditions of their occurrence.

- *Selective coding*: After refining the categories, subcategories and their characteristics and relations, we try to identify one or two main or core categories that could be used to connect the rest of the categories with them, and to build a conceptual frame of the studied phenomenon: characteristics of the mobile phone mathematics learning community.

We used only the first two stages of the grounded theory to arrive at the various characteristics of the mathematics learning using mobile phones perceived by the students.

III.   FINDINGS

We used discourse analysis to describe the students' perceptions at the beginning of the experiment and after carrying out some activities. This analysis was based on the students' comments in the blog. Then, we used open and axial coding to categorize the students' perceptions of the mathematics learning qualities and the characteristics of the learning environment when the learning took place in the mobile phone environment. This coding was done based on the students' comments in the blog and in the interviews.

A.   *Perceptions of the students at the beginning of the experiment*

At the beginning of the experiment, the students perceived various qualities of the mathematics learning using mobile phones as reasons for liking this learning and expecting the success of the experiment. For most of the students, these qualities were equally important for them as reasons to like the learning process and to identify with the experiment. One of the students (let us call her Alaa) was the first to respond to one of the visitors (let us call him Salem) after experiencing the first activity led by the pre-service teacher inside the class:

"Welcome Salem. We thank you for your interest in the experiment.

I'm Alaa from Umelfahem, a member in this experiment and a student in the eighth grade in Elgazali school. A simple description of the experiment: The experiment in which I participate is about learning mathematics using the mobile phone. There are various programs for drawing functions that could be downloaded from the internet to the mobile.

My opinion about this is that I am astonished that we are able to learn mathematics using the mobile phone. I enjoyed this experiment, though we are still at the beginning and though the experiment is a learning one, but we have fun. They also promised us to go in trips so that we discover by ourselves mathematical relations.

We had problems downloading the programs but our overly enthusiasm made us curious and made us want to proceed with the experiment. We downloaded the programs using special cable, and my friend helped me. I operated the graphing programs and liked working with them.

I also had a difficulty writing my remarks in this blog, because I forgot my email, so I had to open a new one.

I will write my remarks in this blog because I am pleased with it."

The first question that could be asked here is: "whom the student is responding to?" To the visitor who requested more information about the experiment, or to the pre-service teacher who requested the students to write feedback about their experience in the blog? It seems that when the student started her response she had the visitor in her mind because she approached him by his name. This could be an indicator that the student feels belonging to the experiment, and she is proud to be part of it. She emphasizes initially that she is a member of the experiment, and then she mentions her school. Identifying herself with the experiment is also obvious in her statement "The experiment in which I participate" which came after the first statement "a member in this experiment". She has not felt proud only, but she was also enthusiastic towards proceeding with the experiment in spite of difficulties that she faced at the beginning of it. It seems that the difficulties she faced downloading the programs were forgotten when she started to discover them. This also encouraged her to proceed with the experiment.

Alaa faced difficulties not only when preparing for the experiment but also in writing remarks about it in the blog. This difficulty too does not make her reluctant to write remarks, because she likes the new experience of writing in a blog. Alaa finds herself in a new learning environment with new means of communication, so she gets enthusiastic and wants to experience this new environment fully, despite some difficulties that come in her way. It is not only the novelty of the tools, which she works with that make her enthusiastic, but also other characteristics of the learning environment. One important characteristic that also other students mentioned is going outside the classroom to study mathematics. Alaa has not experienced yet this outdoor learning, for it is a future one; a promised one. Therefore, what made Alaa enthusiastic about her new learning was not only what she experienced already, but also what she was going to experience - something that was promised. Alaa also enjoys the experiment, though she is aware that it is a learning experiment, as if she did not expect that one could enjoy learning. This enjoyment could be another reason why she is enthusiastic about using mobile phones in mathematics leaning, and why she identifies herself with the experiment. Figure 1 is a relational map illustrating this student's perception.

This student perceives the environment of mathematics learning using mobile phones as having different characteristics than the traditional one. These characteristics enrich her learning and make it more enjoyable. Other students had a similar perception, though the characteristics of the experiment considered by them to be beneficial were partially different. For example, one student identified herself with the experiment because it makes learning mathematics easier, simpler, and collaborative.

It should be noted that we consider the novelty of the experiment to be the main reason for the student's astonishment, because many students mentioned this novelty in the blog.

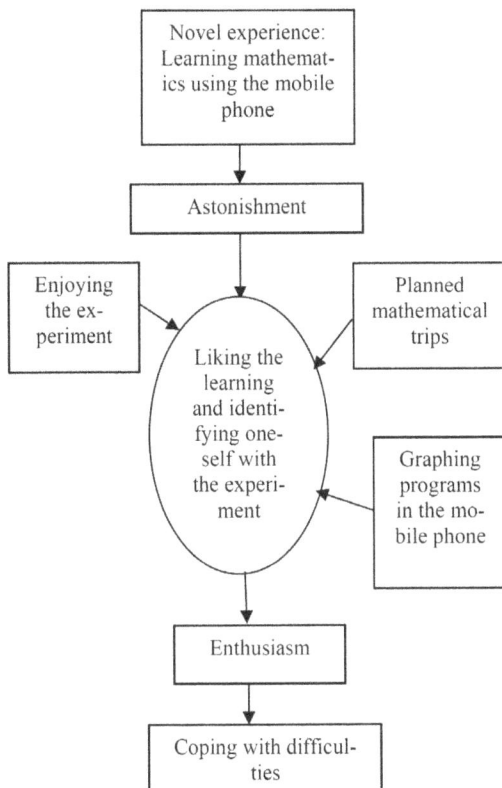

Figure 1. relational map of the student's perception at the beginning of the experiment

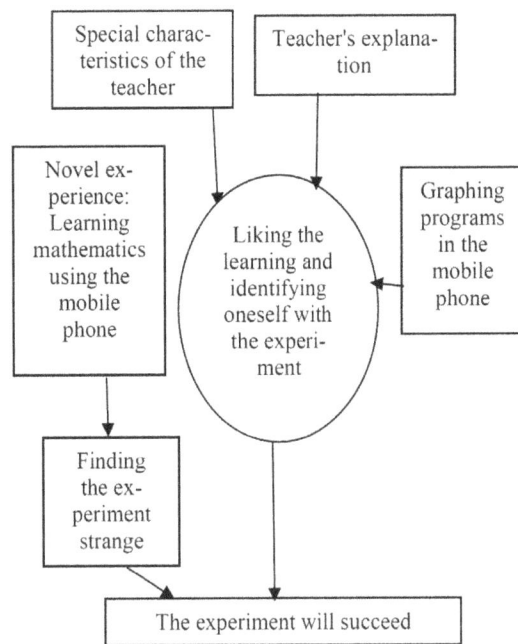

Figure 2. relational map of the teacher's characteristics as influencing the student's perception

B. Perceiving the teacher's characteristics as reasons for liking the experiment and identifying with it

Other students perceived the characteristics of the pre-service teachers and some learning qualities as reasons for liking mathematics learning using mobile phones and identifying with the experiment. One such student, let us call her Sawsan, wrote:

"Hi, my name is Sawsan. I am a student in the eighth grade in Elgazali School, from the town of Umelfahem. I am a member of the experiment too. When the teacher Fatima described the experiment for us, it attracted my attention, because it is strange, because it involves, for the first time, learning mathematics using the mobile phone. I expect that this experiment will succeed because it is new and it is the first time that I hear about it, so we are very eager to do such an experiment. I will be glad if we really learn mathematics through the mobile phone, and graph functions. Fatima encouraged us to participate in the experiment and promised to help us all the way."

Whereas the first student identifies herself with the experiment more than with the school, we see that this student identifies herself with the school more than with the experiment, mentioning first that she belongs to a specific school.

Talking about the experiment, Sawsan mentions only one pre-service teacher of the three managers of the experiment. At the end of the text, she mentions the pre-service teacher by her first name only, which a non-formal approach indicating feeling of closeness towards her teacher. We followed the texts of this student in the blog to find out if and how she mentions the same pre-service teacher. We noted that she admires this pre-service teacher, describing her with very nice words: friendly,

charming, smart, and humorous. Therefore, we can say that this student is influenced by some characteristics of the pre-serve teacher. This student thinks also that the novelty of the experiment is sufficient for making the experiment succeed. Figure 2 is a relational map illustrating this student's perception.

This student perceives the experiment through her relation with her teacher. Though she was aware of some experiment characteristics that contributed to her liking and identification with the experiment, such as its novelty and the use of graphing programs, the main reason for her liking and identification was her closeness to the pre-service teacher.

C. Perceptions of the students after experiencing some activities

Many students perceived the characteristics of the learning environment and qualities of the learning process as reasons for liking mathematics learning using mobile phones and identifying with the experiment. One such student, let us call him Ahmad, wrote:

"At the beginning I hesitated to participate in the experiment because I thought the experiment was a learning one. I was astonished to hear that I would learn mathematics using the mobile phone. At the beginning, I did not conceive this idea because I use the mobile phone for communication, sending messages, hearing music, or playing but not for learning.

I liked the programs that we downloaded from the internet. I explored them by myself, playing with the buttons to discover their functions.

My hopes from the experiment are to spend a good time, to learn new things and to benefit from this experiment. I expect that we will arrive at wonderful things. Today I became confident that the experiment would succeed because we carried out two activities that I was so pleased to carry out. I used the programs in my mobile phone, measured time using it and took pictures. All these actions are fun. This is the first time that I learn in a fun environment."

Ahmad hesitated to join the experiment when he thought it is about learning. It seems that he decided to try it for its innovation. He was astonished to know that there is such a thing like learning using the mobile phone.

Ahmad did not expect that mobile phones could be tools for learning. He perceived them as communication and fun tools only. He did not mention having difficulties downloading the programs, operating them and discovering their functions, on the contrary, he reported that he had fun doing so. This enjoyment was even reported by students who had some difficulties downloading the programs at the beginning, but afterwards found them very interesting and easy to operate.

Ahmad agrees with Alaa in his expectations from the experiment. He expects to enjoy it and to learn new interesting and wonderful things. However, more importantly, Ahmad, as many other students, confirms his expectation immediately after the first two experiences. This confirmation comes because of his positive perception of the qualities of the learning process and the features of the mobile phone used in the learning environment. These qualities and features include easy to use programs (midlets), accessibility to his personal mobile phone, interesting and fun features of the mobile phone. These qualities and features made the environment of learning mathematics using mobile phones a fun one. Figure 3 is a relational map illustrating the students' perceptions when they started to carry out mathematical activities using the mobile phone.

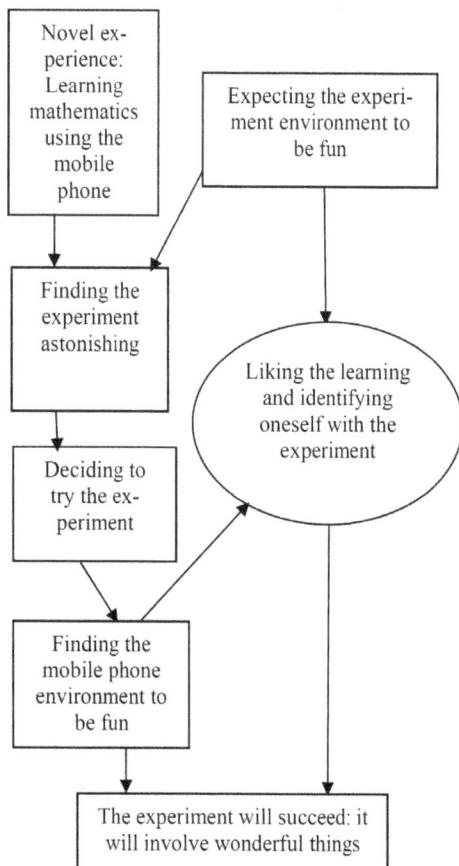

Figure 3. relational map of students' perceptions when they started to carry out mathematical activities using the mobile phone

We can see that the novelty of the experiment is a main factor for the students' participation and involvement. The non-regularity and informality of the experiment made the students curios and consequently they perceived it as strange and astonishing one. It seems that this factor influenced the students' participations and involvement positively. Moreover, the students expected the experiment to be enjoyable and a fun one, and this expectation was realized because of the interesting learning qualities and the mobile phone features. These fun feelings motivated the students making them like the experiment and identify with it, and consequently believe in its success.

D. *Characteristics of the environment of mathematics learning using mobile phones as perceived by the students in the experiment:*

We used open and axial coding to categorize the characteristics of the environment of mathematics learning using mobile phones, which were perceived by the students according to their comments in the blog and in the interviews. We present these characteristics together with selected quotations of the students' comments as evidence for their perceptions.

It introduces novel learning that breaks the routine:
- It makes us break the routine of learning. It is not a learning that is done in the classroom with a chalk and a blackboard.
- It makes us enthusiastic because it is new.
- It makes us use a new technology.

It enables independent learning:
- It enables us to do mathematics alone.
- It enables us to work without the teacher.
- It gives the students an active role over their learning. We decide who does what.

It encourages collaborative learning:
- It encourages collaborative learning.
- It encourages working side by side with our teachers.
- It encourages assisting each other carrying out the activities given to us.

It is a societal and humanistic environment:
- It is a societal experiment: it teaches us patience, competition, initiation and leadership.
- It is a humanistic experiment. It makes the students equal, to the teacher too.
- Working outdoors make us learn and have fun at the same time.
- It makes learning fun because we joke all the time.
- When doing the activities, we felt we belong to the group and want to collaborate to succeed in the activity. When we discussed the results, we felt that all the groups collaborate to explore mathematics together.

It enables exploration of mathematics:
- It enables us to explore mathematics.
- It is interesting to learn the mathematics of our bodies, for example the relation of our weights and heights, the relation between our weights and the time it takes us to go upstairs.

It enables outdoor mathematics:

- It enables us to explore mathematics in the nature, mathematics of trees and rocks and mathematics of walking.
- It makes learning fun because we learn mathematics in the nature.
- It enables us to do applied mathematics, for example, applying mathematics to describe the movement of a thrown ball.

It enables learning mathematics visually or/and dynamically.

- It involves learning mathematics visually when we use the midlets and see the graphs of functions.
- It involves learning mathematics dynamically with the midlets.

It engages the mathematics student in various mathematical actions:

- It enables us to do various learning actions, like measuring, assigning points, drawing functions, making conjectures and modifying them until we arrive at something that seems reasonable.
- We can do different actions in different activities, for example in one activity I do the measuring, while in another activity I work with the midlet to find the mathematical relation. Sometimes I do two different actions in the same activity.

It makes learning mathematics easy and takes less time:

- Working with midlets to learn mathematics is so easy.
- Learning mathematics with midlets is faster than with paper and pen.

It is important to notice that the success of the experiment was dramatically affected by two main factors: taking the decision to participate in the experiment and actively engaging with carrying out the activities. In the following two tables, we describe the characteristics of the environment of mathematics learning using mobile phones that motivated the students to participate in the experiment, and the qualities of this untraditional mathematics learning that kept them engaged with the experiment, as perceived by the students themselves. These characteristics and qualities were mentioned by the students in the blog and the interviews, and were categorized using open and axial coding of the grounded theory approach. The percentages in the tables were calculated using the number of the students that mentioned a specific characteristic or quality out of the 32 students who participated in the experiment.

We see from table 1 that the novelty of the experiment encouraged half of the students to participate in it, and the use of mobile phones was actually the leading characteristic of the environment that motivated the students to participation in the experiment.

We see from table 2 that what attracted the attention of the students mostly was their ability to learn mathematics outdoors. Moreover, they emphasized their collaboration and the dynamic and visualization that the mobile phones provide, as important qualities of the mathematics learning using mobile phones that kept them engaged with carrying out the activities.

TABLE I.
CHARACTERISTICS OF THE ENVIRONMENT OF THE MATHEMATICS LEARNING USING MOBILE PHONES WHICH MOTIVATED THE STUDENTS TO PARTICIPATE IN THE EXPERIMENT (N=32)

The characteristic	%
The novelty of the experiment	50
The use of the mobile phones in the experiment	62.5
The strangeness of the experiment	37.5
The explanations and characteristics of the teachers	25
The enthusiasm towards the experiment	37.5
Breaking the routine	15.63

TABLE II.
QUALITIES OF THE MATHEMATICS LEARNING USING MOBILE PHONES WHICH KEPT THE STUDENTS ENGAGED WITH THE EXPERIMENT (N=32)

The quality	%
Independent learning	15.63
Collaborative learning	50
Learning in a societal and humanistic environment	18.75
Learning mathematics through exploration	15.63
Learning mathematics outdoors	75
Learning mathematics visually or/and dynamically	50
Engaging the mathematics student in various mathematical actions	12.5
Learning mathematics easily and in less time	43.75

IV. DISCUSSION

We notice from the findings that most of the students perceived specific characteristics of the environment of the mathematics learning using mobile phones that motivated them to participate in the experiment. The most common perceived characteristics were: Planned outdoors learning of mathematics, planned use of mobile phones in mathematics learning, the characteristics and explanation of the teacher. These special characteristics attracted the students' attention because they were not used to them in the traditional class. These characteristics made the students consider the experiment novel, untraditional and even strange. Consequently, the students got enthusiastic towards the experiment and took the decision to participate in it. Figure 4 shows the relational map of these characteristics.

We see from figure 4 that the basic properties of the experiment which influenced the students' decision to join it are: (1) Outdoors learning of mathematics (2) Use of mobile phones in mathematics learning (3) The characteristics and explanation of the teacher. Moreover, the novelty of the experiment and the use of the mobile phone in mathematics learning were the most mentioned characteristics by the students as influencing their decision to join the experiment. The influence of the novelty of the technology on students' motivation to use it in the learning process is acknowledged in the literature (for example [16]). On the other hand, the use of the mobile phone is perceived through its playful nature when used as a multifunctional device. Ref. [16] pointed that novelty and playful nature of podcasting could be utilized to foster strong pedagogical principles. We can say the same about using mobile phones for learning mathematics. In fact, this is what happened in the experiment itself, where students discovered and described the benefits of exploring mathematics visually and dynamically, and emphasized the fun of doing so authentically and in the nature.

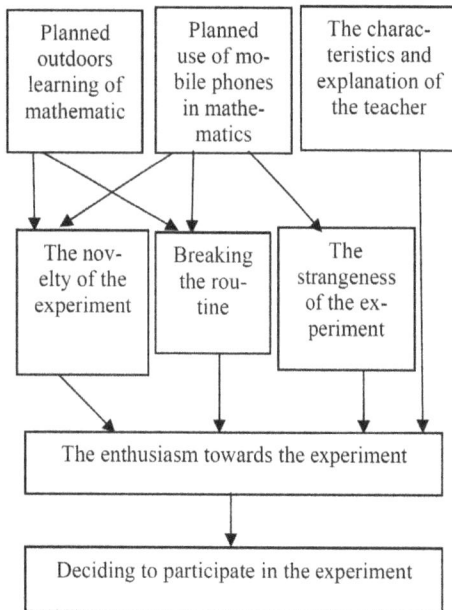

Figure 4. relational map of the factors that influenced the students' decision to participate in the experiment of mathematics learning using mobile phones.

In addition to the playful nature of the mobile phone which made the students enthusiastic to participate in the experiment, another feature of the mobile phone could have influenced the students' enthusiasm: the fact that mobile phones are prohibited in the schools, i.e. prohibited for learning.

Regarding the students' perception of mathematics learning using mobile phones as a result of carrying out the outdoor activities, the students perceived various aspects of their learning experience as presented in table 2. Below we categorize these various aspects into different categories:

- *The relation among the students*: is it a relation of equality? Is it a relation of collaboration? Is it a societal relation? Is it a humanistic relation? Etc.
- *The relation between the students and the teacher*: is it a relation of equality? Is it a relation of dependence? Is it a relation of collaboration? Is it a societal relation? Is it a humanistic relation? Etc.
- *The relation between the students and the learning material*: how does the student learn it? (Does the student read it from a book? Get it from the teacher? Explore it? Etc.) How does the student represent it? (Visually? Dynamically? Etc.) Where does the student learn it? (Indoors? Outdoors?) How is the learning process? (Easy? Simple? Difficult? Etc.) What actions does the student carry out to learn it?

A. The relation among the students:

The students liked that they collaborate to learn mathematics and that this collaboration is on equal terms. These equal terms were possible because the students performed different actions in different activities, sometimes different actions in the same activity. Probably this made them value the need of the different actions and that no one of them can be spared. These equal relations were of competitive nature too, as one of the students wrote: "We work in a group. We help each other and collaborate to carry

out the activity. We discuss the activity … every one listens to the other and respects him, even if he was not right. At the same time every one of us has his own mobile phone and uses it to find the mathematical relation. Sometimes we do not graph the right function, so we try again and again. This teaches us patience. Sometimes we get disappointed but we go on and on till we find the right relation and graph. Carrying the activity teaches us to be leaders too. We compete with each other to be the leader of our group; a leader who leads and helps his group". These different opportunities and capabilities, which the mobile phone provided, attracted the attention of the students and probably made them motivated to be actively engaged in the experiment and enthusiastic to explore mathematical relations. These opportunities can also give the students sense of belonging to the learning group [17]. This is actually what happened in this experiment, for example, the students reported their feeling of belonging to the group when carrying out the activities, and to the whole class when discussing the mathematical relations.

B. The relation between the students and the teacher:

The students were impressed from the fact that the relation between them and their teachers was a social and equal relation. One of the students said: "what impressed me is the collaboration between us and our teachers. We did not consider them as teachers because the environment was that of collaboration and fun". In our opinion, what influenced this relation to be more collaborative and of equal nature is that the mathematical activities were carried out outdoors, i.e. out of the formal order between the teacher and the students in the typical setting of the classroom. We can say that in the mobile phone outdoor setting, the teacher is a facilitator, because facilitating implies equal relations between the teacher and the students [18].

C. The relation between the students and the learning material:

The students valued the fact that they explored mathematics and did not learn it only from the teacher. They knew that they could do so, because they used the mobile phone midlets which enable representing mathematics visually and dynamically. This way of representation of mathematical relations helps the learner to feel the dependencies between the mathematical phenomenon parameters or relations [19]. This feeling also makes the arriving at mathematical relations easier and takes less time.

The students were also interested in where they learn the mathematical material, because: (1) they enjoy learning outdoors (2) outdoor learning answers children's natural curiosity and enthusiasm [20] (3) Learning outdoors was a novel experience for them. Furthermore, learning outdoors is learning with fun and learning about real life, about the nature and about oneself. Students are always enthusiastic to learn about themselves and their environment (see for example [21]). Learning with fun encouraged them to learn, bettered their performance and changed the view that some of them had of mathematics as a difficult science. Ref. [22] reported that playing games to learn mathematics provided a less threatening environment for the students than the typical mathematics classroom environment. The unthreatening environment is one reason for the change in how students look at mathematics. Other reasons for the change in the way students perceive mathematics are related to: (1) the way in which the students learn mathematics, which, in our case, was through exploration (2) the way in which the students represent mathematics, which, in our case, was visually and dynamically (3) the easiness with which the students learn mathematics, which was pointed out by the students in our case. The mobile phone environment provided the students with various opportunities and capabilities that enabled them to perform diversified mathematical actions, thus influencing the way they perceive mathematics.

V. CONCLUSIONS:

- The novelty of the experiment and the use of mobile phones in the mathematics learning were the main characteristics perceived by the students as influencing their decision to join the experiment.

- Regarding the students' perception of mathematics learning using mobile phones as a result of carrying out the outdoor activities, the students perceived the following aspects of their learning experience: The relation among the students, the relation between the students and the teacher, and the relation between the student and the learning material. Regarding each aspect the students perceived different characteristics of it.

- The students perceived various qualities of the mathematics learning that were enabled by the use of mobile phones: (1) exploring mathematics independently (2) learning mathematics through collaboration and team work; where the collaboration is on equal terms (3) learning mathematics in a societal and humanistic environment (4) learning mathematics in authentic real life situations (5) visualizing mathematics and investigating it dynamically (6) performing diversified mathematical actions using new and advanced technologies (7) learning mathematics easily and efficiently.

VI. RECOMMENDATIONS

After experiencing this novel use of mobile phones in mathematics learning, we strongly believe that a lot of opportunities and potentials are yet to be realized. We are still at the beginning of exploring this promising use in the educational environment. In spite of the disruption that these devices could cause in the classrooms, we believe that banning them from schools is not the solution. We should keep studying the pedagogy behind the use of mobile phones in the actual educational environment, and develop appropriate activities that utilize these devices efficiently and profitably in the learning process.

REFERENCES

[1] Chen J. & Kinshuk (2005). Mobile Technology in Educational Services. Journal of Educational Multimedia and Hypermedia, 14(1), 91-109.

[2] Rismark, M., Sølvberg, A. M., Strømme, A., and Hokstad, L. M. (2007). Using Mobile Phones to Prepare for University Lectures: Student's Experiences. The Turkish Online Journal of Educational Technology, 6(4). Retrieved August 11, 2008, from http://www.tojet.net/articles/649.htm.

[3] Cobcroft, R., Towers, S., Smith, J. and Bruns, A. (2006). Mobile learning in review: Opportunities and challenges for learners, teachers, and institutions. In Proceedings of Online Learning and Teaching (OLT) Conference 2006, 21-30, Queensland University of Technology, Brisbane. Retrieved May 09, 2008, from https://olt.qut.edu.au/udf/OLT2006/gen/static/papers/Cobcroft_OLT2006_paper.pdf.

[4] Botzer, G. & Yerushalmy, M. (2007). Mobile Applications for Mobile Learning. Proceedings for "Cognition & Exploratory Learning in Digital Age" (CELDA), Algrave, Portugal.

[5] Genossar, S.; Botzer, G. & Yerushalmy, M. (2008). Learning with Mobile Technology: A Case Study with Students in Mathematics Education. Proceedings of the CHAIS conference, Open University. http://telem-pub.openu.ac.il/users/chais/2008/evening/3_2.pdf

[6] Hartwell, C. L. Lightle, S. S. Maxwell, B. (2005). High School Students' Perceptions of Accounting. CPA JOURNAL, 75 (1), 62-67. http://www.nysscpa.org/cpajournal/2005/105/essentials/p62.htm

[7] O'Malley J. & McGraw, H. (1999). Students' perceptions of distance learning, online learning and the traditional classroom. OnLine Journal of Distance Learning Administration, 2 (1). http://www.westga.edu/~distance/omalley24.html.

[8] Eble, K. (1994). Craft of teaching: A guide to mastering the professor's art (2nd edition), New York: Jossey-Bass.

[9] Quitadamo, I.J. & Brown, A. (2001). Effective teaching styles and instructional design for online learning environments. National Educational Computing Conference, (Chicago, IL). http://confreg.uoregon.edu/NECC2001/program/research_pdf/Quitadamo.pdf.

[10] Duffy, T. M., & Cunningham, D. J. (1996). Constructivism: Implications for the design and delivery of instruction. In D. H. Jonassen (Ed.), Handbook of research for educational communications and technology. New York: Macmillan.

[11] Silander P, Sutinen E & Tarhio J (2004). Mobile Collaborative Concept Mapping - Combining Classroom Activity with Simultaneous Field Exploration. In the Proceedings of The 2nd IEEE International Workshop on Wireless and Mobile Technologies In Education, WMTE 2004. pp. 114-118.

[12] Ting, R.Y.-L. (2007). The Advanced Mobile Learning Practices: Learning Features and Implications. In Proceeding of Advanced Learning Technologies, ICALT 2007. Seventh IEEE International Conference, 718 – 720. http://ieeexplore.ieee.org/iel5/4280926/4280927/04281137.pdf?tp=&isnumber=&arnumber=4281137.

[13] Wetherell, M. & Potter, J. (1988). Discourse analysis and the identification of interpretive repertoires. In Charles Antaki (Ed.), Analysing everyday experience: A casebook of methods (pp. 168-183). London: Sage.

[14] Strauss, A., & Corbin, J., 1998, Basics of qualitative research. Thousands Oaks, CA: Sage Publications.

[15] Talja, S. (1999). Analyzing Qualitative Interview Data: The Discourse Analytic Method. *Library & information science research*, 21 (4), 459-477 http://www.info.uta.fi/talja/LISR%5B1%5D.pdf. (doi:10.1016/S0740-8188(99)00024-9)

[16] Fose, L. & Mehl, M. (2007). Plugging into Students' Digital DNA: Five Myths Prohibiting Proper Podcasting Pedagogy in the New Classroom Domain. *MERLOT Journal of Online Learning and Teaching*, 3 (3) http://jolt.merlot.org/vol3no3/mehl.htm.

[17] Hutchinson L., (2003). ABC of learning and teaching in medicine: Educational environment. *BMJ*, 326, 810-812. http://www.bmj.com/cgi/content/full/326/7393/810. (doi:10.1136/bmj.326.7393.810)

[18] Arnold, R., B.Burke, C.James, D.Martin and B.Thomas. (1991) *Educating for Change*. Toronto: Between the Lines and the Doris Marshall Institute for Education and Action.

[19] Pesonen M. E., Haapasalo L. & Ehmke, T. (2006). Critical look at dynamic sketches when learning mathematics. *The Teaching of Mathematics*, 9 (2), 19-29. http://elib.mi.sanu.ac.yu/files/journals/tm/17/tm922.pdf.

[20] Cleaver, S. (2007). Classrooms are going green: How Green Classrooms are Reconnecting Kids with Nature. *Instructor*. http://www2.scholastic.com/browse/article.jsp?id=3748233.

[21] Spencer, N. (2005). Enticing Queensland Teachers and Students with Real Data. http://www.stat.auckland.ac.nz/~iase/publications/13/Spencer.pdf.

[22] Bragg, L. (2007). Students' Conflicting Attitudes Towards Games as a Vehicle for Learning Mathematics: A Methodological Dilemma. *Mathematics Education Research Journal*, 19 (1), 29-44.

AUTHORS

Nimer Baya'a is a supervisor of computers, ministry of education, Israel, and a senior lecturer in the computer department in Al-Qasemi, academic college of education, in Baka, Israel (e-mail: bayaan@qsm.ac.il).

Wajeeh Daher is the head of the mathematics department and a senior lecturer in Al-Qasemi academic college of education, Baka, Israel (e-mail: wdaher@macam.ac.il).

6

A Framework for Caching Relevant Data Items for Checking Integrity Constraints of Mobile Database

Zarina Dzolkhifli[1], Hamidah Ibrahim2 and Lilly Suriani Affendey[3], Praveen Madiraju[4]

[1,2,3] Universiti Putra Malaysia, Serdang, Selangor, Malaysia
[4] Marquette University, Milwaukee, WI, USA

Abstract—In a mobile environment, due to the various constraints inherited from limitations of wireless communication and mobile devices, checking for integrity constraints to maintain the consistent state of mobile databases is an important issue that needs to be addressed. Hence, in this paper we propose a framework for caching relevant data items needed during the process of checking integrity constraints of mobile databases. This is achieved by analyzing the relationships among the integrity tests (simplified form of integrity constraints) to be evaluated for a given update operation. This improves the checking mechanism by preventing delays during the process of checking constraints and performing the update. Hence, our model speeds up the checking process.

Index Terms—Mobile Database, Integrity Constraints, Integrity Tests, Data Caching.

I. Introduction

Recently, there has been an increasing interest in mobile computing due to the rapid advances in wireless communication and portable computing technologies. Massive research efforts from academia and industry have been put forth to support a new class of mobile applications such as just-in-time stock trading, mobile health services, mobile commerce, and mobile games as well as migrating the normal conventional applications to mobile applications. Users of these applications can access information at any place at any time via mobile computers and devices such as mobile phone, palmtops, laptops, and PDA [10].

While technology has been rapidly advancing, various constraints inherited from limitations of wireless communication and mobile devices remain primary challenges in the design and implementation of mobile systems and applications. These constraints include: limited client capability, limited bandwidth, weak connectivity, and user mobility. In addition, disconnections occur frequently, which may be intentional (e.g., to save battery power) or unintentional (e.g., due to signal interference). These constraints make the wireless and mobile computing environments uniquely different from a conventional wired server/client environment [10].

A general architecture of a mobile database environment is shown in Figure 1 [3, 10]. The architecture consists of base stations (BS) and mobile hosts (MH). The base station is a stationary component in the model and is

responsible for a small geographic area called a cell. They are connected to each other through fixed networks. The mobile host is the mobile component of the model and may move from one cell to another. These mobile hosts communicate with the base stations through wireless networks.

Due to limited storage capabilities, a mobile host is not capable of storing all data items in the network, thus it must share some data item with a database in the fixed network. Data caching technique is used to cache some or most frequently accessed data from the base station into mobile host. By caching the needed data items, it allows mobile host to continue processing without worrying about disconnection.

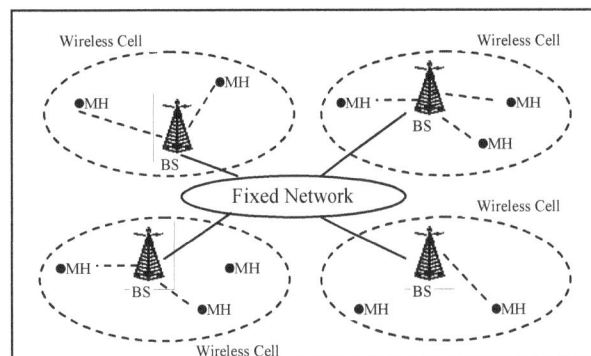

Figure 1. The architecture of a mobile database environment

Another important issue in databases is consistency, which must be maintained whenever an update operation (insert, delete, or modify) or transaction (sequence of updates) occurs at the mobile host. A database state is said to be consistent if the database satisfies a set of statements, called integrity constraints, which specify those configurations of the data that are considered semantically correct. The process of ensuring that the integrity constraints are satisfied by the database after it has been updated is termed constraint checking, which generally involves the execution of integrity tests (query that returns the value true or false). In a mobile environment, checking the integrity constraints to ensure the correctness of the database spans at least the mobile host and one other database (node), and thus the update is no longer local but rather distributed [14]. As mentioned in [14], the major problem in the mobile environment are the unbounded and unpredictable delays can affect not only the update but other updates running at both the mobile and the base stations, which is clearly not acceptable for most applications. With the same intuition as [14], we address the challenge of extending the data consistency maintenance to cover disconnected and mobile operations.

In this paper, a framework is proposed where checking the consistency of mobile databases is performed at the mobile host. This framework is suitable for both intentional (planned) and unintentional (unplanned) disconnection. This framework differs from the approach proposed in [14] since it is intended to cater for the important and frequently used integrity constraints, i.e. those that are used in database application. Mazumdar's approach [14] is restricted to set-based constraints (equality and inequality constraints). In our work, in order not to delay the process of checking constraints during disconnection, a similar concept as proposed in distributed databases [8, 9] is employed, namely localizing integrity checking by adopting sufficient and complete tests. Since sufficient test can only verify if a constraint is satisfied, we propose that the data items required during the checking to be cached at the mobile host during the relocation period. Our approach not only treats the issue of disconnection but also reduces the amount of data items to be cached by analyzing the relationships of the integrity tests to be evaluated. Hence, we achieve speed up in the constraint checking process.

The rest of the paper is organized as follows. In Section II, the previous works related to this research are presented. In Section III, the basic definitions, notations and examples, which are used in the rest of the paper, are set out. Section IV describes the proposed framework, while conclusions are presented in the final section

II. Related Work

Much of the research concerning integrity constraint checking has been conducted in the area of relational database systems. A comprehensive survey on the issues of constraint checking and maintaining in centralized, distributed and parallel databases is provided in [7]. A naïve approach is to perform the update and then check whether the integrity constraints are satisfied in the new database state. This method, termed brute force checking, is very expensive, impractical and can lead to prohibitive processing costs because the evaluation of integrity constraints requires large amounts of data, which are not involved in the database update transition. Hence, improvements to this approach have been reported in many research papers. Many approaches have been proposed for constructing efficient integrity tests, for a given integrity constraint and its relevant update operation, but these approaches are mostly designed for a centralized environment [13, 16, 17]. As centralized environment has only a single site, the approaches concentrate on improving the checking mechanism by minimizing the amount of data to be accessed during the checking process. Hence, these methods are not suitable for mobile environment as the checking process often spans multiple nodes and involves the transfer of data across the network.

Several studies [1, 5, 8, 9, 11] have been conducted to improve the checking mechanism by reducing the amount of data transferred across the network in distributed databases. Nonetheless, they are not suitable for mobile databases. These approaches reformulate the global constraints into local constraints (local tests) with an implicit assumption that all sites are available, which is not true in mobile environment, where a mobile unit may be disconnected for long periods. Even though failure is considered in the distributed environment, none of the approach cater failure at the node where the update is being executed, i.e. disconnection at the target site. Nevertheless, the localization concept proposed in distributed databases is used in our approach.

Other approaches such as [6, 15] focus on the problems of checking integrity constraints in parallel databases. These approaches are not suitable for mobile databases as the intention of their approach is to speed up the checking process by performing the checking concurrently at several nodes.

To the best of our knowledge, PRO-MOTION [14] is the only work that addresses the issues of checking integrity constraints in mobile databases. The difference between our work and the work in [14] has been highlighted in the previous section.

On the other hand, to meet the characteristics of mobile devices (hosts) especially disconnection and limited storage capabilities, many previous works such as [2, 12, 18, 19, 20] have focused on strategies to cache data items into mobile host. These strategies attempt not to delay the mobile operations even during disconnection. However, these works did not focus on strategy to cache relevant data items for the purpose of checking integrity constraints at the mobile host.

III. Preliminaries

Database integrity constraints are expressed in prenex conjunctive normal form with the range restricted property. A conjunct (literal) is an atomic formula of the form $R(u_1, u_2, \ldots, u_k)$ where R is a k-ary relation name and each u_i is either a variable or a constant. A positive atomic formula (positive literal) is denoted by $R(u_1, u_2, \ldots, u_k)$ whilst a negative atomic formula (negative literal) is prefixed by $\neg$. An (in)equality is a conjunct of the form $u_1 \theta u_2$ (prefixed with $\neg$ for inequality) where both u_1 and u_2 can be constants or variables and $\theta \in \{<, \leq, >, \geq, \neq, =\}$.

Integrity tests can be classified into several categories depending on the characteristics of the tests. Three different types of integrity test based on its properties were defined by McCarroll [15], namely: *sufficient tests*, *necessary tests*, and *complete tests*. An integrity test has the sufficiency property if when the test is satisfied, the associated constraint is satisfied and thus the update operation is safe with respect to the constraint. An integrity test has the necessity property if when the test is not satisfied, the associated constraint is violated and thus the update operation is unsafe with respect to the constraint. An integrity test has the completeness property if the test has both the sufficiency and the necessity properties.

Throughout this paper, the following symbols and their intended meaning, which are related to integrity constraints, are used:

- $I^v = \{I_1, I_2, \ldots, I_M\}$, the set of integrity constraints of an application in the whole mobile system.
- $I^{Bi} = \{I^{Bi}_1, I^{Bi}_2, \ldots, I^{Bi}_N\}$, the set of integrity constraints at the base station, i.
- $I^{Mh} = \{I^{Mh}_1, I^{Mh}_2, \ldots, I^{Mh}_O\}$, the set of integrity constraints at the mobile host, h.

From the above, $(\cup^P_{i=1} I^{Bi}) \cup (\cup^Q_{h=1} I^{Mh}) = I^\upsilon$, where P and Q are the number of base stations and mobile hosts, respectively in the mobile system.

Similarly, the following are the symbols and their intended meaning that are related to the data items in the mobile system. Here, data item refers to relation or fragment of relation that appears in the specification of an update operation.

- $R^\upsilon = \{R_1, R_2, ..., R_S\}$, the set of relations or fragments of relations in the mobile system.
- $R^{Bi} = \{R^{Bi}_1, R^{Bi}_2, ..., R^{Bi}_T\}$, the set of relations or fragments of relations at the base station, i.
- $R^{Mh} = \{R^{Mh}_1, R^{Mh}_2, ..., R^{Mh}_U\}$, the set of relations or fragments of relations at the mobile host, h.

From the above, $(\cup^P_{i=1} R^{Bi}) \cup (\cup^Q_{h=1} R^{Mh}) = R^\upsilon$, where P and Q are the number of base stations and mobile hosts, respectively in the mobile system. Also, we assume that for each data item, $R^{Mh}_v \in R^{Mh}$, the same data item appears in one of the base station, i.e. $R^{Mh}_v \in (\cup^P_{i=1} R^{Bi})$ [4].

Update operation in a mobile environment can occur at two different levels:

- $U^{Bi}(R)$, an update operation over the relation R, submitted by a user at the base station, i. This type of update operation is similar to the update operation in distributed databases and thus is not considered in this work. Note that R can also be a fragment of relation.
- $U^{Mh}(R)$, an update operation over the relation R, submitted by a user through his mobile host, h, where R is located at the mobile host. Note that R can also be a fragment of relation.

Throughout this paper the *company* database is used, as given in Figure 2. Table 1 presents some of the integrity tests generated based on the set of integrity constraints given in Figure 2. The derivation of the integrity tests is omitted here since this is not the focus of this paper. Interested readers may refer to [8, 9].

TABLE I.
THE INTEGRITY TESTS DERIVED BASED ON THE INTEGRITY CONSTRAINTS
LISTED IN FIGURE 2

I^υ	Update Template	Integrity Test
I_1	insert(emp(a, b, c, d))	1. $d > 0$[1]
I_2	insert(emp(a, b, c, d))	2. $(\forall x2 \forall y2 \forall z2)(\neg emp(a, x2, y2, z2)$ $\vee [(b = x2) \wedge (c = y2) \wedge (d = z2)])$ [1]
I_3	insert(dept(a, b, c, d))	3. $(\forall x2 \forall y2 \forall z2)(\neg dept(a, x2, y2, z2) \vee [(b = x2) \wedge (c = y2) \wedge (d = z2)])$ [1]
I_4	insert(emp(a, b, c, d))	4. $(\exists x \exists y \exists z)(dept(b, x, y, z))$ [1]
		5. $(\exists t \exists v \exists w)(emp(t, b, v, w))$ [2]
	delete(dept(a, b, c, d))	6. $(\forall t \forall v \forall w)(\neg emp(t, a, v, w))$ [1]
I_5	insert(proj(a, b, c))	7. $(\exists x \exists y \exists z)(emp(a, x, y, z))$ [1]
		8. $(\exists v \exists w)(proj(a, v, w))$ [2]
	delete(emp(a, b, c, d))	9. $(\forall v \forall w)(\neg proj(a, v, w))$ [1]
I_6	insert(proj(a, b, c))	10. $(\exists x \exists y \exists z)(dept(b, x, y, z))$ [1]
		11. $(\exists u \exists w)(proj(u, b, w))$ [2]
	delete(dept(a, b, c, d))	12. $(\forall u \forall w)(\neg proj(u, a, w))$ [1]
I_7	insert(dept(a, b, c, d))	13. $(a \neq 'D1') \vee (d > 4000)$ [1]
I_8	insert(emp(a, b, c, d))	14. $(\forall x \forall y \forall z)(\neg dept(b, x, y, z) \vee (d \leq z))$ [1]
		15. $(\exists t \exists v \exists w)(emp(t, b, v, w) \wedge (w \geq d))$ [2]
I_9	insert(proj(a, b, P1))	16. $(\exists z)(proj(z, b, P2))$ [1]
		17. $(\exists z)(proj(z, b, P1))$ [2]
	delete(proj(a, b, P2))	18. $(\forall x)(\neg proj(x, b, P1))$ [1]
		19. $(\exists z)(proj(z, b, P2) \wedge (z \neq a))$ [2]

Note: a, b, c and d are generic constants; [1]: complete test; and [2]: sufficient test.

IV. THE PROPOSED FRAMEWORK

The proposed framework is illustrated in Figure 3. The framework consists of 4 main components. These components are:

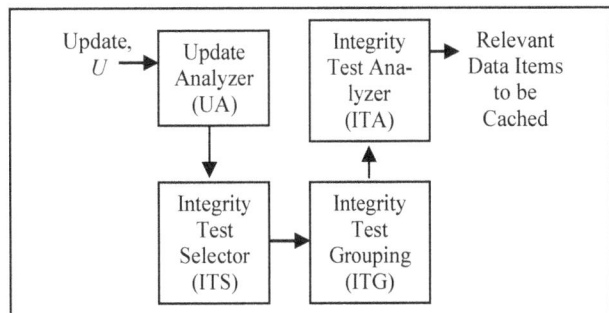

Figure 3. The proposed framework

(a) Update Analyzer (UA): This component accepts an update operation submitted by a user, $U^{Mh}(R)$, and analy-

Schema:
emp(eno, dno, ejob, esal);
dept(dno, dname, mgrno, mgrsal);
proj(eno, dno, pno)

Integrity Constraints:
'A specification of valid salary'
$I_1: (\forall w \forall x \forall y \forall z)(emp(w, x, y, z) \rightarrow (z > 0))$
'Every employee has a unique *eno*'
$I_2: (\forall w \forall x1 \forall x2 \forall y1 \forall y2 \forall z1 \forall z2)(emp(w, x1, y1, z1) \wedge emp(w, x2, y2, z2) \rightarrow (x1 = x2) \wedge (y1 = y2) \wedge (z1 = z2))$
'Every department has a unique *dno*'
$I_3: (\forall w \forall x1 \forall x2 \forall y1 \forall y2 \forall z1 \forall z2)(dept(w, x1, y1, z1) \wedge dept(w, x2, y2, z2) \rightarrow (x1 = x2) \wedge (y1 = y2) \wedge (z1 = z2))$
'The *dno* of every tuple in the *emp* relation exists in the *dept* relation'
$I_4: (\forall t \forall u \forall v \forall w \exists x \exists y \exists z)(emp(t, u, v, w) \rightarrow dept(u, x, y, z))$
'The *eno* of every tuple in the *proj* relation exists in the *emp* relation'
$I_5: (\forall u \forall v \forall w \exists x \exists y \exists z)(proj(u, v, w) \rightarrow emp(u, x, y, z))$
'The *dno* of every tuple in the *proj* relation exists in the *dept* relation'
$I_6: (\forall u \forall v \forall w \exists x \exists y \exists z)(proj(u, v, w) \rightarrow dept(v, x, y, z))$
'Every manager in *dept* 'D1' earns > £4000'
$I_7: (\forall w \forall x \forall y \forall z)(dept(w, x, y, z) \wedge (w = 'D1') \rightarrow (z > 4000))$
'Every employee must earn ≤ to the manager in the same department'
$I_8: (\forall t \forall u \forall v \forall w \forall x \forall y \forall z)(emp(t, u, v, w) \wedge dept(u, x, y, z) \rightarrow (w \leq z))$
'Any department that is working on a project P_1 is also working on project P_2'
$I_9: (\forall x \forall y \exists z)(proj(x, y, P_1) \rightarrow proj(z, y, P_2))$

Figure 2. The Company static integrity constraints

ses the operation to identify the type of update operation (insert, delete, modify), the relation involved, and the set of data values to be inserted/deleted/modified.

(b) Integrity Test Selector (ITS): When a user requests an update, only those constraints that might be violated are selected for evaluation. Based on these constraints, the appropriate tests are selected. Thus, this component selects the integrity tests to be triggered, by comparing the type of update operation and the relation of the user's update operation with the type of update operation and relation of each of the update template stored in the mobile host. For example, if $I^{Mh} = \{I_1, I_2, I_4, I_5, I_8\}$ and $U^{Mh}(R) = $ insert(emp(E20, D1, Analysts, 3400)), then tests 1, 2, 4, 5, 14, and 15 are selected.

(c) Integrity Test Grouping (ITG): This component groups the integrity tests that have been selected by ITS. There are several criteria that can be used for this purpose. For example grouping can be based on the relation specified in the tests, i.e. those tests that will be evaluated over the same relation are grouped together in the same group. Grouping can also be based on the type of tests, i.e. those tests that have the same properties (complete or sufficient) are assigned to the same group. Other criterion that can be used is region, i.e. tests that can be evaluated locally are grouped in the same group. Note also that it is seldom the case that we can select a test from each of the integrity constraint that satisfies the characteristics of the group. Thus, a group may have some tests whose characteristics do not belong to the group but are forced to be the elements of the group, since they are the only tests available for a given integrity constraint. After grouping, this component is also responsible to select one of the group to be evaluated. For the example given in (b) above, the following are some possible groups:

Based on relation (Note, Gi is a label for a group):
    $G1$: {1, 2, 5, 15}
All tests span the *emp* relation except test 1.
    $G2$: {1, 2, 4, 14}
All tests span the *dept* relation except tests 1 and 2. Tests 1 and 2 are selected and grouped in $G2$ as they are the only tests available for I_1 and I_2, respectively.

Based on properties of the tests:
    $G3$: {1, 2, 4, 14}
All tests are complete tests.
    $G4$: {1, 2, 5, 15}
All tests are sufficient tests except tests 1 and 2. Tests 1 and 2 are selected and grouped in $G4$ as they are the only tests available for I_1 and I_2, respectively.

Based on region: Assume that only part of the *emp* relation is located at the mobile host.
    $G5$: {1, 2, 5, 15}
Test 1 is a local test, while tests 2, 5, and 15 have high chances to be evaluated locally.
    $G6$: {1, 2, 4, 14}
Test 1 is a local test, test 2 has high chances to be evaluated locally, while tests 4 and 14 are global tests.

Finally, one of these groups is selected to be evaluated. Decision to select is based on the data items already located at the mobile host. If the tests of the group have more chances to be performed locally at the mobile host, then that group is selected.

(d) Integrity Test Analyzer (ITA): This is the core component of the whole framework that analyses the relationships among the tests that have been grouped and selected to be evaluated by ITG, with the aim to identify the relevant data items to be cached. Here, relevant is defined as the minimum number of data items that needs to be cached given a set of integrity tests to be evaluated. Analysis is performed by comparing the relations, constant values, and equations among the tests. Three main rules are applied as follows:

Rule 1: Test T_i is said to be *redundant* with test T_j if the data item(s) required by both T_i and T_j is the same, i.e. $D_i \cap D_j = D_i$ where D_i and D_j denote the set of data items needed by T_i and T_j, respectively.

Rule 2: Test T_i is said to be *subsumed* by test T_j if the data item(s) required by T_i is part of the data item(s) required by T_j, i.e. $D_i \subseteq D_j$.

Rule 3: Test T_i is said to be *contradicted* with test T_j if the data item(s) required by T_i is not the data item(s) required by T_j although the attribute(s) is the same, i.e. $D_i \cap D_j = \{\}$ and both D_i and D_j are over the same attribute.

The steps performed at this stage are as follows:
1. BEGIN
2. Substitute each test in the group Gi with the actual values as given in the update operation, $U^{Mh}(R)$.
3. Evaluate domain test (if any). If the test is false, then $U^{Mh}(R)$ is aborted. GO TO step 7.
4. For each of the remaining tests in the Gi, identify the data items required by the test.
5. Check for redundancy, subsumption, and contradiction by applying rules 1, 2, and 3.
6. Generate the required relevant data items needed to be cached from base station. For Rule 1, D_i is cached. While for Rule 2, D_j is cached and for Rule 3, both D_i and D_j are cached.
7. END

For example, assume that $G1$ has been selected.

Example 1:
Step 2:
1. $3400 > 0$
2. $(\forall x2 \forall y2 \forall z2)(\neg emp(E20, x2, y2, z2) \vee [(D1 = x2) \wedge (Analysts = y2) \wedge (3400 = z2)])$
5. $(\exists t \exists v \exists w)(emp(t, D1, v, w))$
15. $(\exists t \exists v \exists w)(emp(t, D1, v, w) \wedge (w \geq 3400))$

Step 3:
1. $3400 > 0$ is true.
2. $(\forall x2 \forall y2 \forall z2)(\neg emp(E20, x2, y2, z2) \vee [(D1 = x2) \wedge (Analysts = y2) \wedge (3400 = z2)])$
5. $(\exists t \exists v \exists w)(emp(t, D1, v, w))$
15. $(\exists t \exists v \exists w)(emp(t, D1, v, w) \wedge (w \geq 3400))$

Step 4:

Test	Relation	Attribute	Value
2	*emp*	*eno*	E20
		dno	D1
		ejob	*Analysts*
		esal	3400
5	*emp*	*dno*	D1
15	*emp*	*dno*	D1
		esal	≥ 3400

Step 5:

The data item needed by test 5 is part of the data items required by test 15 (Rule 2).

Step 6:

Thus, the data items to be cached (if and only if the data items are not at the mobile host) are as follows:

Relation	Attribute	Value
emp	*eno*	E20
emp	*dno*	D1
	esal	≥ 3400

As for a second example, consider $I^{Mh} = \{I_5, I_6, I_9\}$ and $U^{Mh}(R) = $ insert($proj(E20, D1, P1)$), then tests 7, 8, 10, 11, 16, and 17 are selected. Assume that the following group has been selected by ITG, $G7 = \{7, 10, 16\}$ (complete tests).

Example 2:

Steps 2 and 3:

7. $(\exists x \exists y \exists z)(emp(E20, x, y, z))$

10. $(\exists x \exists y \exists z)(dept(D1, x, y, z))$

16. $(\exists z)(proj(z, D1, P2))$

Step 4:

Test	Relation	Attribute	Value
7.	*emp*	*eno*	E20
10.	*dept*	*dno*	D1
16.	*proj*	*dno*	D1
		pno	P2

Step 5:

The data item needed by test 10 is part of the data items required by test 16 (Rule 2).

Step 6:

Relation	Attribute	Value
emp	*eno*	E20
proj	*dno*	D1
	pno	P2

We have performed a simple analysis that compares (a) caching the whole data item without analyzing the integrity tests, (b) caching the data items by analyzing the integrity tests individually (i.e. omitting Step 5), and (c) caching the data items by analyzing the relationships be-

tween the integrity tests. For this analysis, we assume the following, the *emp* relation has 500 tuples (2000 data items), *dept* has 10 tuples (40 data items), and *proj* has 100 tuples (300 data items). Note the number of data items is calculated by multiplying the number of tuples with the number of attributes of a relation. Figure 4 illustrates this comparison. From this figure, we can conclude that the number of data items to be cached can be significantly reduced by analyzing the relationships among the integrity tests. Another analysis has been conducted by increasing the number of tuples (records) in each relation. The results are as shown in Figure 5. From Figure 5 we noticed that increasing the number of tuples in each relation has no effect on the number of data items to be cached for both strategies (b) and (c).

The proposed framework improves the constraint checking mechanism mainly by employing an efficient checking strategy, which is achieved through:

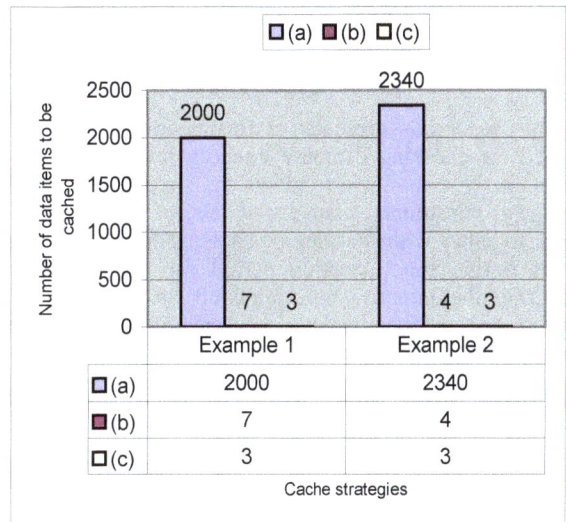

	Example 1	Example 2
(a)	2000	2340
(b)	7	4
(c)	3	3

Cache strategies

Figure 4: Comparison between strategies (a), (b), and (c), with respect to the number of data items to be cached

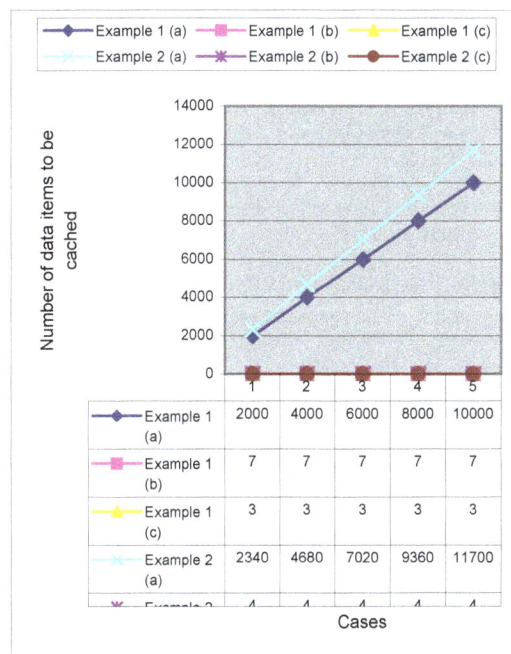

	1	2	3	4	5
Example 1 (a)	2000	4000	6000	8000	10000
Example 1 (b)	7	7	7	7	7
Example 1 (c)	3	3	3	3	3
Example 2 (a)	2340	4680	7020	9360	11700
Example 2	4	4	4	4	4

Cases

Figure5: Comparison between strategies (a), (b), and (c), when the number of tuples in each relation is increased

(i) Caching relevant data items – this has achieved two purposes, namely: (i) upgrading the properties of the tests – by caching the relevant data items it increases the possibility of performing the constraints checking locally at the mobile host, as most of the data required are now available at the mobile host and (ii) the process of checking the integrity constraints at the mobile host can be performed without delay even if the mobile host is disconnected.

(ii) Localizing integrity checking – allow the initial constraints to be validated by accessing data at the mobile host, i.e. at the site where the update is performed. This technique eliminates the cost of accessing remote data, i.e. it minimizes inter-site data communication cost. It also prevents delays during the process of checking constraints and performing the update, especially when the mobile host is disconnected.

(iii) Test filtering – for each update request, only those constraints that may be violated by it are selected for further evaluation.

V. CONCLUSION

This paper has presented a framework, which is designed for checking database integrity in a mobile environment. This framework adopts the simplified forms of integrity constraints, namely: sufficient and complete tests, together with the idea of caching the relevant data items during the relocation period for the purpose of checking the integrity constraints. It has improved the performance of the checking mechanism of mobile databases as delay during the process of checking the integrity constraints and performing the update is reduced.

REFERENCES

[1] Alwan, A.A., Ibrahim, H., and Udzir, N.I., "Local Integrity Checking using Local Information in a Distributed Database", *Proceedings of the 1ˢᵗ Aalborg University IEEE Student Paper Contest 2007 (AISPC'07)*, Aalborg, 2007.

[2] Chan, B.Y., Si, A., and Leong, H.V., "A Framework for Cache Management for Mobile Databases: Design and Evaluation", *Distributed and Parallel Databases*, Vol. 10, 2001, pp. 23-57. (doi:10.1023/A:1019297705159)

[3] Chan, D. and Roddick, J.F., "Context-sensitive Mobile Database Summarization", *Proceedings of the Twenty-Sixth Australian Computer Science Conference (ACSC 2003)*, Adelaide, 2003.

[4] EPFL, Grenoble, U., INRIA-Nancy, INT-Evry, Montpellier, U., Paris, U., and Versailles, U., "Mobile Databases: a Selection of Open Issues and Research Directions", *SIGMOD Record*, Vol. 33, No. 2, 2004, pp. 78-83.

[5] Gupta, A., "Partial Information Based Integrity Constraint Checking", PhD Thesis, Stanford University, USA, 1994.

[6] Hanandeh, F.A.H., "Integrity Constraints Maintenance for Parallel Databases", PhD Thesis, UPM, Malaysia, 2006.

[7] Ibrahim, H., "Checking Integrity Constraints – How it Differs in Centralized, Distributed and Parallel Databases", *Proceedings of the 17ᵗʰ International Conference on Database and Expert Systems Applications – the Second International Workshop on Logical Aspects and Applications of Integrity Constraints (LAAIC'06)*, Krakow, 2006, pp. 563-568.

[8] Ibrahim, H., "A Strategy for Semantic Integrity Checking in Distributed Databases", *Proceedings of the Ninth International Conference on Parallel and Distributed Systems*, IEEE Computer Society, Republic of China, 2002.

[9] Ibrahim, H., Gray, W.A., and Fiddian, N.J., "Optimizing Fragment Constraints – A Performance Evaluation", *International Journal of Intelligent Systems – Verification and Validation Issues in Databases, Knowledge-Based Systems, and Ontologies*, Edited by: Ronald, R., John Wiley & Sons Inc., Vol. 16, No. 3, 2001, pp. 285-306.

[10] Ken, C.K.L, Wang-Chien, L., and Sanjay, M., "Pervasive Data Access in Wireless and Mobile Computing Environments", *Journal of Wireless Communications and Mobile Computing*, 2006.

[11] Madiraju, P. and Sunderraman, R., "A Mobile Agent Approach for Global Database Constraint Checking", *Proceedings of the ACM Symposium on Applied Computing (SAC'04)*, Nicosia, 2004, pp. 679-683.

[12] Madria, S.K., Mohania, M., Bhowmick, S.S., and Bhargava, B., "Mobile Data and Transaction Management, Information Sciences", *Elsevier*, 2002, pp. 279-309.

[13] Martinenghi, D., "Advanced Techniques for Efficient Data Integrity Checking", *PhD Thesis*, Roskilde University, 2005.

[14] Mazumdar. S. and Chrysanthis, P.K., "Localization of Integrity Constraints in Mobile Databases and Specification in PROMOTION", *Proceedings of the Mobile Networks and Applications*, 2004, pp. 481-490.

[15] McCarroll, N.F., "Semantic Integrity Enforcement in Parallel Database Machines", *PhD Thesis*, University of Sheffield, UK, 1995.

[16] McCune, W.W. and Henschen, L.J., "Maintaining State Constraints in Relational Databases: a Proof Theoretic Basis", *Journal of the Association for Computing Machinery*, Vol. 36, No. 1, 1989, pp. 46-68.

[17] Nicolas, J.M., "Logic for Improving Integrity Checking in Relational Data Bases", *Acta Informatica*, Vol. 18, No. 3, 1982, pp. 227-253. (doi:10.1007/BF00263192)

[18] Pitoura, E. and Chrysanthis, P.K., "Caching and Replication in Mobile Data Management", *IEEE Data Engineering*, Bull, 2007, pp. 13-20.

[19] Ren, Q., Dunham, H.M., and Kumar, V., "Semantic Caching and Query Processing", *IEEE Transaction on Knowledge and Data Engineering*, Vol. 15, 2003, pp. 192-210. (doi:10.1109/TKDE.2003.1161590)

[20] Song, H. and Cao, G., "Cache-miss-initiated Prefetch in Mobile Environment", *Science Direct, Computer Communication*, Vol. 28, 2005, pp. 741-753.

AUTHORS

Zarina Dzolkhifli, is a postgraduate student of Department of Computer Science, Faculty of Computer Science and Information Technology, Universiti Putra Malaysia, 43400 Serdang, Selangor, Malaysia. (e-mail: zarinadzolkhifli@yahoo.com.sg).

Hamidah Ibrahim, is an associate professor of the Department of Computer Science, Faculty of Computer Science and Information Technology, Universiti Putra Malaysia, 43400 Serdang, Selangor, Malaysia. (e-mail: hamidah@upm.edu.my).

Lilly Suriani Affendey, is a senior lecturer of the Department of Computer Science, Faculty of Computer Science and Information Technology, Universiti Putra Malaysia, 43400 Serdang, Selangor, Malaysia. (e-mail: suriani@fsktm.upm.edu.my).

Praveen Madiraju, is an assistant professor of the Department of Mathematics, Statistics and Computer Science, Marquette University, Milwaukee WI 53201-1881 USA. (e-mail: praveen@mscs.mu.edu).

7

Investigating Mobile Devices Integration in Higher Education in Cyprus: Faculty Perspectives

N. Eteokleous1, and D. Ktoridou[2]

[1] Frederick University Cyprus, School of Education, Nicosia, Cyprus
[2] University of Nicosia,School of Business, Nicosia, Cyprus.

Abstract—**Mobile devices are everywhere and mobile learning has emerged as a potential educational environment; however it is relatively new to Cyprus educational system. The purpose of this research work is to assess and determine the readiness; and evaluate the viability of integrating mobile technology in Cyprus higher education level. To address the above, a mixed method approach is employed making use of quantitative and qualitative data from faculty members** working in three private universities in Cyprus. **Faculty reactions were mixed with some of them seeing the benefits for mobile learning while others have doubts. The results summarize the technological and pedagogical aspects to be considered prior integrating mobile devices. Additionally, the study supports that one of the major barriers to educators is the lack of understanding regarding mobile devices integration in the teaching and learning process. Finally, there is a need to develop well-defined and well-structured requirements for mobile integration in the classroom.**

Index Terms—**mobile devices, higher education, faculty perspectives, technological & pedagogical aspects**

I. INTRODUCTION AND THEORETICAL FRAMEWORK

The walls of the classrooms are torn down. Computer technology evolution has widened the educational activities for instructors and students in the 90's, removing time and space constraints from instructors as well as from students themselves. With the rapid diffusion of the Internet, computers, and telecommunications; new approaches to learning were created [4]; [12]; [27]. On-line courses appeared as a new method of course delivery. Since then, the interest in the development and use of distance learning in higher education has been steadily increasing [14]. The demands of e-learning on one hand, in connection with the possibilities offered by modern technology (i.e. evolution of mobile devices) on the other hand, pose new opportunities and new challenges to the educational systems [30]; [47].

More specifically, during the past decade every area of education and training has been affected by the introduction and use of technological advancements. E-learning (electronic learning) originated from D-learning (distance learning) and now, the follow-up is M-learning (mobile learning). As latest models of mobile phones combine PDA functions with cameras, video and MP3 players, m-learning becomes more convenient and exciting. The process of adding mobility to interactivity transformed the role of the Internet and has set new beginnings to innovations, services and applications. This new learning environment supports collaborative and accessible learning experiences for both instructors and learners that are integrated anytime and anywhere beyond the classroom [30]; [47]

A. M –Learning

M -learning can be defined as e-learning using mobile devices and handheld IT devices, such as PDAs or Personal Digital Assistants (e.g. Palm, Pocket PC), mobile phones, laptops, tablet PC technologies and smart phones (e.g. Blackberry, iPhone) with wireless networks; digital media players (e.g. iPods, MP3 players), intelligent active badges and portable game devices [3]; [20]; [27]; [42]. Some view e-learning as the immediate ancestor of m-learning. E-learning is defined as learning supported by digital electronic tools and media, and, by analogy, m-learning is defined as e-learning that uses wireless transmission and mobile devices such as PDAs, mobile phones, laptops and tablet PCs [32]; [41]; [47]. Along the same lines, the eLearning Guild Report supports that "mobile learning is a subset of e-Learning, yet it has its own distinctive, expensive, and restrictive qualities" [33, p. 1]. In agreement, Evans mentions that "m-learning inherits the advantage of e-learning, but extends their reach making use of portable wireless technologies" [20, p.492]. In Quinn's study it is defined simply as learning that takes place with the help of mobile devices [38]. Turunen, Syvaenen and Ahonen view mobile devices as a pervasive medium that may assist us in combining work, studying and leisure time in meaningful ways [45]. Polsani considers these definitions 'restrictive' and proposes instead the term 'network learning' (or 'n-learning'). He defines mobile learning as a form of education whose site of production, circulation, and consumption is the network [37].

Mobile technology integration has already started. Mobile learning (m-learning) has emerged as a potential educational environment to support learning as well as to increase the quality of learning with proper use [27]. Mobile tools immerse in the learning context and surround the educators, the students and the environment they operate. Some are purely assistive and supportive in nature and others are becoming increasingly intelligent. These tools can be directly integrated in classroom activities in order to enhance and promote new ways of teaching and learning. Mobile learning allows instructional designers and instructors to utilize the strengths of mobile platforms to bring a variety of new applications to the learning environment [3]; [27]; [29]; [35]; [48].

B. Mobile Devices Integration

Mobile devices are everywhere, and various innovative practices and experiments took place in all levels of education from elementary to higher education (formal education) as well as to informal and non-formal educational settings [3]; [29]. Research studies appear to report positive and encouraging results [2]; [3]; [15]; [16];[20];[22];[27];[30];[35];[42]; [44]; [47]. Examples of mobile devices integration can be found in various countries across the globe, mainly in colleges and universities in Europe, and the USA [27];[33];[47]. As the eLearning Guild Report supports "colleges and universities have taken the lead in mobile learning adoption" and "have been among the early adopters of mobile learning" [33, p. 5].

Mobile wireless computers, mobile wireless phones and PDAs are the ones mostly used in higher education [27];[47]. Based on various research studies, we suggest that mobile use can be categorized based on the following parameters: 1) educational level (e.g. primary, secondary or higher level); 2) educational setting (e.g. informal, formal); 3) purpose of use (e.g. learning and teaching purposes, organization and administration purposes, research purposes); 4) person using the device (e.g. the educator/instructor, the student; 5) location of use (e.g. in classroom, at the library, outdoor activities - arts and science centers, museum and field trips) [3];[20];[27];[29]; [35];[47]. Each mobile use has various characteristics, for example it could be used at the secondary level, by the teacher for administration purposes in classroom. Based on the category "person using the device", find below a summary of how students and educators use mobile devices for numerous purposes. First of all students, use mobile devices for notes taking, and homework completion; conducting research and class-work anywhere anytime; podcasting revision; collecting, organizing, and exchanging data (e.g. send and receive library data, exchange e-books through beaming).

Additionally, they use mobile devices as communication tools, graphing calculators, and mapping concepts creators [3];[20];[27];[29]; [35];[47]. Educators take advantage of mobile devices mainly for three purposes: teaching and learning; administration and organization; and research purposes. Regarding teaching and learning purposes, mobile devices are used for quizzes, in-class tests, and language instruction; to share syllabi, lectures' schedules, lesson materials, notes, images and photos; and provide web-based curriculums. As far as it concerns administration and organization purposes, mobile devices uses are summarized as follows: manage data such as classes schedule, students' grades and attendance' organize courses, lecture material, student assignments, and exams as well as access central school data. Finally, for research purposes mobile devices are used as performance and decision support tools, to gather and analyze data as well as to manage research results and information [3];[20];[27];[29]; [35];[47].

Several advantages and educational benefits were reported regarding mobile devices integration. The above can be summarized as mobility, easy of movement and use, flexibility, functionality, convenience, simplicity, speed, affordability, economic/ cost benefits, information management capacity, and share information instantly. Additionally, mobile devices use improves efficiency and effectiveness in teaching and learning, enables one to one learning, enhances private and self-constructed/ self-paced learning and independent work, improves communication and collaboration among faculty members and students, promotes inquiry-based investigation, improves interpersonal communications and social interactions, accommodates students' needs, and finally increases productivity, performance, and learners' access [3];[20];[27];[29];[33];[35];[47]. The major concerns or in other words drawbacks of m-learning mentioned in various studies are the following: security, limited bandwidth and capacity, small screen size, technical and design obstacles, as well as pedagogical concerns such as the complexity in how mobile devices support meaningful learning [3];[27].

Based on the above, the pictorial in Figure 1 represents and explains a learning environment where mobile devices are integrated, as well as the relationship among the learner, the instructor and the academic community. The learner and the instructor are considered to be the center of this m-learning environment within the academic community. All three parties are influenced by the following key issues:

- *Underlying principles for implementing mobile learning technologies*: the major issue for a successful implementation is to develop well-defined and well-structured requirements.
- *Effects of ubiquitous access to a WLAN change on insight and outside the classroom learning*: insight the classroom, mobile learning offers lecturers and learners increased flexibility and new opportunities for interaction. M-Learning allows a collaborative and accessible learning experience that integrates with the world beyond the classroom.
- *Best practices for using mobile learning*: developing and delivering mobile learning is a new practice, yet is a question how much of what people are doing constitutes best practice. The focus and attention must be on the development, distribution, and adoption of best-practices for mobile learning [33].
- *Provision of best end-user support for mobile learning*: The issue of end-user support is of major importance for m-learning integration. The lecturer as well as the student must be provided with the best support. The academic society must be well equipped to handle any questions and problems from an extensive peer-to-peer on campus support group.

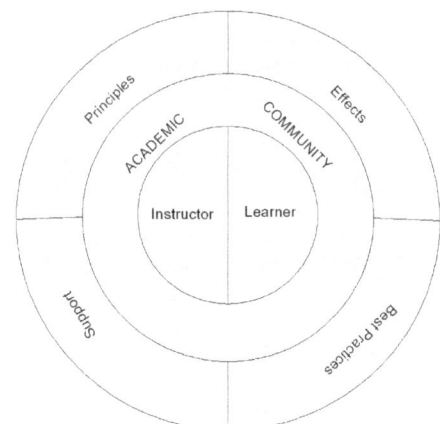

Figure 1.   Learning environment

C. Current Situation in Cyprus Higher Education

Cyprus is the third biggest island in the Mediterranean sea, situated in the crossroads of three continents: Europe, Asia and Africa [18];[46]. The context of this study is Cyprus higher education level, that can be characterized as a newly established field, since Cyprus did not have a public university until 1992 [18];[46] even though other higher education private and public institutions have been established and operating in the island for too many years [13]; [18]. More specifically, the development of Cyprus higher education officially started in the late 80s "where the establishment of the Department of Higher Education within the Ministry of Education in 1984 gave the momentum needed for the field to develop" [18, p. 3]. The Department of Higher and Tertiary Education of the Ministry of Education and Culture (MOEC) is responsible for Third-level education. More specifically, Cyprus educational system is centralized and the highest authority for educational policy making is the Council of Ministers [18];[25].

As of now higher education is provided in three different types of institutions, i.e. public and private, universities, public tertiary institutions, and private higher institutions (colleges). The private universities (previously were operating as private colleges) began their operation in September 2007. Overall, in Cyprus there are three public and three private universities, there are eight public higher institutions, and there are also twenty-three private third-level education schools, colleges and institutes [10];[18];[25];[36].

D. Technological Advancements and Technology Integration in Cyprus Educational System

Cyprus is regarded as a developing nation and not that technologically advanced nation [46]. However, in the past couple of years, various technological developments and expansions in the areas of telecommunications, broadcasting, third generation (3G) networking technology, broadband wireless networking, GSM network and satellite technology; provided the appropriate foundation and infrastructure for the enhancement and promotion of technology integration in education.

The use of technology in Cyprus educational system was limited during the years of 1960s to 1980s, "to the use of traditional audiovisual equipment and some government produced educational radio and television programming" [46; p. 125]. The official and at the same time advanced technology integration in Cyprus educational system took place in the early 1990s, mainly in primary education. The launch of an ICT policy by the Cypriot MOEC, gave the momentum to technology integration to flourish in education. Some primary schools were equipped with computers at an experimental level. Also, a Departmental IT group was created as a part of the Department for Programs Development of the MOEC, while the governmental Pedagogical Institute started offering at the end of the 1990s an optional training program for teachers. 'Evagoras' (1999) was the first formal ICT policy document, and describes the action plan for the embedding of new technologies in primary education from 2000 to 2005 [17]. Schools were connected to the Internet and equipped with computer labs; and classrooms were equipped with few computers. Regarding secondary education, there is no clear-cut policy to integrate technology as a tool in the classroom.

However, access to technology was given to all secondary schools by being connected to the Internet and equipped with computers labs. Additionally, information technology and computers classes were introduced in the secondary education curriculum. The infrastructure and equipment give the opportunity to the teachers to use computers in all areas of the curriculum [39]. As far as it concerns higher education, even though it is under the control of the MOEC (see above), the institutions have the flexibility to develop their own agenda regarding technology use. High speed Internet connection, network capabilities, high-end computer labs, provision of desktop computers and laptops to faculty members as well as in some cases to students (i.e. Frederick University Cyprus) highlight the situation in Higher education institutions.

Various empirical studies examined computer technology integration in Cyprus educational system in all levels; primary, secondary and higher education [1]; [17]; [23]; [26]; [46]. The same does not exist regarding mobile devices integration since attempts to integrate them mainly focuses on primary and secondary education. An example of the above is called *HandLearn* and it "aimed at investigating the use of handheld computers within the context of elementary school science" [3, p. 358]. Although, the global interest has been growing and proceed to various attempts in integrating mobile devices in higher educational settings [16];[20]; [27]; [35]; [47], there were no attempts to integrate mobile devices in Cyprus higher education.

More than 90% of public universities and 80% of private universities in the US are using at some level wireless technologies [42]. Given the above, one of the most important challenges that Cyprus higher education has to face is to effectively and efficiently integrate mobile devices in its practices in order to follow the global trend. As Kim and Holmes mention "the movement of mobile devices wireless technologies in education is a recent trend, and it is now becoming the hottest technology in higher education" [27, p.78].

Many of the studies that examined mobile devices integration in various educational systems and countries [2]; [3];[15];[16];[20];[22];[27];[29]; [35];[42]44];[47] appear to report positive and encouraging results. However, the successful stories and the innovations occurred at a localized level having minimal impact [21] and consequently they have not been scaling up. Why the above happens? Are we moving too fast towards integrating mobile devices into current educational setting and activities? [19] Has the readiness of the educational systems to accept and support mobile devices integration been addressed? Is faculty willing to integrate mobile devices in their classroom practice? Does faculty posses the necessary technological literacy, skills and abilities in order to integrate mobile devices in their educational practices? What organizational, educational and personal factors might influence mobile technology integration in classrooms? Harnessing mobile technologies and putting them in the true service of learning is not going to be an easy task. Through critical consideration and evaluation of experiences gained during the computer technology integration, online and e-learning, it is crucial to be proactive, make educated and conscious decisions, avoid mistakes of the past and take full advantage of the experiences, lessons learned and effective practice guidelines [33];[47].

Given the above, the literature lacks of academic research on assessing the needs and demands from technical and pedagogical perspectives before integrating mobile devices in the educational settings [30]. Even though mobile devices integration has been examined mainly in developed countries, the literature lacks on studies from developing countries, without much experience using mobile devices for higher education, which although have the appropriate equipment and infrastructure, such as Cyprus.

In order to be able to meet and follow the pressures of global competition in a knowledge-based, net-centric new economy, as well as apply the new trends and innovations in education, and experience these innovations in our system; Cyprus's educational leaders should increasingly consider mobile devices integration in their practices. This paper focuses on examining the readiness of higher education context in integrating mobile devices. Specifically, the perceptions of higher education lecturers as well as the challenges that should be taken into consideration by the higher educational institutions for the realization of this integration into their educational practices [7]; [43].

II. MAIN AIM AND RESEARCH OBJECTIVES

The purpose of this research work is to assess and determine the readiness; and evaluate the viability of integrating mobile technology into the teaching and learning processes in higher education in Cyprus. More specifically, it explores faculty perceptions regarding mobile technology integration and identifies their willingness to integrate mobile devices in their classroom practices. In addition, it attempts to identify the factors such as organizational, educational, and personal that influence mobile technology integration in higher education. Finally, the proposed study aims in defining and delineating the technological and pedagogical aspects to be taken into consideration for successful mobile technology integration. Examining the innovation through faculty perspectives is a good starting point since they are one of the most important "players" in mobile devices integration. Since, mobile integration has not yet started in Cyprus it is extremely valuable to consider various parameters in advance in order to achieve a successful integration of mobile devices.

The paper has both practical and theoretical significance and contribution. The practical significance of the paper lies in the fact that it can provide guidelines and recommendations to countries that are interested or planning to implement mobile learning. The practical value of the paper goes beyond the borders of Cyprus since it aims to develop a set of directions and guidelines of mobile learning in an educational system. More specifically, the paper is of great importance for the design and planning of mobile learning integration in developing countries, such as Cyprus, where the technological infrastructure exists, although lacking experience on this kind of integration. Besides the above, it could be valuable for educational system and countries that already using mobile devices to take corrective actions and address any problems appeared. Furthermore, the study adds to the current body of literature since it assesses the readiness of an educational system, to integrate mobile learning; and literature lacks of this kind of studies [33]. Along the same lines, the paper differs from other studies conducted

related to mobile learning, since those investigated specific mobile practices and applications (e.g. how mobile devices have been integrated, their effectiveness, and contribution to the learning environment). However, this study is conducted on a broader level besides the boundaries of a classroom or an institution. Additionally, the results of the study provide the foundation of examining the feasibility of mobile integration in other formal, non-formal or informal educational settings, countries and levels. There are not yet many studies that investigate the feasibility of mobile devices integration through identifying the requirements needed for successful integration. Consequently, it stimulates further research to be conducted in the field. Besides the above, this field is relatively new and studies regarding mobile integration are still rare in Cyprus. Also, the mixed results of the paper could promote debates in various educational settings and levels. Finally, it consists of the foundation of the development of a framework of guidelines and suggestions in systemically integrating mobile devices integration in an educational system.

III. RESEARCH METHODOLOGY

To address the above a mixed method approach was employed [11]; [28]. The study made use of both quantitative and qualitative data. To better 'use' the data gathered the study applied a sequential explanatory strategy where first the quantitative data were collected and analyzed and then the qualitative data collection and analysis followed. The two methods were integrated during the interpretation phase of the study.

The research population consists of private universities faculty members. The quantitative component was addressed through a survey administered to a sample of private college faculty members. A survey was considered the most appropriate method since it enabled the researchers to collect data from a wide range of faculty. Additionally, the survey provided the opportunity to capture an initial picture of faculty perspectives regarding mobile technology integration in education as well as identify the factors that might influence mobile devices integration.

Quantitative data were collected through questionnaires. The survey sample selection was based on purposive sampling in an attempt to get faculty members with various characteristics such as age, fields of teaching, educational background, as well as computer and mobile literacy level. The questionnaires were circulated among 200 faculty members in three private universities in Cyprus. The response rate was 60% since, 120 faculty members completed and returned the questionnaires. The current research work does not aim to generalize the findings rather than evaluate the feasibility of mobile devices integration by identifying the aspects needed for successful integration from faculty members' point of view.

The phenomenological approach was applied in order to address the qualitative part of the current study. Phenomenological research was used to understand the experiences of other people and the meaning they make of that experience [34]. It requires that researchers put aside their opinions and egos and focus on the worth of the stories of those individuals who we are interviewing. This approach assisted in the construction of semi-structured, open ended questions that encouraged the participants to use their own terminology to describe their own experiences and perceptions on the subject under investigation.

Purposive sampling was used in an attempt to draw the subjects for the interviews. More specifically, a sub-sample of those faculty surveyed were selected to participated in the interviews. Twenty participants were chosen to be interviewed (See Table 1) based on their fields and their responses at the questionnaires. The interviews provided the opportunity to explore faculty perception on various parameters related to mobile devices integration. More specifically, the subsequent interview protocols, aimed at providing a deeper level of data that were used to evaluate, confirm, complement and/or better understand the survey findings [31]; [40]. Interviews took place as soon as the quantitative analysis ended.

TABLE I.
INTERVIEW PARTICIPANTS FROM EACH FIELD

Field	N	%
Sciences and Engineering	8	40%
Business	3	15%
Arts and Languages	4	20%
Education	5	25%
Total	**20**	**100%**

IV. DATA ANALYSIS & RESEARCH OUTCOMES

A. Quantitative Data Analysis

The questionnaire was divided into the following four parts: a) background information, b) technology and mobile learning literacy, c) current mobile devices integration in teaching and learning process, and d) potential for future integration of mobile devices. The purpose of the first part of the questionnaire was to identify the demographic information and the academic background of the respondents. This information served as the basis for the analysis of the replies to the other parts of the questionnaire. Based on the responses to the questions in the first part of the questionnaire four broad categories of faculties were revealed: sciences and engineering, business, arts and languages, and education.

Concerning the general knowledge on mobile learning, it has been observed that faculty of sciences and engineering were more knowledgeable of mobile technology in general and, more specifically, of mobile learning. The group with the least degree of awareness in mobile technology and learning was that of business. The findings are summarized in Table 2.

TABLE II.
FACULTY'S GENERAL KNOWLEDGE ON MOBILE TECHNOLOGIES AND LEARNING

Faculties	Yes	No
Sciences and Engineering	63%	37%
Business	17%	83%
Arts and Languages	20%	80%
Education	50%	50%

It was also revealed that even though the participants had a general idea on mobile technologies and learning, they lacked the expertise to implement it as a tool for their classes. When it comes to using such technology, the responses indicated that none of the respondents is utilizing this tool.

As far as the potential of future mobile integration in higher education learning is concerned, a broad quantification of the results led to the findings summarized in Table 3. Obviously, the potential was higher for faculty in Sciences and Engineering which was not surprising as they were more comfortable with the advancements of new technologies.

TABLE III.
POTENTIAL FOR FUTURE MOBILE DEVICES INTEGRATION

Faculties	Low	Medium	High
Sciences and Engineering			X
Business		X	
Arts and Languages	X		
Education		X	

The very last part of the questionnaire engaged in identifying the factors that might influence educators' usage of mobile devices in the classroom. As shown in Table 3, more important appeared to be 1) the need for professional development and training in integrating mobile devices into classroom activities and 2) the lack of knowledge and skills in integrating mobile devices into classroom activities. Similarly, less important factors proved to be the lack of campus-wide wireless network, and the lack of time to study for integration (See Table 4).

TABLE IV.
RANKING OF IMPORTANCE OF VARIOUS FACTORS THAT MIGHT INFLUENCE EDUCATORS' USAGE OF MOBILE DEVICES

Factor	Ranking in terms of importance
Personal attitude towards mobile technology integration regarding their use in the classroom activities	4
Level of mobile devices literacy - knowledge and skills in using mobile devices for any purposes (not only educational/in classroom)	3
Lack of knowledge and skills in integrating mobile devices into classroom activities	1
Need for professional development and training in integrating mobile devices into classroom activities	1
Lack of time to study for integration	5
Lack of campus-wide wireless network	5
Lack of technological resources (software and hardware)	2
Use of mobile technology may sometimes lead to pitfalls	3

B. Qualitative Data Analysis

1) Wireless and mobile technologies in education: Technological aspects

This section addresses the types of *technologies currently available for m-learning, and discusses the benefits and limitations of their integration.* The authors presented and explained the following available mobile technology devices to the participants. It was necessary to gain some initial knowledge of what the mobile devices look like and what they can offer in order to be able to further discuss the technological and pedagogical aspects to be considered:

SMS: Short Message Service allows users to send/receive messages of up to 160 characters between mobile phones (text messaging).

MMS: Multmedia Messaging Service serves the same purpose as SMS but allows the inclusion of graphics. *Mobile Learning software:* Specifically designed learning modules using m-learning software.

Wireless access points WAPs: There are two popular wireless standards, wireless fidelity - Wi-fi, also known and Bluetooth. Wi-fi is used primarily for Internet access.

GPRS: (General Packet Radio Service): In Wikepedia article this mobile data service is available to users of specific phone types. It can be used for WAP service, SMS, MMS, email, and access to the World Wide Web.

Bluetooth: a short-range wireless connection between PCs, handhelds, PDAs, mobile phones, camera phones, printers, digital cameras, e.t.c. It uses Radio Frequency (RF) for communication with the possibility of secure communication between multiple devices within a 30-foot range.. Bluetooth technology uses a globally available frequency band (2.4GHz) for worldwide compatibility.

3G and 4G phones: 3G technologies enable network operators to offer users services: wireless voice telephony, video calls, and broadband wireless data, with data transmission capabilities able to deliver speeds up to 14.4Mbit/s on the downlink and 5.8Mbit/s on the uplink. By the end of the decade 4G (4th Generation mobile phones) will provide up to 100 megabits per second transmissions adequate for multimedia.

PDAs: Personal Digital Assistants, usually called a pocket PC, which can store documents, spreadsheets, calendar entries, games, databases, and lots of other resources normally associated with a laptop or desktop computer using the Palm OS or MS Pocket PC operating system. PDA's are relatively inexpensive, highly portable, and are designed to utilize small, low-bandwidth files and applications.

MP3s: Audio file format that efficiently compresses files and enables them to be shared.

CAMs: Video cameras now embedded into mobile phoneand PDAs.

Given the above descriptions and presentations, the participants were asked to comment on the benefits and limitations of mobile devices integration in education, mainly from a technological aspect of view. The limitations of m-learning devices given by the participants are delineated below. The following limitations were mainly discussed by the faculties from the Sciences and Engineering fields, without implying that the rest of them did not contribute to this part. The small screens of mobile phones and PDAs, the limited storage capacities in PDAs and the battery life/charge were the three limitations discussed by almost all of the participants. The lack of common operating system and common hardware platform makes it difficult to develop content for all. As all of the technology devices can become out of date quickly and for some of them there is limited potential for expansion with some devices. Another two limitations reported were the difficulty that still exists in using graphics and the difficulty with printing, unless connected to a network. Finally, the limited wireless bandwidth that

may degrade with a larger number of users and the concerns about security issues were also reported from the participants.

The benefits reported were grouped in two categories the technology-oriented and the pedagogy-oriented ones. The most important technological benefits reported were the following.

1. *Portability*: The ability for the lecturer and the learner easily to move with a device inside a learning environment or to different learning environments.
2. *Cost Effectiveness*: Handhelds are becoming more affordable and therefore accessible to students. Additionally, the majority of the students own devices like that.
3. *Accessibility:* Easy access to mobile devices, even though students might be in different locations as well as equal access for learners with disabilities.
4. *Convenience*: Students can have access to content including theoretical information, quizzes, journal entries, balance sheets, learning games from anywhere.
5. *Ownership - Increase motivation:* Since students own their devices they are motivated to use them and learning from it.

The pedagogical benefits were summarized in three main categories: students' collaboration, interaction, and increased involvement:

1. *Collaboration:* Students can learn best when they share with each other and get immediate instructions and feedback
2. *Interaction:* Student can interact with instructors and among each other effectively and easily.
3. *Increased involvement:* The new generation likes mobile devices such as PDAs, phones and games devices, so they feel more motivated and interested in using them.

2) Pedagogical aspects - Learning with Mobile Technologies

This part of the paper discusses how mobile technologies can be used, and the various issues to be taken into consideration before integrating mobile devices as tools in the educational practices.

Two approaches of mobile devices integration were mostly discussed: 1) as a supportive tool; and 2) as an instructional tool. The approaches discussed below were mostly recommended and argued by the educators from the engineering and computer sciences fields; without implying that the rest of the participants did not address this question. Two of the educators reported having the experience of using mobile devices in their teaching practices. The educators argued that as supportive tools mobile devices allow the recording and maintenance of the lessons, the instructional procedures, the type of mentoring and the pedagogical approach, the role of the teacher and students. Additionally, the educators supported that mobile devices can facilitate communication between faculty members and students. More specifically, one of them mentioned that the above can be achieved "...through file sharing capabilities, built-in networking and a friendly interface with on-line discussion and e-mail options". Additionally, the educators reported that mobile devices can be used as instructional tools to support constructive learning [48]. Educators can provide students with electronic books, school-specific context, internet reference sites, graphing calculator, dictionary, and thesaurus etc. Finally, "...electronic quizzes and tests can be taken through mobile devices", as one of the participants argued. Educators from the Business and Arts and Languages Field revealed to be sceptical towards using mobile technologies as instructional tools.

Concerns and issues to be taken into account were discussed by all of the participants, include: the lack of knowledge regarding mobile devices integration, the need for skills development in successfully implementing mobile devices in teaching practices, the need for appropriate learning materials, instructional approaches and strategies to be applied, and educators' professional development programs specifically designed for this purpose. The above issues are further discussed below.

All of the participants seemed to agree that the new mobile learning arena imposes significant new design requirements of the curriculum per se. These requirements are not limited to the ways in which it is delivered and received but moreover in the ways the curriculum is structured and the ways in which it is maintained. A participant discussed the need for the "...curriculum units to be project-based, including a well-defined pedagogical and technological angle". The majority of them mentioned that the activities within the curriculum can be designed to take place in classroom (deskwork) or mainly outside the classroom (fieldwork).

Given the fact that the participants got an initial idea of what mobile devices are and what capabilities they have, participants were asked to think of examples of using mobile technology in the courses they were teaching during that time. They were instructed that by completing this activity the students should reach a number of educational objectives. Amazingly, all of them managed to briefly describe a lesson from the courses they teach where they could integrate mobile technology. Even the sceptical ones managed to describe either a game or an activity to take place in their lessons. One of them also argued that... "It is unrealistic to support that mobile devices could be used for all classroom activities". Another one came up with the idea of using mobile technology to evaluate students learning as well as assess students' attitudes to learning.

Some of the participants from the Arts and Languages and Education Fields expressed their concerns that m-learning technologies might support individualism. On the other hand some others said that it can facilitate the application of constructivist techniques where collaboration and team work can be enhanced and promoted [48].

All of the participants strongly commented on the need of professional development training programs to be designed for this specific purpose. "We all use mobile devices for calls, to send messages, for calendar and reminder purposes, but it is a totally other issue using them for teaching and learning purposes", one of them said. The participants agreed that they have to be trained on how to use mobile technology as an instructional tool.

All of the participants strongly argued that they should be involved throughout the entire process of designing, developing and implementing mobile technology integration. Being directly and actively involved, the participants argued that it would be easier for them to 'accept' and 'embrace' this innovation and successfully integrate it in their teaching practices. One of the participants referred to the power that educators have in boycotting innovations and ideas. Educators' feelings have to be considered regarding this innovation. Positive and negative reactions are expected to emerge. Finally, the majority of the participants positively commented on the need of students' involvement in the process of mobile devices integration. Some of the participants strongly supported that students need to have direct input on the process and the features to be developed. "I believe that they can provide valuable suggestions since they view mobile integration through a different perspective", one of the respondents commented.

Finally, the participants firmly discussed the importance of collaboration among various stakeholders. The collaboration of various stakeholders such as educators, students/ learners, computer scientists and engineers, is a critical element to successful mobile devices integration in education.

V. DISCUSSION

Some of the technical limitations mentioned by the participants, such as the limited storage capacities in PDAs and the battery life/charge, the lack of common operating system and common hardware platform makes, the small screen, etc; are not out of the control of a higher educational system. These are design issues that companies and designers need to consider. Regarding the limited wireless bandwidth, higher educational institutions could take care of that given the fact that in Cyprus there are the possibilities of today's cutting-edge third generation (3G) networking technology, increasing the capacity, improving quality, and allowing the use of advanced services over the existing, Cyprus Telecommunication Authority (CYTA), GSM network. With the new 3G mobile devices higher education will a digital, connected learning environment will emerge and information will be provided in a compact and convenient format. Learners will have a remote and instant access to a range of people and resources as well as the ability to process data [8].

Regarding the pedagogy-oriented limitations such as the lack of knowledge regarding mobile devices integration, the need for skills development in successfully implementing mobile devices in teaching practices, the need for appropriate learning materials, instructional approaches and strategies to be applied, and educators' professional development programs; are possible to be addressed. As new technologies emerge, lecturers should be encouraged and reinforced to integrate them without hesitation nor fear. What could be done to help lecturers overcome their fears so as to absorb and utilize new technologies? In order to overcome the above mentioned limitations, the higher educational institutions could organize information- and training-days to develop the appropriate technical and pedagogical knowledge and understanding needed. Initially training seminars could be offered presenting lecturers the potential advantages that these technologies have on student learning. Further training with a series of hands-on experience training will reinforce lecturers with knowledge, skills, time and accessibility to mobile technologies integration.

The integration of mobile devices in several sectors of everyday life facilitates our way of living. In Cyprus, however, particularly in the area of education, we have still not benefited as the responses received show. This preliminary study has revealed that educators, who play the major role in the implementation of mobile devices in the teaching and learning process, appeared to have limited experience and knowledge on the subject. On the other hand, faculty with technical background seemed to have the basic knowledge for the technology behind mobile devices, even though they lack capabilities to integrate it into the curriculum. Furthermore, some of those who lacked basic knowledge on the subject expressed interest in learning more. Organizational, educational, and personal factors seem to influence mobile technology integration in higher education. At the moment, faculty members find it impossible to implement this new educational tool in their classes. This is mainly due to the lack of training, knowledge and skills, and, less importantly, due to the lack of the necessary infrastructure and time to learn how to integrate them. Besides all the above, it seems to be great amount of potential to integrate mobile devices in Cyprus higher education.

As previously supported there are various technological and pedagogical aspects to be considered regarding m-learning. The new mobile learning arena imposes significant new requirements not only for the technological support and implementation but also for the educational perspective. Talking about technological challenge we mean that we must find ways to create and set up highly supportive environments which could provide support to contribute to different kind of learning settings. A technological opportunity is the fundamental transformation from the existing online learning using the advantages of 3G and 4G mobile phones and wireless communication networking.

Along the same lines, the pedagogical challenge related to m-learning is to find ways on how mobile devices can be integrated into classroom activities as well as successfully address all the parameters related to and influence mobile devices integration in education. A pedagogical opportunity is that the m-learning widens the educational horizons of students as well as enhances the educational options for educators. The categories of pedagogical aspects to be considered for successful mobile devices integration as revealed through the current study are discussed below:

1) Mobile devices applications in higher education classrooms

Mobile devices can be used by the students and the educators for professional, personal, and educational purposes. They can also be used as supportive as well as instructional tools. Mobile devices can be treated as tools to help students execute their tasks and promote the balanced development of their mental abilities by functioning as intellectual partners to the instructor and the learner.

2) Curriculum and learning materials development

Along the same lines, the need to produce innovative material that maintains a clear perspective on the learning goal is addressed [9]. As Carboni et al. mention, it is a complementary approach to the classic classroom lessons [6]. It might not be able to deliver three hour course on a PDA but is it feasible to deliver small learning activities and a number of documents, and exercises. To produce materials and design the content to be appropriate to stimulate and support the learner, knowledge of the technological constraints should exist as well. Consequently, to produce acceptable learning materials for mobile devices there is a need for educators, engineers, and computer scientists to collaborate and coordinate their actions and activities.

3) Appropriate contents for mobile technology to be used

The contents which mobile devices can be applied vary. Research so far shows that the experiments took place in various fields such as: Business and specifically MBA classes, Accounting, English, Social Studies, Mathematics, Science and Geography classes etc. Other activities include innovative games, exploring museums and exhibitions. Additionally, mobile learning devices can be used for evaluation and assessment purposes. For example, evaluate students' learning as well as assess students' attitudes to learning. Cyprus higher education (private and public) offers a variety of fields of study where mobile devices could be integrated. Educators are advised to embed and apply mobile devices in the context of teaching and learning in various contents and through various activities.

4) Pedagogical methods and instructional approaches

The authors support that there is a need for a shared, progressive pedagogy for mobile learning that will provide the scientific basis for networked and collaborative learning in both a virtual and a virtual-augmented environment. It must accommodate different teacher- and learner perspectives, promote learner-centered environments and collaboration among learners and between learners and educators. Finally, the new pedagogy must support ambient learning.

5) Educators' training

Additionally, educators need to be trained on how to apply mobile devices in their practices. To integrate computers in classroom practices, researchers were addressing the need that educators should be computer literate; in this case they have to be mobile literate. This is a great challenge because they have to deal with various types of equipment (hardware) and software. Additionally, the role of the educators needs to move towards facilitation and not teaching. With the ongoing technological advancements lecturers should gain the necessary skills through training, information and access, to overcome their fears and understand the advantages and opportunities these new technologies have to offer to their professional and personal lives.

6) Collaboration among various stakeholders: educators, students/ learners, engineers, computer scientists

Adopting an innovation is a risky process. But in order to minimize that risk and increase the success probabilities, it is important to be proactive and apply a systemic, holistic approach to mobile technology integration. The systemic approach to an innovation implies the involvement and participation of different parties in the design, development and implementation of the innovation. The systemic approach suggests the involvement and contribution of various stakeholders such as educators, students/ learners, engineers, computer scientists, community members, parents. The above stakeholders need to communicate, coordinate their actions, transfer and share their knowledge and experiences, as well as align their needs and goals. Educators need the help, support and knowledge of engineers and computer scientists and vice versa. It is not feasible to achieve m-learning without the coordination and knowledge integration of the above fields. Specifically, the role of the educators and the students in the design, development and implementation of the innovation is a necessity. Educators and students should be taken into consideration about the innovation and be actively involved in any steps and actions related to the integration [Kim and Holmes, 2006] . Having in mind the obstacles, failures, and problems faced when computer technology was first attempted to enter the classrooms and be established, it is extremely important to involve educators and students in this process.

VI. CONCLUSION

Mobile technologies offer learning experiences which can effectively engage and educate contemporary learners. Although mobile technologies can be considered as computers, lecturers are hesitant to integrate them into their everyday curriculum. This research indicated that the lack of understanding and the fear about where to start in a relatively new boundary of education prevents lecturers to integrate this new tool. Lecturers need to realize the potential of mobile devices to add a new dimension to their classrooms due to their personal and portable configuration and their type of interactions they can support with other learners and the environment [33]. The challenge for lecturers is to design mobile learning opportunities that properly utilize the power, convenience, contextualization, mobility, portability, connectivity and personalization of mobile devices.

It is reasonable and expected that some researchers, educators and practitioners are wondering and trying to understand what the educational benefits from m-learning are. This concern is even more "stronger" in Cyprus since there have been made no attempts so far to integrate mobile devices in higher education. However, research showed that through mobile devices reluctant learners can be motivated, hard-to-reach learners can be reached, various skills can be developed and improved as well as better communication among learners and between learners and instructors can be achieved [2]; [9]. Consequently, there is a need for some experiments to take place in order to examine the integration of mobile devices and their effects on various parameters such as students' learning, performance, and behavior, before moving further.

In order to take full advantage and materialize the potentials that mobile integration provide us, this research proved that, before making any further attempts to integrate mobile devices in education, it is essential to develop well-defined and well-structured requirements for mobile integration in the classroom. In order to meet these requirements we need to consider parameters such as faculty and students' role, technological resources, organizational, structural and technological issues, and

content design and delivery based on instructional methodologies and pedagogy. Mobile devices with advanced computing features are proliferating. They are even entering classrooms, and, before we realize it, they will probably find their ways in many educational settings just like computers did some decades ago [19].

The results of the current research work provide the basis for development and expansion of mobile devices integration in education by identifying the requirements needed for successful integration in educational systems. This work will contribute mostly to countries who still lack behind m-learning and they are willing to use this "new vehicle" for delivering education to today's learners via mobile phones PDAs, tablet PCs, etc. Higher institutions as well as educators, interested in transforming their traditional teaching/learning environment in a mobile will be introduced to challenges of mobile learning and some solutions to these challenges for a successful integration. Additionally, the results of the study prevent the danger of experiencing analogous obstacles and problems, as with computer technology integration. Since this danger is high, it is extremely important to be proactive, make wise decisions, and take appropriate actions [19]. As Wagner mentions "mobile learning represents the next step in a long tradition of technology mediated-learning" [47, p.44].

The feasibility of mobile devices integration could be further examined by focusing on other stakeholders such as students, parents, policymakers. Having evaluated the feasibility from various stakeholders, a complete set of requirements could be developed on how to successfully integrate mobile devices in higher education.

Finally, we could further expand this study by designing and implementing classroom experiments. Through experiments mobile and traditional teaching and learning processes could be compared based on various parameters such as student interaction, communication with the instructor, student performance and satisfaction.

REFERENCES

[1] Angeli, C., & Valanides, N. (2005) A socio-technical analysis of the factors affecting the integration of ICT in primary and secondary education. In L. T. W. Hin & R. Subramaniam (eds.) *Literacy in Technology at the K-12 Level: Issues and Challenges.* Heshey, PA: Education Media International.

[2] Attewell J., "Mobile learning: reaching hard-to-reach learners and bridging the digital divide," In G. Chiazzese, M. Allegra, A. Chifari, S. Ottaviano, ed., Methods and Technologies for Learning, WIT Press, Southampton, 2005.

[3] Avraamidou, L. (2008). Prospects for the use of mobile technologies in science education. *AACE Journal, 16*(3), 347-365.

[4] Berge L.Z., and Collins, M.P. "Computer mediated communication and the online classroom," eds. *Distance learning: Volume III.* Cresskill, NJ: Hampton Press, 1995.

[5] Brown, Doug (2006). Personalised learning – the technology challenge. Global Summit 2006 Technology Connected Futures. Retrieved on November 3rd, 2008 from: http://www.groups. edna.edu.au/file.php/1030/GS2006_BROWN.pdf.

[6] Carboni, et al. "Mobile Lessons and GPSWeb: mobile classrooms with goerefernced information," *In:* G. Chiazzese, M. Allegra, A. Chifari, S. Ottaviano, ed. *Methods and Technologies for Leaning.* Southampton: WIT Press, 2005, 349-353.

[7] Carr-Chellman, A.A. "Systemic Change: Critically Reviewing the literature," *Educational Research and Evaluation,* Vol. 4, No. 4, 1998, pp 369-394. (doi:10.1076/edre.4.4.369.6952)

[8] Clark, J. D. (2007). *Learning and Teaching in the Mobile Learning Environment of the Twenty-First Century. Retrieved on*

July 15th, 2008: http://www.austincc.edu/jdclark/mobilelearningenables.pdf

[9] Colley, J. & Stead, G. "Take a bite: producing accessible learning materials for mobile devices," *In:* J. Attewell, C. Savill-Smith, ed. *Learning with mobile devices, research and development.*UK: Learning and Skills Development Agency, 2003, 43-47.

[10] Council of Europe. (2004). *Global education in Cyprus.* The European Global Education Peer Review Process. National Report on Cyprus.

[11] Cresswell J. W., "Research Design: Qualitative, Quantitative, and Mixed Methods Approaches," 2nd ed., Sage Publications, California, 2003

[12] Crosta,. L. "Beyond the use of new technologies in adult distance courses: an ethical approach," *International Journal on E-Learning,* Vol. 3, No. 1, 2004, pp 48-61.

[13] Cyprus Ministry of Education and Culture. (2004). Department of Higher and Tertiary Education. Retrieved May 20, 2006, from www.moec.gov.cy.

[14] Dabbagh, N., & Kitsantas, A. Supporting self-regulation in student-centered web-based learning environments.*International Journal on E-Learning.* Vol. 3, No. 1, 2004, pp 40-48.

[15] Dawabi, et al. "Using mobile devices for the classroom of the future," *In:* J. Attewell, C. Savill-Smith, ed. *Learning with mobile devices, research and development.*UK: Learning and Skills Development Agency, 2004, 55-60.

[16] Duncan-Howell, J. & Lee, K.T. (2007). M-learning: Finding a place for mobile technologies within tertiary educational settings. In *ICT: Providing choices for learners and learning. Proceedings ascilite Singapore 2007.* Retrieved at November 10, 2008: http://www.ascilite.org.au/conferences/singapore07/procs/duncan-howell.pdf

[17] Eteokleous, N. (2008). Evaluating Computer Technology Integration in a Centralized Educational System. *Computers and Education Journal* (2007), doi:10.1016/j.compedu.2007.07.004

[18] Eteokleous, N., & Ierodiakonou, C. (2006). Leading change in academic institutions. Published in the Proceedings of the The Commonwealth Council for Educational Administration and Management (CCEAM) Conference: Recreating Linkages between Theory and Praxis in Educational Leadership.

[19] Eteokleous, N., & Laouris, Y. (2005). Are we Moving to Fast in Integrating Mobile Devices into Educational Practices?. *In: K. Nyriri, Communications in the 21st Century – The Mobile Information Society,* 197-205.

[20] Evans, C. (2008). The effectiveness of m-learning in the form of podcast revision lectures in higher education. *Computers and Education,* 50, 491-498. (doi:10.1016/j.compedu.2007.09.016)

[21] Fishman, B., Soloway, E., Krajcik, J., Marx, R. & Blumenfeld, P. (2001, April). *Creating scalable and systemic technology innovations for urban education.* Paper presented at the Annual Meeting of the American Educational Research Association (AERA), Seattle, Washington.

[22] Giroux, S., et al. "Mobile Lessons: Lessons Based on Geo-Referenced Information," Proceedings of E-Learn 2002 Conference, World Conference on E-Learning in Corporate, Government, Healthcare and Higher Education, 2002, pp. 331-338.

[23] Hadjithoma, C. (2007) New Technologies: New Schools? Embedding ICT in Primary Education: Exploring the Implementation Process in Relation to the Context and Teachers' Work (in Cyprus). PhD thesis, Graduate School of Education, University of Bristol, UK.

[24] Hewitt, J. S. M. "Design Principles for the Support of Distributed Processes," *Educational Psychology Review,* Vol.10, No. 1, 1998, pp 75-96. (doi:10.1023/A:1022810231840)

[25] International Bureau of Education. (2001). *The development of education: National report of Cyprus.* The Ministry of Education and Culture.

[26] Karagiorgi, Y. (2000) *The Introduction of Educational Technology into Elementary Schoolsin Cyprus: A Critical Analysis of the Implementation of an Innovation.* PhD thesis, Institute of Education, University of London, UK.

[27] Kim, S.H., Mims, C., & Holmes, K.P. (2006). An introduction to current trends and benefits of mobile wireless technology use in higher education. *Association for the Advancement of Computing in Education Journal, 14*(1), 77-100.

[28] Krathwohl R. D., "Methods of Educational and Social Science Research: An Integrated Approach," 2nd ed., Longman, 1997.

[29] Kukulska-Hulme, A. (2005). Introduction. In A. Kukulska-Hulme & J. Traxler (Eds), *Mobile learning: A handbook for educators and trainers* (pp. 1-8). London: Routledge.

[30] Kurubacak, G. (2007). Identifying Research Priorities and Needs in Mobile Learning Technologies for Distance Education: A Delphi Study. *International Journal of Teaching and Learning in Higher Education,* 19 (3), 216-227.

[31] Kvale, S. "InterViews: An Introduction to qualitative research interviewing," Thousand Oaks, CA, Sage, 1996.

[32] Milrad M. "Mobile learning: challenges, perspectives, and reality," In K. Nyiri (Ed.): Mobile learning essays on philosophy, psychology and education, Passagen Verlag, Vienna, 2003.

[33] The eLearning Guild. (2006). *"Mobile Learning Research Report",* Retrieved on November 3rd. 2-8 from: http://www.elearningguild.com/pdf/1/july_2006_-_mobilelearning.pdf

[34] Moustakas C. "Phenomenological Research Methods," Sage, Thousands Oaks, 2004.

[35] Naismith, L., Lonsdale, P., Vavoula, G., & Sharples, M. (2005) *Literature review in mobile technologies and learning.* A report for NESTA futurelab. Retrieved October 30, 2008, from http://www.futurelab.org.uk/

[36] Pneumatikos, T. & Michael, E. (2005). *Towards the European higher education era: Bologna process – National Reports 2004-2005.* Department of Higher and Tertiary Education. Ministry of Education and Culture.

[37] Polsani P. "Network learning," In K. Nyiri K. (Ed.): Mobile learning essays on philosophy, psychology and education, Passagen Verlag, Vienna, 2003.

[38] Quinn, C., M-Learning: Mobile, Wireless, In-Your-Pocket Learning, LiNE Zine, 2000

[39] Republic of Cyprus (1999) *Annual report on education 1999,* Ministry of Education and Culture, Nicosia.

[40] Rist R. D., "On the application of ethnographic inquiry to education: Procedures and Possibilities", Journal of Research in Science Teaching, 1982, vol. 19, pp. 439-450. (doi:10.1002/tea.3660190602)

[41] Seppala, P., & Alamaki, H. (2003). Mobile learning in teacher training. *Journal of Computer Assisted Learning,* 19(3), 330-335. (doi:10.1046/j.0266-4909.2003.00034.x)

[42] Swett, C. (2002, October). College students' use of mobile wireless-internet connections becomes more common. *Knight Ridder Tribune Business News,* Washington, DC.

[43] Trifonova, A. and Ronchetti, M. "Prepare for a bilingualism exam with a PDA in your hands," *In:* G. Chiazzese, M. Allegra, A. Chifari, S. Ottaviano, ed. *Methods and Technologies for Leaning.* Southampton: WIT Press, 2005, 343-347.

[44] Turban E., McLean E., and Wetherbe J. "Information Technology For Management: Transforming Organizations in the Digital Economy," John Wiley& Sons, inc, 2004.

[45] Turunen H., Syvaenen A., Ahonen M., Supporting observation tasks in a primary school with the help of mobile devices, In K. Nyvri (Ed.), Mobile learning: essays on philosophy, psychology and education, Communications in the 21st Century, Passagen Verlag, Vienna, 2003.

[46] Vrasidas, C. (2002). Educational technology in Cyprus and strategies for higher education. *Education Media International,* 39, 123-131. (doi:10.1080/09523980210153462)

[47] Wagner, E. D. (2005). *Enabling mobile learning.* EDUCASE, May/June 2005, 41-52.

[48] Zurita, G., & Nussbaum, M. (2004). A constructivist mobile learning environment supported by a wireless handheld network. *Journal of Computer Assisted Learning,* 20, 235-243. (doi:10.1111/j.1365-2729.2004.00089.x)

AUTHORS

N. Eteokleous is with the Frederick University Cyprus, Yianni Frederickou, Str., POBox 24729, Nicosia 1303, Cyprus (e-mail: nikleia@cytanet.com.cy).

D. Ktoridou is with the *University of Nicosia,* 46, Makedonitissas Ave., P.O. Box 24005 1700 Nicosia, Cyprus (e-mail: ktoridou@cytanet.com.cy).

8

The MOBO City: A Mobile Game Package for Technical Language Learning

F. Fotouhi-Ghazvini[1, 2], R. Earnshaw[2], D. Robison[2] and P. Excell[3]

[1] Qom University, Qom, Iran
[2] University of Bradford, Bradford, UK
[3] Glyndŵr University, Wrexham, UK

Abstract—In this research we produced a mobile language learning game that is designed within a technical context. After conceptual analysis of the subject matter i.e. computer's motherboard, the game was designed. The action within the game is consistent to the theme. There is a story, simplifying and exaggerating real life. Elements of control, feedback and sense of danger are incorporated into our game. By producing an engaging learning experience, vocabularies were learned incidentally. Deliberate vocabulary learning games were also added to our package to help students solve their common errors.

Index Terms—Mobile learning, Educational games, Language learning, Vocabulary learning

I. INTRODUCTION

Mobile learning offers the potential for radical change, not only in learning itself [1], but also in relation to the digital enfranchisement of previously-excluded populations in the developing world. In Iran, as a typical developing world country, there is massive pressure for educational development, in order to underpin industrial and economic development. The penetration of cabled access is low, reaching only a modest proportion of the population. Ownership of personal computers is also very limited, economic factors being compounded by problems of unreliable electricity supplies. On the other hand, access through the mobile phone system has the potential to be far more propitious, since the terminal hardware is far cheaper than a PC, has a more reliable power supply, avoids cabled connections, is already very widely deployed and has the great virtue of familiarity to a substantial proportion of the target audience. Following a study of opportunities for mobile learning in Iran it was recognized that most students in urban areas possess mobile phones and the only mode of communication available for rural students is that of mobile phones [2]. In the survey it was obvious that mobile phones are facilitating access to information amongst Iranian students. However, it is still not recognized as a training tool. In this paper we examine possibilities of using mobile games in language learning classes. References [3] and [4] developed several innovative projects using mobile phones to teach English by SMS. Unfortunately, problems with SMS teaching include its limited features and lack of students' engagement. It is not possible to keep students motivated. According to reference [2] 89% of students had WAP 2 enabled mobile phone and the use of Java mobile games is growing in Iran. Using games that increase students' interest and simultaneously increasing their amount of practice of language is very appealing. Game-based learning can be viewed as a particular form of incidental learning where the learner is engaged in an activity that may not be directly tied to the task at hand. A detailed analysis of game playing and digital games for education has been provided in references [5] and [6]. Game-based learning has being proposed for Higher Education to motivate assignments, curricula, and undergraduate research [7]. A number of games, however, have been proposed and used for teaching English as a second language [8]. Mobile language learning games have been also designed, for example crossword game [9]. The research in this paper revolves around a Language Learning project using game for Iranian University students. The authors demonstrate that it can facilitate students' learning motivation and engagement in the interactive learning environment.

II. THEORY

Modern linguistic theories of and instructions for second language acquisition emphasize greatly on the use of language for *meaningful* communication [10]. They argue, students can usefully be taught some non-language related subjects, such as history or computing, in a second language. The assumption is that the learners *would* acquire the second language simply by using it to learn the subject matter content, without the second language being the focus of explicit instruction. In this research we produced a scheme that pursues this line of theory. To teach technical English vocabulary to students, we chose to teach a technical subject (motherboard components), and we introduced necessary vocabulary indirectly during our instruction. Concept maps were used as a kind of *template* or *scaffold* to help us to organize computing knowledge. As a result the subject was divided into small units of interacting concept and propositional frameworks. This follows from Novak, who believes concept maps facilitate meaningful learning and the creation of powerful knowledge frameworks [11]. It seems evident from diverse sources of research that our brain works to organize knowledge in hierarchical frameworks and that learning approaches that facilitate this process significantly enhance the learning capability of all learners [12], [13].

Obviously, our brains store more than concepts and propositions. While the latter are the principal elements that make up our knowledge structures and form our cognitive structure in the brain, other forms of learning exist such as *iconic learning*. This involves the storage of images of scenes we encounter, people we meet, photos, and a host of other images. These are also referred to as *iconic* memories [14]. While the alphanumeric images Sperling used in his studies were quickly forgotten, other kinds of images are retained for much longer periods. Our brains have a remarkable capacity for acquiring and retaining visual images of people or places, but soon forget the details [15]. To teach about computing, we integrated various kinds of images into our concept maps. The idea was to enhance iconic memory via conceptual frameworking. Therefore we used both visual and verbal mental imagery to relate a word to be memorized. By this stage the context is created in an organized structure but what is missing is the motivation of our students to use our materials. Most m-learning content tends to consist of summarized PowerPoint files, PDFs, WAP sites "gussied up" with graphics, photos and in advanced cases, videos. Providing some verbal or visual material to students without keeping them immersed and interested in context is not as useful as it could be. If we make the educational content more interesting, we can make it more effective. At its best, learning should be a wildly enjoyable experience. There should be joyous discoveries, satisfied completions and sudden recognitions. We need to address the emotional side of learning as well as the knowledge side. The other important factor in instruction is the "normal flow of learning". Several researchers defined flow as "*the state in which we are so involved in something that nothing else matters*" [16]. Cleverly designed educational games can provide such a flow for an individual learner and keep them simulated throughout instruction. Interaction is another important aspect of educational games, which is proposed in game learning theories and models [17], [18] and [19]. The learner's interaction with game is essential as it defines how could the learners control the game and learn from it. The interaction element adopted in our game uses the Interaction Cycle suggested by Barendregt and Bekker where the interaction between a user and a computer game happens in terms of cognitive and physical user actions. [18]. At the first stage, learners understand the rules and goals of each task in the game then they decide the actions to be taken to accomplish the task. After taking appropriate actions to complete the task, the game provides feedback to learners. Based on the feedback, learners evaluate whether the task is completed successfully or not. This is useful in deciding whether the learners have conceived the correct information. Learners will then repeat the interaction cycle on the same task or proceed to the next task. Motivation here is a key aspect for effective learning and is sustained through feedback responses, reflection and active involvement in order for designed learning to take place. Other factors that has an impact upon learners' motivation in educational games relates to sense of challenge, game realism, opportunities to explore or discover new information. According to the researchers these motivational variables should be considered in game development and use [20], [21], [22] and [17]. These elements could greatly facilitate the learning process and has been incorporated in our game.

III. METHOD

Our mobile game was designed to teach technical English to Iranian students in the University of Qom. The package was produced as a platform-independent application. The chosen development environment was Java 2 MicroEdition (J2ME). After organizing the sources about motherboard components into concept maps, their corresponding environments and characters were designed and then the necessary vocabulary was inserted inside the game as a kind of verbal feedback and guide throughout the game.

MOBO city stands for Motherboard City because the game's main theme is that of a motherboard. The metaphor is of a city where at different locations, electronic components are located, just like in a real motherboard. The complex task for the motherboard is to move data in the right order and right manner, to the right recipient. Our main characters are a red bus and its driver (Fig. 1 and Fig. 3). The bus represents a 'motherboard bus' whose its duty is to transfer information across the relevant components. The bus driver's name is OS which is abbreviation for operating system. An operating system manages hardware and guides the flow of data; hence what drives and directs the bus of data is called OS. In the game the OS will do the paper work for data bus passengers in each station (Fig. 3 and Fig. 4).

Our game is an adventure where the player assumes the role of a character within a world of fantasy (Mobo City). The player can control his character and thereby cause the character to move about in the fantasy world, investigate and interact with whatever is encountered in the world. The character can, for example carry out dialogues with other characters in the fantasy, for example security men. The story begins when our red bus receives a new task, for example "There is some data just arrived from scanner ship, pick them up from USB port and take them to monitor theatre, where they have to perform a show that has been organized by Viewscan Corporation" which illustrates how scanned data is shown in a monitor, using Viewscan software. There is a clear goal that the player will be trying to achieve, i.e. to successfully send data to its destination, the monitor. This goal will provide a motivation for the action and a metric for attainment. The questions normally asked in the game are related to the computer's common processes, such as accepting input, executing instructions, generating output and displaying or storing results. However it must be in accord with the game story, providing a task for the city bus. The bus has to move data through MOBO City passing through different components in the right order. For each question a flow diagram is produced.

The bus must reach its destination in order for a player to win a game. The game displays pertinent information related to the state of the game such as life points, which is initially 5, and the score, which is initially 0, in the information bar displayed along the top of the game screen (Fig. 1 and Fig. 2). The game finishes when the player loses all their life points. The game is not static; other characters such as virus ships can move about and act on their own. The bus must be aware of virus ships at all times, because they try to destroy the data bus (Figure 1). If hit, the player looses 1 life point. The game consists of a network of distinct physical *contexts* such as the rooms of an office or bus stations of a city. On the motherboard, next to each main component, there is a bus station or bus stop. A congratulatory message will be shown each time the bus passes through a correct bus stop, the player scores 5 and the bus moves on. Some components have the sign of a station next to them which mean that some tasks which need to be done (Fig. 4). If the bus passes through the wrong component a 'you are not allowed!' message will be shown and the player looses 2 points. If the player passes through a component too early, a 'come back later!' message will be shown. Passing through a correct station produces a congratulatory message and the player gains 5 points (Fig. 1 and Fig. 2)

When a message page appears all virus ships will stop. We limited the message pages to a few seconds in order to avoid the players loosing the game flow. In each different bus station, the physical feature such as background and characters are changed according to the kind of work that particular component involves. We made a concept map for each component, questioning: What parts does it consist of and what do they do? The graphic inside each component is produced according to its concept map. The bus driver OS moves inside the component. The component consists of rooms inside which there are different characters, each responsible for different kinds of jobs. The driver has to meet all of them, but in the correct order. The security men in front of each room help the player to get an idea of what job characters in each room are involved in. However, the player is constantly being followed by spy viruses that have escaped from the security men, in front of the station. If he is attacked by one of viruses, the player looses 5 points. Each time the player meets the correct character, he or she scores one point. The environment inside a CPU and the dialogue where the bus driver gets involved with different characters is shown in Fig. 3.

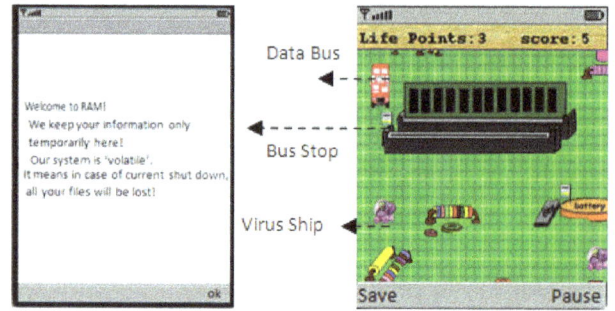

Figure 1.   The game shows an appropriate message when the bus arrives at RAM bus stop.

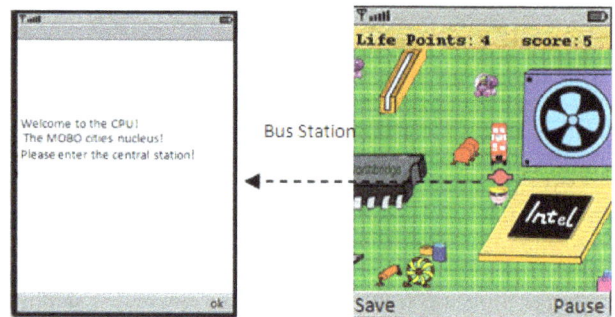

Figure 2.   The Game shows an appropriate message when the bus arrives at CPU bus station.

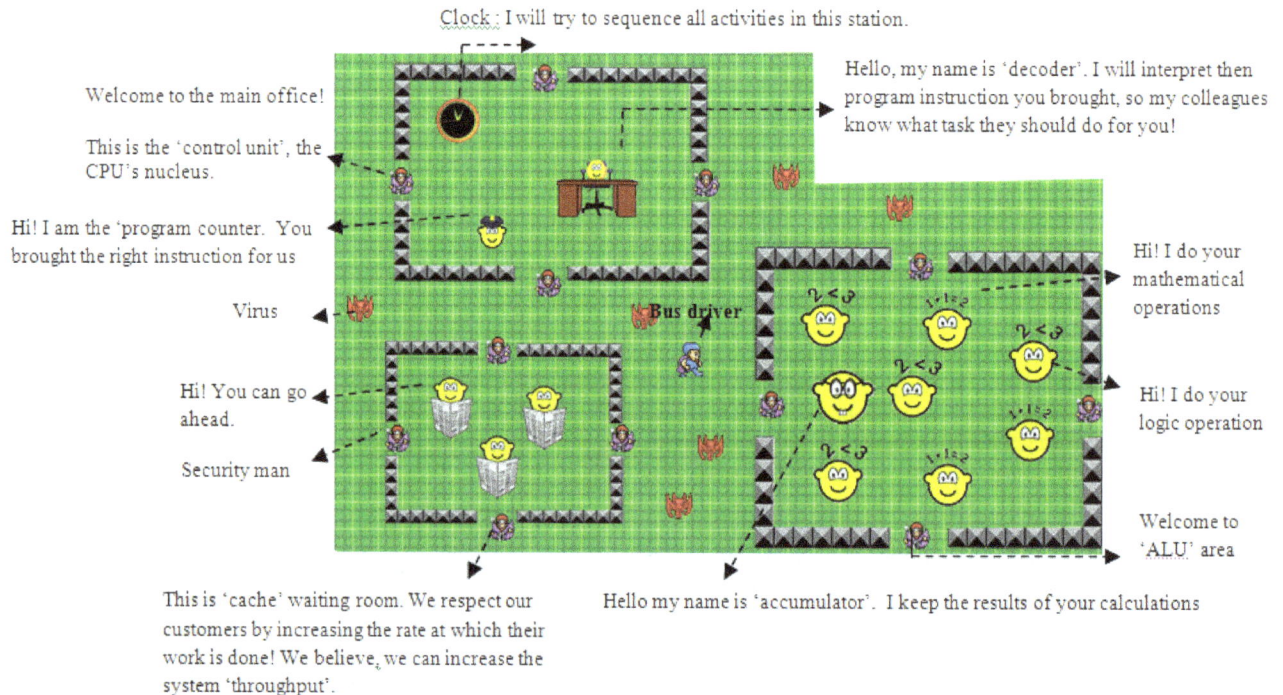

Figure 3.   The game background inside the CPU

IV. EVALUATION

Technical English course units are very important for computer engineering students because relevant textbooks are mainly written in English. Due to the nature of their field of studies, they have to extensively use the Internet, which is also dominated by English. The role of English in passing MSc and PhD entrance examinations is significant. Despite its importance, it is offered for only one semester during the student's four years of study and its lecture time is very limited, only 2 hours per week. On the other hand there are few lecturers that are both competent in English Linguistics and Computer Engineering. In addition to all this, the students' attitude towards these classes is that of weariness and it has very little appeal for them. 15 students from the third year of computing Engineering whose levels of English were as equal as possible were selected. We divided them into three groups, the first 5 tried to read a comprehension describing motherboard components without using a dictionary. The second 5 were asked to use the dictionary and to memorize a list of vocabularies and the last 5 were asked to play with our MOBO city game (Table I). Then, a list of vocabularies that all three groups have encountered in their tasks was presented to them and their level of vocabulary understanding together with spelling was examined.

The results show that students reading skill is very low, using the dictionary does help (Table 1). However it has two main drawbacks; first, it is a tedious job and students are often reluctant on using it; secondly they often only learn the first meaning and consequently it produces out of context learning. For example in this study the word 'nucleus' referred to 'a central part about which other parts are grouped or gathered' but in some dictionaries it is described as 'usually spherical mass of protoplasm encased in a double membrane' in other dictionaries as 'the positively charged mass within an atom' and others as 'a mass of nerve cells in the brain'. This is the result when the vocabulary is taught out of context. Computer students in our evaluation not only had a problem with second language learning, they also lacked some technical concepts of the motherboard and this made understanding the related comprehension very difficult for them. For example they were not familiar with vocabularies such as: AGP, PCI, Bios, expansion slot. MOBO city, by helping them learn the subject matter, helped them greatly in learning technical vocabularies. Below are listed the remarks of students' observations of the game:

"The objective of the game was clear, we knew what was expected from us." A basic rule of instructional design put forward by Gagne in 1965 seems far too commonsense but still, inordinately, relevant today: to inform learners of objectives and goals [23]. These goals need to be presented early [20], need to be clearly stated and should be personally meaningful, obvious and easily generated [20].

"The storyline was both fun and educational." Designers need to tune the educational message to the content, in other words, game goals which are fun and learning goals need to be in harmony with one another.

"It felt real, we got feedback for what we did and we should have been ready for dangerous surprises from viruses."

In the real world we can control events and get perceptual feedback concerning what we have done, we must constantly be ready for dangerous surprises. Perhaps, when this sense of vulnerability in usual online learning is absent, our whole experience is sensed as unreal. By means of educational games we can invent virtual worlds that simulate this sense of reality for students.

"The learning experience produced by game was neither too difficult nor too easy!" The degree of difficulty is an important feature in games; for players to enjoy playing, the game must be neither too difficult nor too easy [24].

"When playing the game I lost track of time." What makes game learning so distinctive from other types of learning is its essence of flow, context, control and in brief immersion and engagement in learning, which is difficult to achieve with other types of learning.

"After playing a game I could easily produce its concept maps, the learning really sticks in." On the other hand, when the learners' 'heart and mind' are captured they are cognitively and effectively connected to a learning experience.

" For the first time I was exposed to such a vast number of vocabulary at once, I really got to learn many words" In a first language, as the learner encounters most words on a frequent basis in a wide range of contexts, the words are often learned incidentally in an incremental way. In a short space of time, a large number of words are thus learned and this lexical repertoire then forms the basis for learning other new words. In this research we tried to simulate the process of implicit vocabulary learning. A selected number of high frequency words were chosen and integrated in the game to help students learn incidentally. They were exposed to a large quantity of input, a condition that otherwise was impossible to achieve for non-native speakers. We provide a cumulative learning environment; different vocabularies were continuously encountered to allow the learning of each word to become stronger and to enrich the knowledge of each word.

TABLE I.
RESULT OF HOW TO SPELL...? AND WHAT DOES... MEAN? FROM 46 VOCABULARIES.

	Group using reading comprehension					Group using dictionary					Group playing games				
Students No.	1	2	3	4	5	1	2	3	4	5	1	2	3	4	5
Spelling Results	22	28	24	25	25	26	26	25	27	30	26	25	29	23	22
Meaning Results (answers in context)	9	15	16	19	13	30	25	12	20	35	41	35	43	37	31
Meaning Results (answers out of context)	-	-	-	-	-	15	12	10	12	10	-	-	-	-	-

However, they also showed that the learners continued to make certain persistent spelling errors, even after playing quite a few times (Table 1). In other words, a communicative approach helped learners to become fluent but was insufficient to ensure comparable levels of accuracy. It seems as if a certain amount of explicit instruction focusing on language form may be necessary as well. As a further refinement to our package we added a few word games and dictionaries to the package, focusing only on technical and non-technical vocabularies that students encountered in MOBO city games.

1. A word search game that comprised seemingly random letters arranged in a rectangular grid. A list of 10 hidden words is provided. The object of the game is to find and mark all of the words hidden in the grid. The first letter of each word provided the cue which was accessible by the command 'Cue' at the bottom of the screen.

2. A 'Butterfly shooter' game was created where a butterfly represented the player. The meaning of a random word was given at the bottom of the screen. By shooting the red bullet towards the rolling balls, one could choose one's desired letter. Each time a bullet hit the wrong letter, the butterfly lost a wing. On losing all its 6 wings, the game terminates. A bilingual single-field dictionary was produced that narrowly covered computer engineering terms, in Farsi and English.

V. CONCLUSION

Effective learning cannot be achieved by only introducing different modes of learning, but also requires increasing students' motivation and keeping them motivated and engaged until the educational goals are achieved. In educational games, the learner will be immersed in the context which is achieved by designing a suitable theme appropriate to learners. Elements of control and a sense of danger were incorporated into our game and thematic feedback was delivered as appropriate. For example, if the learner has made an incorrect choice, dialogue from a character provided a feedback mechanism. All this produced an immersive experience that helped students learn vocabulary incidentally, as they were having fun. Deliberate vocabulary learning was also added to our package to help students solve their common errors. In this game we illustrated a simplistic illustration of components and different processes of computers. It is possible to elaborate on extra features and expand this game for the future. For example other kinds of buses such as control bus, address bus and power bus could be described. The motherboard has different components, and some motherboards are designed for multiprogramming and multiprocessing. There are many other examples that can be thought of. When adding new features and new components, new environments and characters must be designed. In the process, different storylines will evolve. Further learning content such as this can help to enrich the game further and to teach students even more vocabulary. Future work on this project will attempt to provide social interaction between students while playing. By adding Bluetooth capabilities, we are aiming to produce multiplayer version of this game. The shared experiences could greatly increase the appeal and longevity of the game and works as a powerful motivator to engage girls within the educational content.

REFERENCES

[1] B. Eschenbrenner and F.H. Nah. "Mobile technology in education: uses and benefits", International Journal of Mobile Learning and Organisation, Vol. 1, No. 2, pp.159-183, 2007. (doi:10.1504/IJMLO.2007.012676)

[2] F. Fotouhi-Ghazvini, P.S. Excell, A. Moeini and D.J. Robison "A Psycho-Pedagogical Approach to M-learning in a Developing World Context", International Journal of Mobile Learning and Organisation (IJMLO). Volume 2, No. 1, pp 62-80, 2008. (doi:10.1504/IJMLO.2008.018718)

[3] P. Thornton and C. Houser, "Using mobile phones in English Education in Japan", Journal of Computer Assisted Learning, Vol. 21, pp. 217-228, 2005. (doi:10.1111/j.1365-2729.2005.00129.x)

[4] M. Levy and C. Kennedy, Learning Italian via mobile SMS, In A. Kukulska-Hulme and J. Traxler (Eds.), Mobile Learning: A Handbook for Educators and Trainers. London: Taylor and Francis, 2005.

[5] M. Prensky, Digital Game-Based Learning. McGraw-Hill Trade, 2001.

[6] K. Salen and E. Zimmerman, Rules of Play – Game Design Fundamentals, MIT Press, Cambridge, Mass (2003).

[7] E.Clua, B. Feijó, J. Schwartz, M. Graças, K. Perlin, R. Tori, T. Barnes, "Games and Interactivity in Computer Science Education", Panel at SIGGRAPH, Boston, MA, August 2006.

[8] EslFlow-2006 http://www.eslflow.com/games1.html, July 2008.

[9] H. C. Hung and S.C. Young, "Constructing the game-based learning environment on handheld devices to facilitate English vocabulary building" Seventh IEEE International Conference on Advanced Learning Technologies , ICALT (2007), pp. 348-350, 2007.

[10] W. Littlewood, Communicative Language Teaching: An introduction. Cambridge: Cambridge University Press, 1981.

[11] J. D. Novak and J. Wandersee, "Coeditors, special issue on concept mapping", Journal of Research in Science Teaching, Vol. 28, No. 10, 1991.

[12] J.Bransford, A. L. Brown and R. R. Cocking, (Eds.) How people learn: Brain, mind, experience, and school, Washington, D.C.: National Academy Press, 1999.

[13] J. Z. Tsien, "The Memory Code", Scientific American Magazine, Pp. 52-59, July 2007.

[14] G. Sperling, "A model for visual memory tasks", Human Factors, Vol. 5, pp. 19-31, 1963.

[15] R. N. Shepard, "Recognition memory for words, sentences, and pictures", Journal of Verbal Learning and Verbal Behavior, Vol. 6, pp. 156-163, 1967. (doi:10.1016/S0022-5371(67)80067-7)

[16] M. Csikszentmihalyi, Flow: The Psychology of Optimal Experience. New York: Harper & Row, 1990.

[17] A. Amory and R. Seagram, "Educational Game Models: Conceptualization and Evaluation", South African Journal of Higher Education, vol. 17, No. 2, p. 206-217, 2003.

[18] W. Barendregt, and M.M. Bekker, "Towards a Framework for Design Guidelines for Young Children's Computer Games", In Proceedings of the 2004 ICEC Conference, Eindhoven, The Netherlands: Springer, 2006.

[19] N.S. Said, "An engaging multimedia design model" In Proceeding of the 2004 conference on Interaction design and children: building a community, Maryland: ACM Press, 2004.

[20] T. W. Malone, "Toward a Theory in Intrinsically Motivating Instruction", Journal of Cognitive Science, Vol. 5, Issue 4, Pp 293-388, 1981

[21] L. P. Rieber, Seriously considering play: Designing interactive learning environments based on the blending of microworlds. Simulations and games. Educational Technology, Educational Technology Research and Development, Vol. V44, No. 2, pp. 43-58, 1996. (doi:10.1007/BF02300540)

[22] P. Thomas and R. Macredie, "Games and the design of human–computer interfaces" Educational Technology, Vol. 31, 134–142, 1994.

[23] R. Gagne, The Conditions of Learning. New York: Holt, Rinehart and Winston, 1965.

[24] A. McFarlane, A. Sparrowhawk and Y. Heald, "Report on the Educational Use of Games", TEEM (Teachers Evaluating Educational Multimedia): (2002). www.teem.org.uk, last accessed July 2008.

AUTHORS

F. Fotouhi-Ghazvini is a Lecturer in Mobile Communication at the University of Qom, 37165, Iran and PhD student at the University of Bradford, UK, BD7 1DP, ffotouhi@bradford.ac.uk

R. Earnshaw is Pro Vice-Chancellor (Strategic Systems Development) and Professor of Electronic Imaging and Media Communications at the University of Bradford, UK, BD7 1DP, R.A.Earnshaw@Bradford.ac.uk,

D. Robison is a Lecturer in Media and Mobile Media at the University of Bradford, UK, BD7 1DP, d.robison@Bradford.ac.uk, Tel: +44 (0)1274-235465.

P. Excell is a Professor of Communications and Head of the School of Computing & Communications Technology at the University of Glyndŵr, Wrexham, LL11 2AW, p.excell@newi.ac.uk

9

The Barcelona Mobile Cluster:
Actors, Contents and Trends

C. A. Scolari[1], H. Navarro Güere[1], I. García[1], H. Pardo Kuklinski[1] and J. Soriano[2]

[1] Universitat de Vic, Vic, Spain
[2] Universitat Autónoma de Barcelona, Barcelona, Spain

Abstract—Communication mediated by mobile devices is one of the most dynamic sectors of the global economy and is transforming different aspects of our lives, including our ways of relating and our cultural production, distribution and consumption models. Media studies should not ignore these transformations. This paper presents the results of a study carried out during 2008 to determine the actors, contents produced and trends of the mobile communication companies in Barcelona. The study outlines an initial map of the situation, proposes a series of analysis categories and lays the foundations for more specific future studies on *mCommunication*.

Index Terms—mCommunication, mobile communication, mobile marketing, mobile gaming, Barcelona

I. INTRODUCTION

The web has brought new challenges to traditional mass communication studies. Mass communication theories, founded in the broadcasting logic, have undergone an upheaval since the new forms of interactive communication, that combine different media, formats and languages, arrived [1]. When these processes started to become part of research agendas, and the web was just starting to take off, a new medium began to forge a path in the communication ecosystem: mobile devices.

It is now possible to say that the digitalization of the content production and distribution processes and worldwide diffusion of the Internet was only the first stage of an evolution that is far more extensive and complex. The second phase of communication digitalization has begun with two very clear trends:

- The traditional model of communication media, based on the diffusion concept (one-to-many) has been challenged by the appearance of new collaborative logics (many-to-many) [2] [3].
- The diffusion of portable communication devices able to connect to the Internet as terminals and to receive and transmit all kinds of digital contents has opened the doors to what is now called the 'mobile Internet' [4] [5].

From the 90s the mobile telephone has undergone a remarkable transition in its evolution process: it has stopped being an instrument of interpersonal communication for an elite group of professionals to become a multifunctional product of the masses that connects to the Internet as one more terminal. This leads us to consider the appearance of a new form of communication: *mobile communication (mCommunication)*.

A. The Relevance of mCommunication

According to the studies carried out by the International Telecommunications Union in September 2007 there were more than 3,300 million mobile telephones in the world (in 2000 this figure was only 800 thousand) [6]. In Spain there are 50,548,312 mobile telephone lines including those associated with computers (+ 6.2% compared to April 2007). For every 100 inhabitants, 44.5% have landlines but 109.1% have mobiles [7]. No other communication medium or device has reached this level of penetration in society.

This diffusion of mobile devices has caused a repositioning of the large economic and technologic actors. All of the Spanish companies involved in the sector, from the telecommunication operators to those who produce contents, consider mCommunication as a new frontier for their business activities [8] [9] [10]. In this extremely dynamic, global context, the city of Barcelona has become a very relevant location. The Mobile World Congress (ex 3GSM World Congress) has been held in this city for four consecutive years (2006-2009), which has favored the public visibility of many local companies and their integration into international projects and networks. The Mobile World Congress is one of the most important conferences in the world of mobile phones. It is a huge commercial and networking initiative created by the GSM Association, an organization that brings together more than 700 GSM operators (Global System for Mobile Communications) in 220 countries with more than 170 associated members.

Every day there is an increasing number of scientific books and articles produced on mCommunication, mainly in the United States, Northern Europe and Asia, and it is evident that the Spanish academic world is lagging behind. Among the work on mobile communication in Spain we can mention the pioneer analysis on the impact of new media [11] or more recent studies from the perspective of media studies [12] [13] [14]. In Catalonia, there are few research works on mCommunication from the perspective of media studies; among these it is worth noting the analysis of journalism via MMS that uses the service provided by the newspaper La Vanguardia as a case study [15]. The growing importance of Barcelona and Catalonia in mCommunication demands a more in-depth study of the business and communication dynamics within this geographic area.

B. Towards a Definition of mCommunication

mCommunication is a social practice of content production/consume and technological appropriation carried out through the massive diffusion of multifunctional wireless devices. The diversification of the technology (mobile communication devices now incorporate an increasing number of functions, from cameras to music players, web navigators or mini consoles for videogames) and the extended range of terminals (mobile telephones, palmtops, smartphones, iPods, etc.) has generated the support for mCommunication.

The telephone has evolved from being a one-to-one communication device, and has adopted new forms that are already used in the Internet (one-to-many, many-to-many, etc.). The third generation mobile devices incorporate different communication modes, from the most massive and public (receiving television) to the most personal (sending and receiving SMS messages). Between these two modes there are a wide range of possible communication and exchange forms that no other technical device offers in the same way. To some degree, the new generation of mobile devices is closer to personal computers than traditional telephones; this is why they can also be considered a *metamedium*, which is a concept that certain researchers have used to refer to the web [16].

In synthesis we can say that mCommunication is a phenomenon within the confluence of a series of properties and functions:

- Ubiquity and the ability to be carried (communication anywhere, anytime)
- Convergence of functions, media and languages (metamedium)
- Integration of communication models (broadcasting, unicasting, multicasting, etc.)
- Bidirectionality (consumption and production of contents)
- Contents and services designed according to localization.

II. OBJECTIVES AND RESEARCH METHODOLOGY

There is no need to discuss the importance of mCommunication: mobiles are now part of the media system and will soon be considered simply 'one more medium' like television or radio. Until recently, research into the mobile device sector has mainly taken the form of applied technical studies (studies based on the development of hardware or software for the functioning of these devices), sociological studies (investigating the uses and the social impact) or economic analyses (studies of the mobile telephone market, commercial strategies and business models). There is also research that analyzes mCommunication from the media studies perspective [17]; however, in Catalonia there is practically no research into this sector from this point of view. The general objective of this study was to analyze mobile communication in Catalonia from the media studies perspective. Therefore, an initial map of the economic actors, contents, services and trends of the mobile communication sector in Catalonia was constructed. As all the companies are located in and around the city of Barcelona we prefer to define this reality as the 'Barcelona Cluster'.

A. Methodology

For some time now there have been scientific observations, long-term studies, longitudinal surveys, registers and periodic audits of the mass communication media in Catalonia that evidence the size and evolution of the sec-

tor; however, in the case of content production for mobiles there are few data sources of this style. For this research an initial exploratory empirical study was carried out and information was gathered directly from the sector. This implied becoming familiar with a very little known phenomenon and therefore the research team drew up a flexible work plan in order to deal with the different aspects that could arise during the study [18].

The research team's first objective was to identify the actors who could be reliable information sources in order to outline the map of the sector in Catalonia. The Chamber of Commerce and specialized publications were consulted although the main source of information was the Mobile World Congress. When the work began in October 2007 we had a list of 84 companies, among which were the most outstanding content and services producers in Catalonia.

The sampling carried out to select the interviewees combined elements of both quota and snowball sampling, neither of which are probabilistic. The sample size, which was finally 23 interviewees, was determined by the type of content produced (journalism, publicity, etc.). Each of these production sectors had a different number of interviews depending on the number of businesses in the market and the information saturation point of the interviews:

- Marketing and publicity contents (7 companies: Ubiqua, Frog Mobile Services, Tempos 21, Mobbiz Communication, CPM Telecom Publitono, AdsMedia, Daem Interactive)
- Education and services contents (6 companies or institutions: Ajuntament de Barcelona [The Barcelona Council], Digital Work Force, e-movilia, Childtopia, Bluethchannel, Ta with You)
- Audiovisual and videogame contents (5 companies: Digital Legend Entertainment, Kailab, Lechill Mobile, Cromosoma, Microjocs)
- Journalistic contents (4 companies: CCRTV Interactiva, La Vanguardia/Grup Godó, Agència Catalana de Notícies, Vilaweb)
- User-generated contents (1 company: Ready People).

The requisite for the interviewees was that they were in charge of content production.

The study was enriched by subjective information from experts in the sector, their opinions, experiences and expectations by combining structured and semistructured interview modes in a single questionnaire that included different aspects (open and closed questions and objective data on the company). The interviews lasted between 30 and 90 minutes.

The analytic process was based on the intersection between the productive sectors and analytic categories in order to outline descriptions that cover the entire sector. This allowed us to observe the behavior of the categories beyond their productive specialties. This information process was complemented with an exhaustive study of bibliographic sources, statistics on the mCommunication market and information from specialized publications.

III. THE mCOMMUNICATION ACTORS IN CATALONIA

In this section we present the results obtained in the research on the content and service producers for mobile communication in Catalonia.

A. Company Profiles

Most of the companies analyzed are private initiatives. The exceptions are three cases of public journalism institutions or companies that later incorporated content and service production for mobile communication (Ajuntament de Barcelona, CCRTV Interactiva and the Agència Catalana de Notícies [Catalan News Agency]).

Most of the people interviewed have a university degree in engineering and/or multimedia. The professional origin of the actors differed depending on the production sector. For example, in the case of mobile marketing (*mMarketing*) there were publicists, copywriters and media planning experts. The role of journalists is decisive in the creation of information contents for mobile devices and therefore the task of programmers is relegated to second place; however, in the videogame sector, programmers play a very important part.

B. The Size of the Companies

The highest number of permanent workers in these companies fluctuated between 20 and 25 people. However, in the case of companies that were already established before they entered the mobile market and in which content production for mobiles is only a part of the main activity (for example, journalistic companies), there is generally no more than 10 professionals dedicated to mCommunication.

In other sectors, like the production of low-cost videogames or contents for mMarketing, it is also not strange to find companies with no more than 10 employees. Having a small number of permanent personnel gives these small companies a great capacity to adapt to the needs of the market. The only exceptions to this model are the 'fusions' with or acquisitions by large companies.

C. The Size of the Projects

On one hand there are companies that carry out simple short-term projects (from one to six months), for example low-cost videogames or mMarketing platforms. Due to the characteristics of there markets they need to respond immediately to the client's needs, especially publicity agents or operators; some small companies have been able situate themselves very well in this type of project. On the other hand, there are companies that specialize in complex, long-term projects (from 12 to 18 months) that require a group of very specialized professionals and a high level of coordination.

How do the companies manage these projects? Flexibility seems to be the key concept for understanding their capacity to adapt to the market's rhythm. The companies that carry out different short-term projects need to work together in parallel. Often these companies are precisely those with the fewest employees, and therefore the situation is always critical, as they may carry out up to six products at the same time.

D. The Fusion Myth

In such a dynamic market as that of mCommunication it is possible that a company with few employees can be absorbed by a large company with diversified business models on a global scale. Many young enterprises wish to the bought by a larger company. This discourse, which could be defined as the *myth of fusion*, emerged in many of the interviews we carried out, mainly for the mMarket-

ing and videogame companies. Why do we say 'myth'? Because many business people, although they vindicate their independence and capacity to adapt quickly to the client's demands as a result of their small size, really aspire to fuse with a larger company.

One of the paradigmatic cases of fusion, widely referred to by the interviewees, is the case of Digital Chocolate buying the company Microjocs. The global expansion of Digital Chocolate has not been limited to Catalonia, they have also bought a studio in Helsinki (Finland) and a programming company in Bangalore (India). These fusions, that really have little to do with 'fusing' as the large company buys the small one, have not led to any radical changes in the production routines of the bought company: the company that buys the other company is looking for precisely this, a particular way of designing and producing contents.

This model reproduces a organizational form already tested successfully in many other sectors of the economy (automobile industry, electrical appliance industry, etc.) in which the business headquarters is located in the United States, the design is carried out in Europe and the production in countries with low labor costs.

E. Multiskilling and Outsourcing

In the companies with few employees, which is the majority in the sample studied, the employees assume various tasks. One person interviewed defined it as an 'adaptable, amorphous structure'. This is not a new situation: multiskilling is common in digital communication companies in Catalonia [19] [20].

Unlike other production centers, Barcelona has little outsourcing: Catalan companies prefer to concentrate their production in their own territory. However, in certain sectors such as the mMarketing sector there are some companies who work for large foreign publicity agencies. These are global or national companies that need a videogame or contents specifically for mobiles as part of a multimedia publicity strategy. Some small companies from Barcelona have successfully found a niche in this area, making Catalonia a receiver rather than a generator of outsourcing.

F. Company Models, Business Models

Two different company profiles became clear during the research:

- Companies of recent origin created after 2000 that were formed exclusively for mobile communication (*native companies*);
- Consolidated companies that incorporated the mobile media into their traditional products after the year 2000 (*migrant companies*).

Native companies predominate in the sample studied here. Although there are few migrant companies, they contribute significantly to the content production. Native companies tend to be formed by a group of young professionals who decided to work together, or as a new company within a larger one that is consolidated in another sector (computing, audiovisual, etc.). Most of the companies that operate in the mMarketing area belong to the first group, and there is one example of the second type in the audiovisual sector.

Migrant companies are companies that come from other markets, for example audiovisual production, journalism or the Internet, which have included contents for mobile communication devices in the products they offer. Journalistic companies predominate in this group, as they use this new channel as another platform for spreading information content. In most cases the production for mobiles is an appendage to the contents designed for the Internet, and in this sense these companies still do not have a mature business model.

Both models are found in the videogame sector, from companies created for designing games for consoles and computers that migrate to the mobile sector to companies created for producing interactive contents for mobile communication devices.

In the context of conforming to a new market, it stands out that the contents offered by many new, independent, private producers is based on a single product (*monoproduct* phase). Nearly all of the enterprises interviewed were created with this format, generally based on SMS messages, and they then broadened their services, continually adapting to the constant changes of the market (MMS, advergames, clips, etc.) (*diversification* phase).

G. The Company Network

In saturated markets, for example the editorial or television markets, it is very difficult for new actors to break in. However, the opposite is true for the mobile communications sector: the difficulty here lies in the fact that there is no consolidated market with established actors with shared game rules. This is why it is very difficult to find a place in an ecosystem that is still in its initial construction phase. Both the native and migrant companies have original business proposals that involve a high degree of experimentation and commercial uncertainty. It is common for small companies to establish periodic tactical alliances in accordance with the demands of the moment to carry out more ambitious projects and/or offer more services to their clients.

The model of the company network that Castells [21] describes fits this dynamic. It is an:

'[...] organizational form constructed around a business project, which results from the cooperation between different components of various companies that operate in a network while a certain business project lasts; the networks are reconfigured to complete each project' (pp. 84).

In the interviews carried out various enterprises from the videogame and marketing sectors highlighted this aspect of their production dynamic. The flexibility to grow without losing control of the production process is one of the most essential characteristics of Castells' model and the work method of the companies interviewed, especially the native companies. Identifying this small swarm of multiskilled, flexible companies that can adapt to the permanently mutating market has been one of the most notable discoveries of this study.

H. An Operator-Dependant Model

Another aspect that stands out and which conditions the functioning of these companies is the central role of the mobile network operators (Movistar, Vodafone, Orange, etc.) through the construction of what are known as *walled gardens*. The interviewees complained of the operators' excessive influence on the market and the way in which

this conditions the evolution of their products. The great majority of the current business models are *operator-dependent,* which leaves the small companies at the mercy of the large telecommunication companies. This generates great uncertainty for the companies interviewed.

The way to escape from this is to diversify clients and products. The companies that find themselves tied to mobile network operators know that the solution is to develop different types of contents and services for other buyers. For example, a company that worked exclusively for Nokia developed one of the first videogames for Apple's iPhone 3G; other companies diversify their clients and attempt to produce for the television industry, editorial groups or larger companies in the mobile sector. This situation is starting to change with the appearance of 3G devices and Mobile Internet, since in these new scenarios the content producers do not depend exclusively on the operators to offer services to the users.

I. Economic Actors: General Situation

Most of the interviewees recognized that they are new actors with little experience in the mCommunication business and that they operate in an expanding industry in its initial stages. The large migrant companies that come from the journalism and audiovisual sectors and who have recently entered mCommunication are also in this situation. The 'new medium', both technologically and culturally, is currently in a process of construction in which constant experimentation dominates most of the business practices.

However, not all the areas within the mobile communication sector in Barcelona are in the same economic situation. While some recognize the relative immaturity of the market and note the limited amount of business, others feel that there are more and more possibilities every day. For example, the videogame industry is now in a very good moment, with a very competitive professional environment, clients who bring a large amount of business and enormous possibilities to continue growing. After the videogame industry comes mMarketing, a sector that is becoming profitable and whose actors are confident will become a very lucrative business very soon given that the market is opening up quickly. Further behind we find the currently loss-making journalism market. The audiovisual sector and the sector of user-generated contents (mobile web 2.0), except for some exceptions of contents for an adult public, have almost inexistent economic turnover in Catalonia.

For the companies interviewed, the fact that the Mobile World Congress is held in Barcelona is essential for promoting the visibility of their companies and constructing international networks. This event is a commercial beacon for launching many Catalan companies and products and an enormous opportunity that other Spanish or European communities do not have.

IV. mCOMMUNICATION CONTENTS AND SERVICES IN CATALONIA

A. Journalism Contents and Services

Since the year 2000 the main Catalan journalistic companies provide information services through mobile communication devices. The expectations for the new medium have led these companies to create contents, but always as an appendage to the contents produced for the Internet. In most cases this has inhibited the creation of structures exclusively for content production for mobiles and therefore has multiplied the tasks of ciberjournalists.

The contents chosen for mobile devices can be defined perfectly as service journalism, as the journalistic content is adapted to the way mobile users consume information. The main contents are weather forecasts, sports results, political events, etc.. Interpretive journalistic practices and opinions are not included in information production.

Among the formats most used are SMS alerts sent to subscribers (based on push technology) and the contents that are available online 24 hours a day that users can consult when they wish (pull technology). Together with some pioneering experiences in the video sector (see 4.4.), the alert services and WAP portals are the only journalistic content created specifically for mobile devices in Catalonia.

The alerts are brief and immediate, written in a telegram style like news headlines. The alert format blurs the boundary between journalistic information and service information. They are sent to subscribers who have signed up to receive this information periodically. The communication groups analyzed in this study who offer alert services send about 30 messages every month. For the managers interviewed, the main added value of this format is its immediacy. That is, the users, normally influential people in the economic and political worlds, can access current news items as soon as they happen, and can therefore act on this information immediately. At the moment this is the only service that generates revenue, as there are no profits from advertising in this sector.

The importance and speed with which the alert news is spread can be represented by a pyramidal hierarchy: first for mobiles (quick, instantaneous, short, spread quickly), then for the web (where the journalist has to spend more time writing up the news item) and lastly in the traditional media such as newspapers (where journalists write a more detailed, elaborate article).

The WAP format has become bogged down after its initial burst of activity at the end of the 90s. This is due to the high hopes producers have for the 3G and 4G technologies, as they offer better connectivity and more features.

The main companies in this sector -CCRTV Interactiva, Vilaweb and La Vanguardia- are also important actors in the diffusion of news content over the Internet, which confirms the subordinate nature of content production for mobile devices in relation to the web, which itself was subordinate, even until very recently, to the traditional media (printed press, television, etc.).

B. mMarketing: Products and Services for Advertisement

mMarketing is defined as the use of mobile platforms for sending messages (SMS, MMS), downloading applications or surfing the web for interactive publicity purposes. mMarketing is included in publicity campaigns as it favors client loyalty, improves the brand's image, encourages repeated buying, directs buyers to the selling point and establishes a new communication channel between the user and the brand. Among the contents most diffused in the media are company logos, wallpapers, audio publicity jingles or files, SMS, MMS, watermarks, advergames and brandgames, WAP portals, etc. [22] [23] [24].

All the companies studied produce their own contents and have their own platforms to manage and carry out this activity. This way the companies offer their products and services to the clients but it is the clients who manage the platform and follow the campaigns.

Some companies produce content exclusively for mobiles (SMS, MMS, advergames, etc.) while others are publicity or consultancy agencies that produce content for mobiles and other channels (polifunctional agencies).

Almost all the companies studied agree that currently SMS is the leader in publicity campaigns. SMS has become the killer application of mMarketing thanks to its low cost and technological simplicity as all mobile devices on the market can send and receive this type of message. If the rates are changed it is probable that SMS will be replaced by MMS, but at the moment the cost of these messages in Spain depends on the success or failure of this type of advertising. In this context, Bluetooth technology is increasingly used as it allows the user to access contents and download them for free, as well as produce contents and services based on the user's location (events, train stations, airports, etc.).

C. Mobile Gaming

Mobile Gaming (mGaming) is the result of a technological development which combines consolidated experiences and product lines. The first pre-installed game in a mobile phone was Snake, a classic for portable consoles which Nokia introduced in some of its models in 1998. Today the market has diversified and it is possible to access videogames from a wide range of wireless communication devices (PSP, Nintendo DS, iPods, smartphones, etc.). Catalan companies are very dynamic actors in this sector. Entertainment products designed in Barcelona have gained international recognition on various occasions.

There are two types of production: on one hand, there are the companies that produce low-cost games over a short time period (less than six months), and therefore the quality of the graphics and plot is not very high; and on the other hand, there are the companies that design more elaborate, high-quality games – for example, using 3D and motion capture systems – which can take up to 18 months to produce. According to the interviewees the success of the game does not depend so much on the graphic quality or the time employed to produce it but on other factors such as playability, being culturally in tune with the public, and the publicity strategies the operators apply.

The speed of production depends on different factors, from the complexity of the game to the structure of the company, as well as the number of versions (not forgetting the market's many different standards) and deadlines. As in the mMarketing sector, videogame producers are very creative and in most cases the work-teams are small and flexible and can adapt to the clients' demands and the current trends. Except for some specific cases, each game is an independent production and not a version of a famous game designed previously for wireless consoles or home technologies.

The people interviewed during this study debate between designing high-quality games and producing games quickly that end up feeding the operators' content offer. In this second sector videogame and marketing platforms overlap. Examples of this kind of hybrid product are *Girl Finder* and *Lynx FX* designed by Kailab for the *Lynx* campaigns in the United Kingdom. These types of campaigns include spots, advertisements in the press, videogames and contents for mobiles, videos in Youtube, etc. and are an example of transmedia narrative [25] applied to marketing.

One of the most sophisticated videogames produced in Catalonia is *One*, designed by Digital Legends for Nokia's N-Gage device. This fighting game was created using an innovative 3D technology with real combatants and motioncapture processes, as well as the possibility to participate in a virtual arena with players from all over the world. Other companies such as Microjocs-Digital Chocolate are based in retrogaming, that is, they convert classic videogames into a format for mobiles, among the most successful are *Movi Domino* (2006) and *Movi Futbolín* (2006). Other popular, widespread videogames for mobiles produced in Catalonia are *The Lord of the Rings* (Microjocs), *Fernando Alonso Racing* (Gaelco Móviles),and *Rafa Nadal Tennis* (Virtual Toys).

D. Mobile Television

According to the North American researcher Amanda Lotz [26] television is expanding 'outside the box'. Kumar [27] considers that mobile television:

> '[...] is emerging as the killer application of the 21st century [...] Mobile TV, the newest addition to the mobile services portfolio, is a sunrise technology with a potential user base of over 200 million by 2011' (pp. xiiii).

However, it is a peculiar fact that people talk a lot about mobile television (*mTV*), and construct hypotheses about its impact [28], but very few people actually watch it. Why? Because the offer of audiovisual content specifically for mobile devices is in its initial stages and is very under developed compared to the production of videogame, journalistic and publicity content.

In Catalonia the audiovisual content for mobiles, for example in mobisodes, trailers, and fictional and non-fictional clips produced especially for mobiles, is very limited and only works well for certain genre, such as content for adults. Contradictory attitudes to mTV emerged in the interviews: some companies keep it in mind, attentively follow its evolution and produce some contents in order to understand the dynamics of this new channel; however, other companies are distrustful and even resistant to the idea.

A Catalan company holds a predominant place in the generation of contents for adults, which is adapted from other media, generally television or cinema. The adaptation work consists in reducing the productions to short clips of about three minutes and reframing the scenes to improve their visibility on the small mobile screens.

Another emerging sector in Catalonia is the production of clips adapted from television. The CCRTV experience called *TV3minutes* is a pioneer in this area. The Catalan public entity, through CCRTV Interactiva, produces short, three-minute videos that summarize the television news *Telenotícies*, weather forecast or some of the main fictions such as *Ventdelplà* or *El Cor de la Ciutat*.

The child public is absent from the strategies of the companies interviewed. Catalonia has always been at the forefront of this communication area, from editorial production to audiovisual and interactive content; however, this sector has characteristics that limit specific contents being produced. The contents produced for children are usually not in tune with the image the mobile phone has in schools and the family. The companies that produce contents for children want to protect their reputation by not producing for this 'new' medium, as mobile devices could end up sharing some of the negative connotations that have always been associated with television.

E. Institutional Information Contents

Most of the Catalan public institutions that took part in this study do not create contents specifically for mobile devices: they generally limit themselves to adapting the contents that they produce for other formats (web or paper) or that they have bought from external companies. The contents for mobile devices in this sector, which are similar in some ways to journalistic contents, can be text, photographs, audiovisual and may even be interactive. Currently, public organizations carry out experimental and educational work so that users become used to using this channel and discover the advantages it offers.

Like in mMarketing and journalism, SMS is currently the format used most, as it costs little and has few technical problems. Examples are educational content for university students designed by one of the companies interviewed, public service contents, the Ajuntament de Barcelona's information service, for example information on public transport (timetables, delays, agenda, directory, BCN guide, How to get to, etc.), and cartographical services that use GPS technology and maps. These contents and their corresponding distribution platforms are bought from third parties; within Barcelona Council the team that generates the information services for mobile devices is the same that designs the contents for the Internet.

F. User-Generated Contents

The convergence between technical aspects and consumption practices has led to the hybridization of mobile devices and the web 2.0 [2] [3] to generate a new space called mobile web 2.0 [29] [30]. If the mobile Internet defines using the Internet on mobile devices, the mobile web 2.0 refers to using and producing content via mobile devices in social networks and in all Internet applications in which the contents are generated by the users.

In Catalonia this market is still in its initial stages; there are currently only three companies in this sector. The predominant business model is based on constructing virtual communities, for example for businesspeople, members of the university community [31] or for users interested in entertainment uses. However, these developments are still in a very early phase and there is no relevant economic turnover.

G. Contents and Services: General Situation

As the research advanced a clear taxonomy of contents for mobile devices began to appear: *specific, adapted* and *non-adapted* contents. Specific contents are created especially for mobile devices, while adapted contents, which generally come from the web or television, are transformed in order to be distributed through this new channel; finally, non-adapted contents arrive directly to the mobile device without having been transformed.

If we apply this classification to the production of contents for mobile devices in Catalonia, we find the following panorama: there is extensive production of videogames and specific mMarketing contents, while journalistic and audiovisual production falls into the category of adapted contents (for example, CCRTV Interactiva's fictional and non-fictional productions and the videos for adults). None of the companies interviewed produces non-adapted contents.

Before ending this section we would like to highlight the recombination of different genre and the appearance of hybrid products. In many cases the boundary between a videogame and a marketing product is blurry. Many marketing strategies are decidedly transmediatic and include videogames for mobiles, MMS, videos on Youtube, traditional advertisements and television spots.

V. CONCLUSION

The objective of this work was to create a situation map of the cluster of actors, contents and trends in mobile communication in Catalonia. Considering the transformations that this sector is undergoing, this initial map provides a general panorama that should be extended in more detail with more specific studies and/or comparisons with other realities.

Although the current situation in Spain trails behind other societies in which mCommunication is extremely widespread (i.e. South Korea, Japan, Finland, etc.), Barcelona, thanks to the Mobile World Conference and the development of a network of small companies that create contents and services, is slowly entering the world map of mobile communication. Like the other countries that are more advanced in this sector, Catalonia is in a dynamic and exciting moment in history.

From the perspective of communication studies mobile devices can be considered the 'new' new media, with all the theoretical, methodological and epistemological consequences that derive from this. Due to its newness, recent mobile communication is now entering the media studies agenda; however, it is necessary for it to become a more integral part of the research into traditional or interactive media. In this context, this study of mobile communication in Catalonia is in its initial stages.

Various possible directions for future work have emerged during this research. We would like to point out two aspects that we consider relevant:

1) To carry out their projects, the companies, especially the small ones, need to adopt flexible work forms that promote multiskilling and interchanging roles. In certain sectors such as mGaming or mMarketing the smallest companies usually create collaborative networks (Castells' 'company network') depending on the projects they are undertaking. This work model deserves to be examined in more detail in future studies.

2) A more in-depth study of the new hybrid formats that goes beyond the classification proposed here (specific contents, adapted and non-adapted contents) is necessary. In this study we observed that sometimes the boundary between a videogame and a marketing product is blurry. Many communication strategies are decidedly transmediatic and include videogames for mobiles, MMS, videos on Youtube, traditional advertisements and television spots. Future research should examine in more depth the

narrative structures and the multimodal meaning strategies that are being generated.

Research into mCommunication from the media studies perspective has only just begun and has a long way to go. To follow this path it is necessary to understand that a new communication medium has entered the cultural industry, a medium with its own business models, grammar, production practices and consumption dynamics. The aim of this study was to make an initial approach to the production of contents and services for mobile devices in Barcelona. With this objective we have elaborated a series of analytic categories (native company/migrant company, monoproduct phase/diversified phase, adapted/non-adapted/specific contents, etc.) that allow us to begin to understand the dynamics of this sector of the cultural industry.

This first exploratory study should be followed by others that examine some of the open research lines and incorporate others. Unlike other historic moments, for example the birth of cinema, media studies already has the theoretical, methodological and analytical elements for studying the appearance and development of a new 'species' within the media system. The consequences of this apparition will be felt in the entire communication system, which makes it necessary to include mobile communication in the agendas of researchers and any other related area, from consultant agencies to the entities that establish the sector's policies.

REFERENCES

[1] C. Scolari, *Hipermediaciones. Elementos para una teoría de la comunicación digital interactiva,* Barcelona: Gedisa, 2008.

[2] T. O'Reilly, *What Is Web 2.0 Design Patterns and Business Models for the Next Generation of Software*, 2005. Online version: http://www.oreillynet.com/pub/a/oreilly/tim/news/2005/09/30/what-is-web-20.html

[3] C. Cobo Romaní and H. Pardo Kuklinski, *Planeta Web 2.0. Inteligencia colectiva o medios fast food*, Barcelona / México DF: Grup de Recerca d'Interaccions Digitals, UVic / Flacso México, 2007.

[4] D. Steinbock, *Wireless Horizon. Strategy and Competition in the Worldwide Mobile Marketplace*, New York: Amacom, 2003.

[5] D. Steinbock, *The Mobile Revolution*, London/Philadelphia: Kogan Page, 2005.

[6] International Telecommunications Union, *Mobile Cellular Suscribers*, 2007. Online version: http://www.itu.int/ITU-D/icteye/DisplayCountry.aspx?countryId=73

[7] CMT - Comisión del Mercado de las Telecomunicaciones, *Nota Mensual – Abril 2008*. Online version: http://www.cmt.es/

[8] GAPTEL - Grupo de Análisis y Prospectiva del Sector de las Telecomunicaciones, *Contenidos Digitales. Nuevos Modelos de Distribución Online*. Madrid: Red.es, 2006. Online version: http://observatorio.red.es/estudios/documentos/ContenidosDigitales_final.pdf

[9] Fundación Telefónica, *La sociedad de la información en España 2007 – Resumen Ejecutivo*. Madrid: Fundación Telefónica, 2007. Online version:http://sie07.telefonica.es/

[10] V. Reding "La tecnología de la información y la comunicación: motor de la economía moderna" in *Nuevo paradigma de los medios de comunicación en España*. Madrid: Nueva Economía Forum / Forum Europa, 2007.

[11] A. De Miguel and R. Barbeito, *El impacto de la telefonía móvil en la sociedad española*. Madrid; Tabula Ikonika, 1997.

[12] D. Parra, "Internet móvil como nuevo canal de información especializado" in M. Cebrián Herreros and D. Bartolomé Crespo (eds.) *Investigación sobre medios de comunicación/ Medienforschung*, Seminarios Internacionales Complutense, pp. 147-55, Madrid: Servicio de Publicaciones UCM, 2002.

[13] J. M. Aguado and I. Martínez, "El proceso de mediatización de la telefonía móvil: de la interacción al consumo cultural" in *Zer*, 20, 2006, pp. 319-43.

[14] F. Vacas Aguilar (2007) "Telefonía móvil: la cuarta ventana" in *Zer*, 23, 2007, pp. 199-217.

[15] M. Avià and E. Castelló "Periodisme i Internet mòbil: un cas de servei informatiu via MMS: 'El día en cinco imágenes' de 'La Vanguardia'" in *Anàlisi: Quaderns de comunicació i cultura*, 31, pp. 123-48, 2004.

[16] F. Colombo, "La comunicazione sintetica" in G. Bettetini and F. Colombo (eds.) *Le nuove tecnologie della comunicazione*, Milan: Bompiani, 1996, pp. 265-97.

[17] J. Groebel, E. Noam and V. Feldmann, *Mobile Media. Content and Services for Wireless Communication*, Mahwah, NJ: Lawrence Erlaboum Associates, 2006.

[18] A. Laramée and B. Vallée, *La recherche en communication. Éléments de méthodologie,* Québec: Presses de l'Université de Québec, 1991.

[19] C. Scolari, J. Micó, H. Navarro and H. Pardo Kuklinski, *Nous perfils professionals de l'actual panorama informatiu, audiovisual i multimèdia de Catalunya*, Vic (Spain): Eumogràfic, 2006.

[20] C. Scolari, J. Micó, H. Navarro and H. Pardo Kuklinski, "El periodista polivalente. Transformaciones en el perfil del periodista a partir de la digitalización de los medios audiovisuales catalanes" in *Zer*, 25, 2008, pp. 37-60.

[21] M. Castells, *La galaxia Internet: Reflexiones sobre Internet, empresa y sociedad*, Barcelona: Plaza y Janés, 2001.

[22] T. Ahonen, T. Kasper and S. Melkko, *3G Marketing: Communities and Strategic Partnerships,* Chichester: John Wiley and Sons, 2004.

[23] R. Mathieson, Branding *Unbound; The Future of Advertising, Sales, and the Brand Experience in the Wireless Age,* New York: Amacom, 2005.

[24] T. Weiss, *Mobile Strategies: Understanding Wireless Business Models, MVNOs and the Growth of Mobile Content*, London: Futuretext, 2006.

[25] H. Jenkins, *Convergence Culture: Where Old and New Media Collide,* New York: New York University Press, 2006.

[26] A. Lotz *The Television Will Be Revolutionized,* New York: New York University Press, 2007.

[27] A. Kumar *Mobile TV: DVB-H, DMB, 3G Systems and Rich Media Applications*, Burlington, MA: Elsevier, 2007.

[28] S. Orgad, *This Box Was Made for Walking... How will mobile television transform viewers' experience and change advertising?*, London: London School of Economics and Political Science/Nokia, 2006. Online version: http://www.lse.ac.uk/collectio ns/pressAndInformationOffice/PDF/Mobile_TV_Report_Orgad.p df

[29] A. Jaokar and T. Fish, *Mobile Web 2.0: The Innovator's Guide to Developing and Marketing Next Generation Wireless/Mobile Applications*, London: Futuretext, 2006.

[30] H. Pardo Kuklinski, J. Brandt and J. Puerta, "Mobile Web 2.0. Marco teórico y tendencias de desarrollo en la industria de la comunicación móvil', *V Colóquio Brasil-Espanha de Ciências da Comunicação*, Brasilia, Brasil, 28-30 August 2008,

[31] H. Pardo Kuklinski and J. Brandt, "Campus Móvil: designing a mobile Web 2.0 startup for higher education uses", *5th International Conference on Social Software – BlogTalk Conference*, Cork, Ireland, 3-4 March 2008.

10

M-Learning: The Usage of WAP Technology in E-Learning

J. Al-Sadi, B. Abu-Shawar

Arab Open University, Amman, Jordan

Abstract—This paper presents the experience of Arab Open University; AOU for short; on using WAP technology in mobile learning; m-Learning. The goal is to enhance e-Learning aspects of the existing learning management systems (LMS); e.g. Moodle. In addition to presenting technical aspects of the WAP, we also introduce the advantages and disadvantages of using WAP technology in the learning process. Furthermore, a description of AOU's learning management system with the outcomes of the tutors and students is introduced. This paper discusses also the suitability and feasibility of using WAP technology devices for distance learning in real-time.

Index Terms—e-Learning, m-Learning, WAP, CMC, LMS, Open source.

I. INTRODUCTION

As technology rapidly advances, new ideas for learning will also emerge. Computer-mediated communication (CMC) facilitates the development of such learning communities. In the broadest sense, CMC refers to any communication via computers; common applications include e-mail, online bulletin board, and online chat. The most important educational affordance of CMC lies in its connectivity: it connects learners at different geographical locations beyond the boundaries of classrooms; allows learners to exchange information within a short period of time synchronously or asynchronously; and provides the flexibility for one-to-one, one-to-many and many-to-many communication. Connectivity among learners is critical for a learning community, without which social construction of collective knowledge would be greatly impeded.

The growth of Internet-based technology have brought new opportunities and methodologies to education and teaching representing in e-learning, online learning, distance learning, and open learning. These approaches are typically use in place of traditional methods and mean that students deliver their knowledge though the web rather than face-to-face tutoring.

e-Learning is a new trend of education system, where students deliver their materials through the web. e-Learning is the "use of internet technology for the creation, management, making available, security, selection and use of educational content to store information about those who learn and to monitor those who learn, and to make communication and cooperation possible." [1].

Kevin [2] addressed the benefits of e-learning for both parties: organization and learners. Advantages of organizers are reducing the cost in terms of money and time. Learning time reduced as well, the retention is increased, and the contents are delivered consistently. On another hand, learners are able to find the materials online regardless of the time and the place; it reduces the stress for slow or quick learners and increases users' satisfaction; increases learners' confidence; and more encourages students' participations.

New technologies such as Wireless Application Protocol (WAP), General Packet Radio Service (GPRS) and 3G (3rd Generation) technologies further augment the educational potential of CMC by allowing learners and tutors access to the Internet, anywhere and anytime, via the micro browser equipped mobile phone.

WAP technology and the concept of e-learning is an evolving trend in education at all levels. Even classrooms with younger children are benefiting from the use of hand-held mobiles and laptop computers.

Recent advancement in mobile and wireless technology as basic requirements of WAP applications has helped to improve commerce [3, 4, 5, and 6] and services [7, 8]. Mobile technology is strategic to many organizations and activities [9, 10]. Education is no exception! The use of mobile technology has not only extended desktop-based online learning environment into the mobile and wireless channel but also enabled education to take place anytime, anywhere [11]. Educational materials can be delivered to students through mobile devices. Mobile technology can also be integrated into learning management system to improve interactivity in the classroom and also in distance learning.

Research into new WAP technologies that enhance instructional activities will continue to support the educational community as it embraces the idea of anytime, anywhere learning. E-learning is not just about readily accessible information; it opens up the possibility for the delivery of multimedia information, interactive learning and assessment, and real-time distance collaboration.

Applications of WAP technology in education can provide benefits to both students and educators. WAP technology provides greater flexibility in student learning. Students can have access to educational materials through their WAP enabled mobile devices, which enable them to learn as and when the need arises and when the time is right for them, no matter where they are, even when they are on the move. With mobile devices, educational materials are not only readily available to students but they can also be delivered to students based on their needs and preferences [12, 13].

Sharma and Kitchens stated that M- learning offers a unique opportunity for teachers and students in different kinds of learning environment settings. The unique feature of this mode of learning is that it enhances flexibility for students; however, it demands new pedagogies, and new approaches to deliver a course. If appropriately facilitated, m-learning helps learners in a great way by providing virtual classrooms on their mobile devices. Teachers will ultimately spend more time for course-delivery and follow-up as compared to traditional classroom method. In addition, teachers will have to provide a rich learning resource and environment, which in turn, contributes to the quality of learning [14].

In this paper, the technical overview of WAP technology is presented in section two. A full description of the Arab Open University (AOU) LMS is presented in section three. Sections four and five present the using of WAP technology including its requirements and benefits. Sections 6 concludes this paper and section 7 presents the future work.

II. TECHNICAL OVERVIEW

Wireless Application Protocol (WAP) is a secure specification that allows users to access information instantly via handheld wireless devices such as mobile phones, pagers, two-way radios, Smart phone and communicators.

WAP is designed to be user-friendly and innovative data applications for mobile phones easily. There are three types of terminals have been defined:

- Feature phones, which offer high voice quality with the capability of text messaging and Internet browsing;
- Smart phones, with similar functionality but with larger display.
- The communicator, which is an advanced terminal designed with the mobile professional in mind, similar in size to a palm-top with a large display.

WAPs that use displays and access the Internet run what are called micro browsers; browsers with small file sizes that can accommodate the low memory constraints of handheld devices and the low-bandwidth constraints of a wireless-handheld network.

WAP uses Wireless Markup Language (WML), which includes the Handheld Device Markup Language (HDML) developed by Phone.com. WML can also trace its roots to eXtensible Markup Language (XML). A markup language is a way of adding information to your content that tells the device receiving the content and what to do with it. The best known markup language is Hypertext Markup Language (HTML). Unlike HTML, WML is considered a Meta language. Basically, this means that in addition to providing predefined tags, WML lets you design your own markup language components. WAP also allows the use of standard Internet protocols such as UDP, IP and XML.

Although WAP supports HTML and XML, the WML language (an XML application) is specifically devised for small screens and one-hand navigation without a keyboard. WML is scalable from two-line text displays up through graphic screens found on items such as smart phones and communicators.

WAP also supports WML Script. It is similar to JavaScript, but makes minimal demands on memory and CPU power because it does not contain many of the unnecessary functions found in other scripting languages. Because WAP is fairly new, it is not a formal standard yet. It is still an initiative that was started by Unwired Planet, Motorola, Nokia, and Ericsson.

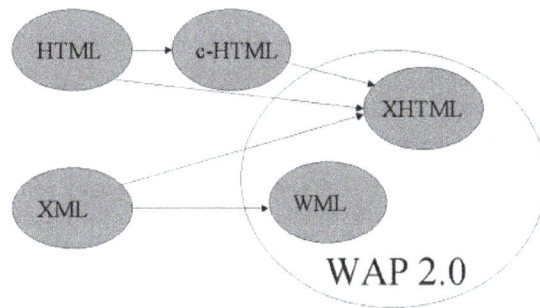

Figure 1. Migration of Markup language

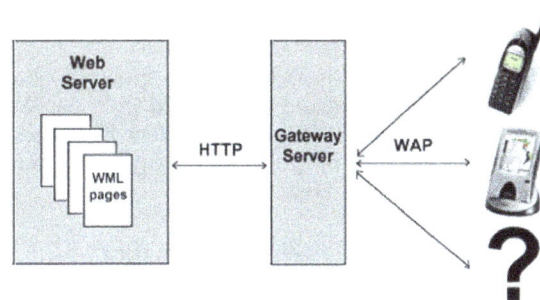

Figure 2. WAP Technology Infrastructure

There are three main reasons why wireless Internet needs the Wireless Application Protocol:

- Transfer speed: most cell phones and Web-enabled PDAs have data transfer rates of 14.4 Kbps or less. Compare this to a typical modem, a cable modem or a DSL connection. Most Web pages today are full of graphics that would take an unbearably long time to download at 14.4 Kbps. In order to minimize this problem, wireless Internet content is typically text-based in most cases.

- Size and readability: the relatively small size of the LCD on a cell phone or PDA presents another challenge. Most Web pages are designed for a resolution of 640x480 pixels, which is fine if you are reading on a desktop or a laptop. The page simply does not fit on a wireless device's display, which might be 150x150 pixels. Also, the majority of wireless devices use monochrome screens. Pages are harder to read when font and background colors become similar shades of gray.

- Navigation: navigation is another issue. You make your way through a Web page with points and clicks using a mouse; but if you are using a wireless device, you often use one hand to scroll keys.

WAP takes each of these limitations into account and provides a way to work with a typical wireless device.

Here's what happens when you access a Web site using a WAP-enabled device:

- You turn on the device and open the mini-browser.
- The device sends out a radio signal, searching for service.
- A connection is made with your service provider.
- You select a Web site that you wish to view.

- A request is sent to a gateway server using WAP.
- The gateway server retrieves the information via HTTP from the Web site.
- The gateway server encodes the HTTP data as WML.
- The WML-encoded data is sent to your device.
- You see the wireless Internet version of the Web page you selected.

Although WML is well suited to most mundane content delivery tasks, it falls short of being useful for database integration or extremely dynamic content. PHP fills this gap quite nicely-integrating into most databases and other Web structures and languages. It's possible to "cross-breed" mime types in Apache to enable PHP to deliver WML content. WML pages are often called "decks". A deck contains a set of cards. A card element can contain text, markup, links, input-fields, tasks, images and more. Cards can be related to each other with links.

When a WML page is accessed from a mobile phone, all the cards in the page are downloaded from the WAP server. Navigation between the cards is done by the phone computer (inside the phone) without any extra access communications to the server.

III. THE E-LEARNING PLATFORM OF THE AOU

Arab Open University was established in 2002 in the Arabic region, and adopted the open learning approach. AOU has partnerships with the United Kingdom Open University (UKOU) and other national educational institutes, such as MoHE, and international institutions, including UNISCO, to help ensure a high quality of teaching.

An open learning system is defined as "a program offering access to individuals without the traditional constraints related to location, timetabling, entry qualifications." [15].

The aim of AOU is to attract large number of students who can not attend traditional universities because of work, age, financial reasons and other circumstances. The "open" terminology in this context means the freedom from many restrictions or constraints imposed by regular higher education institutions which include the time, space and content delivery methods.

Freed et al. [16] claimed that the "interaction between instructors and students and students to students remained as the biggest barrier to the success of educational media". The amount of interaction plays a great role in course effectiveness [17]. For this purpose and to reduce the gap between distance learning and regular learning, the AOU requires student to attend weekly tutorials. Some may argue that it is not open in this sense; however the amount of attendance is relatively low in comparison with regular institutions. For example, 3 hours modules which require 48 hours attendance in regular universities, is reduced to 12 hours attendance in the AOU.

At the beginning the AOU used the FirstClass system as a computer mediated communication (CMC) tool to achieve a good quality of interaction. The FirstClass tool provides email, newsgroups and conferences as possible mediums of communication between tutors, tutors and their students, and finally between students themselves. The most important reason behind using FirstClass was the tutor marked assignment (TMA) handling services that FirstClass provided. However, the main servers are located in the UKOU which influences the control process, causes delays, and totally depends on the support in UKOU for batch feeds to the FirstClass system [18].

To overcome these problems, AOU use Moodle nowadays as an electronic platform. Moodle is an open-source course management system (CMS) and learning management system (LMS) used by educational institutes, business, and even individual instructors to facilitate web technology to their courses. A course management system is "often internet-based, software allowing instructors to manage materials distribution, assignments, communications and other aspects of instructions for their courses."[19]. CMS and LMS or virtual learning environments (VLE), are web applications, meaning they run on a server and are accessed by using a web browser. Both students and tutors can access the system from anywhere with an Internet connection. Moodle provides many learning tools and activities such as forums, chats, quizzes, surveys, gather and review assignments, and recording grades as components of its LMS.

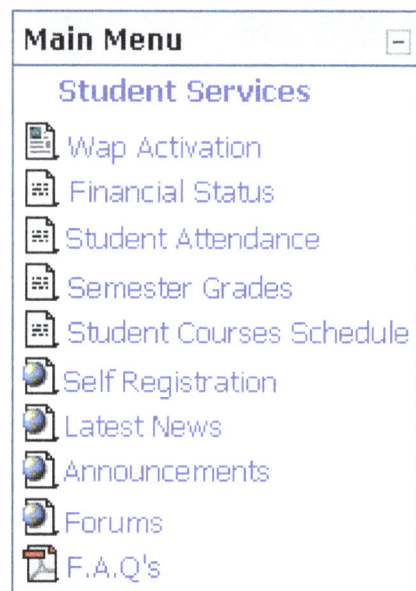

Figure 3.   AOU-LMS integrated tools

Figure 4.   The unified image of the AOU e-learning systems

Moodle was used in AOU mainly to design a well formed learning management system which facilitates the interaction among all parties in the teaching process, students and tutors, and more over to integrate the LMS with the student information system (SIS).

In addition that Moodle is easy to learn and use, and that it is popular with large user community and development bodies. Moodle is flexible in terms of:

- Multi-language interface,
- Customization (site, profiles),
- Separate group features, and pedagogy.

The unified image of the e-learning platform of the AOU from the starting web page shown in Fig. 4, the users will be able to:

- Connect to the SIS, where they could do online registration, seeing their grades and averages.
- Perform learning activities through the LMS, such as submitting assignments, do online quizzes, etc.
- Retrieve resources through AOU digital library subscriptions.

To fit the AOU requirements and specification, a number of modifications and customizations were made (see Fig. 4), including:

- Log records. Logs are replicated into other isolated tables, to increase performance, and to keep track records for long period, while removing these log records from original tables.
- Some facilities and activities are added.
- Student's attendance and absences sheets are provided.
- Grades customizations (fractions) excel sheets are available.
- Randomly captured assignments for quality assurances purposes.
- WAP (wireless application protocol) services (grades, schedule, financial issues, and news) are presented.

IV. WAP TECHNOLOGY USED IN AOU-LMS

AOU has a pioneer experience of developing and integrating the well known learning management system Moodle.

One of our recent integrations of our LMS is the use of WAP technology to mediate the system learning tools such as forums, chat rooms, and announcements. In the next paragraph, we will discuss integration opportunities and limitations of WAP technology for supporting a learning community. Recently, El-Seoud et al. [20] presented a framework for a mobile interface for moodle which utilize proper moodle activities over mobiles.

The main goal is to widen the usage of technology in education to make is reachable, feasible, and reachable to the students anytime, and anywhere. Lee [21] explored the potential of WAP technology as an effective online communicating tool when coupled with other tools.

As a component tool of AOU-LMS courses websites, students have the opportunity of using the integrating instructional technology effectively into their classroom practices. There are approximately 25% of face-to-face tutorial lessons for each course and the remainder of 75% of the course time is delivered using online e-learning activities. AOU has adopted Moodle, an online learning delivery and management system that allows students to learn independently and tutors to customize the online learning package according to their students' needs. Students are equipped with WAP enabled mobile phones to participate in the e-learning activities.

A mobile telecommunication service provider provides the mobile phones and the WAP services subscriptions. Students have the choice of using the e-learning tools via Moodle from their personal computers or via their WAP enabled mobiles using the WAP technology. All e-discussions took place at the class level, consisting of students and a tutor. In-house software developers developed a WAP-based e-discussion application to allow the students in the study to access the e-discussion forums via their WAP-enabled phones.

In this application, the tutor was able to do the administrative function, namely to manage the forums and his group, and to carry the discussion with his students via the WAP-enabled phones. Students logged on to the e-discussion via their phones to discuss with other classmates and the tutor. They created new threads, viewed threads, replied to and deleted messages. There are at least three forums for each course; extra forums can be added as required. One forum of a particular course is on case studies.

During a study of the WAP services, it was observed that students experienced difficulties in using the WAP-enabled phones as a result of slow transmission speed, navigational problems and short battery life. On the other hand, WAP technology offers opportunities for e-discussions in a learning community. It has been observed that WAP technology had helped to build a learning community. We believe that WAP technology have mediated the formation of a closely-knit group where everyone is able to participate and learn from one another. Its mobility is offering the tutor and students anytime-anywhere participation in the e-discussion forums. This opportunity for anytime-anywhere participation might motivate students to participate in the e-discussion forums.

The main limitation of WAP technology is the technical problems of using the WAP-enabled phones. These problems include:

- Short life span of the phone battery,
- Difficulties in logging into the WAP-based forums,
- Slow transmission, failure to send and the need to resend messages,
- Navigational problems, it is difficult to read and browse the messages on the small screen of the phone.
- Difficulties to key in messages using the WAP-enabled phones.

Technical problems are not the most critical, but the first, hurdle that must be crossed. Improving the technical capabilities of the WAP-enabled phones would encourage more participation in the WAP-based forums, and might even enhance the quality of WAP-based messages.

When comparing the mobile network to the fixed network there seems to be many limitations, as already discussed. The mobile network also provides unique advantages or features such as the position or location of the device and personalization (both user preferences and device capabilities). The WAP language supports these features of Positioning and Personalization. A WAP language component supports User Profiles which contains information on the user preferences and the device capabilities.

Furthermore, current WAP technology makes it best suited to particular aspects of e-learning courses, such as:

- Quick reminders and alerts
- Communication with peers and managers
- Multiple-choice quizzes with immediate feedback
- Daily tips
- Glossary information
- Browsing e-learning course material
- Searching for specific information within a topic
- Links to WAP sites
- Course registration

Tutors and students interact via forums as shown in Fig.5. There are mainly four types of forums as follows:

- News forum is used to announce important dates, such as final exams time table, and other news related to AOU.
- Course forum, is used by the tutor to send lecture notes to all students who are registered in a specific module.
- The tutor group forum as shown in Fig. 5 is used as an interacting media between tutors and students. Any message send by the tutor in the tutor forum could be accessed by all students registered in this module with this tutor. At the same time, if student send a message to the tutor or to another student, this message could be read by all.
- Dialogue forum, is used to add some privacy in student/tutor relationship, in which messages in these case will be not be accessed by all, only between both parties.

The new activity to be added is to ask a student if he/she wants to receive the tutor group forum messages through their mobile devices, if yes, then the sent messages will be delivered to the system as well as the students' mobiles.

In the same manner, the chatting is done through the website; it could be done using the mobile.

Moreover, any announcement via the LMS can be directly transmitted to the intended receivers' mobiles to guarantee a full awareness of such announcements.

V. REQUIREMENTS AND BENEFITS OF ADOPTING WAP TECHNOLOGY IN EDUCATION

Computer-based learning environments are increasingly becoming commonplace at colleges and universities. Myers and Talley [22] classify the computer-based technologies into two categories:

1. Online course management systems and specialized collaborative environments that allow electronic communication of classroom instruction materials to students, among students and between student and instructor;

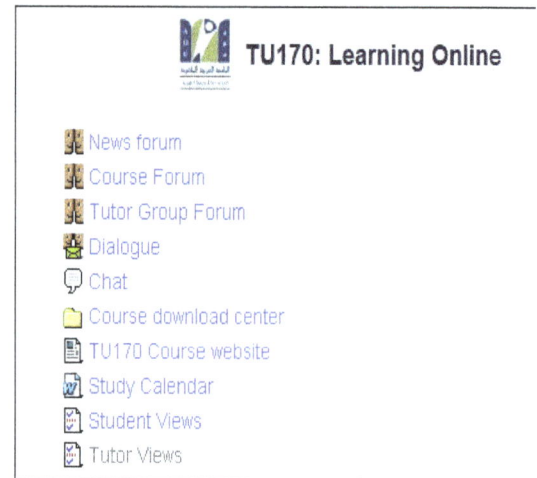

Figure 5. TU170 course (AOU) with its forums

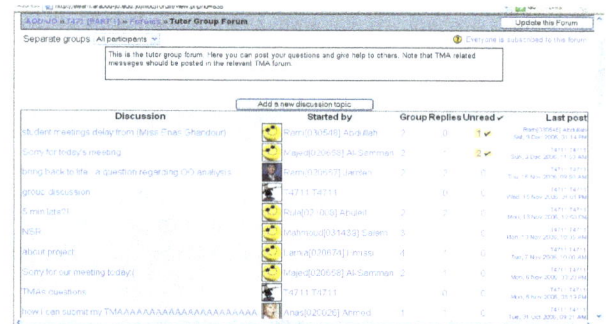

Figure 6. Tutors group forum view

2. Wireless network systems that allow faculty and students with mobile computers to communicate with their campus networks.

Mobile and wireless technology has been served in many domains such as commerce, and education. Mobile learning "m-Learning" is e-learning delivered through mobile computational devices. So mobile learning can provide online learners with capabilities to get instant notification on e-mail, access learning sites, report data from the field, and collaborate with learning colleagues. [1].

M-learning does not replace traditional learning; it just represents another way of learning. Siau and Nah [9] listed the usefulness of mobile technology for both students and tutors. The students' benefits are:

1. The education materials are available for students regardless of where they are;
2. Education materials can be delivered to students based on their needs and preferences.
3. Students can communicate and interact with pee students and educators in real time.

The benefits for tutors are presented by providing a new means of education delivery as well as adding a new dimension for student-tutor interaction. For example, wireless classroom response systems can be integrated into classroom instructions to gather students' responses and provide instantaneous feedback to students on their performance. This could improve classroom interactivity, enhance teaching effectiveness, and promote student learning.

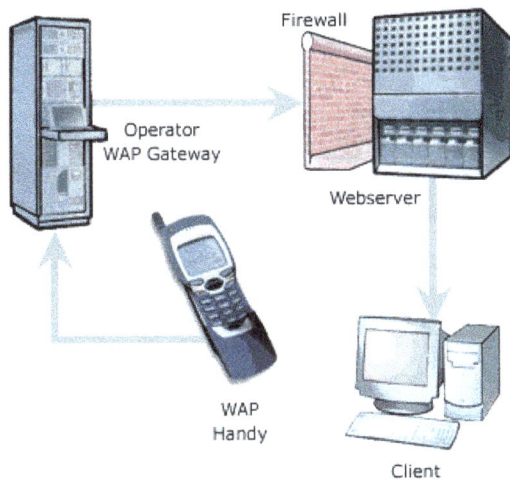

Figure 7.   The infra structure of enhancing the mobile technology

The attributes of mobile devices are operating system, large graphical display, touch screens, connectivity, memory, programmability, and personal information manager (PIM) functionality [23]. Recent evolutions on laptop computing have led to what is called TabletPCs, which is digital pen-based mobile PCs that use digital ink enabled software, and it is used in teaching [24], the general structure of enhancing the mobile technology with e-learning is shown in Fig. 7.

However, the availability, mobility, and performance of wireless technology depend on five major areas [25]:

4. Platform, the majority of PDAs and cellular phones being produced have embedded in them some type of Web browsing technology.

5. Connectivity, the true wireless connectivity is wireless Radio Frequency (RF). The wireless communication categories are Wireless Local Area Network (WLAN), and Wireless Wide Area Network (WWW). Training sites at corporations and students at universities use wireless connectivity to facilitate access to information, information exchanges and learning [26].

6. Wireless Middleware, provides services specific to the world of wireless and handheld computing. Wireless middleware services are secure communication management, synchronization, message processing and management tools [23].

7. Back-End System, handheld and wireless computing extends the reach of corporate data and corporate transaction engines. The data stored on a Web site, mainframe, UNIX server, or an Oracle database.

8. Security, in a wireless world, security includes communication links, integrity of the channel, and accuracy of transactions.

VI. CONCLUSION

This paper has explored the opportunities and limitations of WAP technology in mediating e-learning tools to build a learning community. To make WAP-based discussions more successful online communication technology for education, we must continuously refine our research plans and explore future areas of research. More studies may be conducted to explore the possible ways to make WAP technology a successful social and cognitive tool for facilitating individual learning and enhancing the social

construction of knowledge. The technical aspects of the WAP-enabled mobiles may also need to be considered seriously in the future development of WAP technology with respect to e-discussions and/or learning.

In conclusion, WAP is suitable for the creation of mobile learning training course material. The optimization of WAP and the handling of the design challenges make it feasible to use mobile handheld devices for distance learning in real-time. The application developer must always be aware of the user and take into account the usability issues if the application is to be a success. The experience of AOU on enhancing its LMS with WAP technology for a move toward m-learning has been introduced.

VII. FUTURE WORK

Our future trends are to keep enhancing our m-learning system and its integration with other online system used in the educational institutes. We will also do a complete comparison of infrastructure technologies used as a backbone of m-learning including WAP, 3G, and GPRS. Furthermore, we will do a performance analysis to measure the suitability of using m-learning in education and compare it with other learning technological platforms.

REFERENCES

[1]   F. Mikic, L. Anido, "Towards a standard for mobile technology", *Proceedings of the International Conference on Networking, Systems, Mobile Communications and Learning Technologies ICNICONSMCL'06* , vol. 0, pp. 217-222, 2006.

[2]   K. Kruse, "The benefits of e-learning", http://www.executivewomen.org/pdf/benefits_elearning.pdf, 2003

[3]   K. Siau, E. Lim, Z. Shen, "Mobile Commerce – Promises, Challenges, and Research Agenda," *Journal of Database Management*, vol. 12, no. 3, pp. 4-13, 2001.

[4]   J. Krogstie, K. Lyytinen, A. Opdahl, B. Pernici, K. Siau, K. Smolander, "Mobile Information Systems - Research Challenges on the Conceptual and Logical Level," *Lecture Notes in Computer Science – Advanced Conceptual Modeling Techniques*, vol. 2784, pp. 124-135, 2003.

[5]   K. Siau, Z. Shen, "Building Customer Trust in Mobile Commerce," Communications of the ACM, vol. 46, no. 4, pp. 91-94, 2003. (doi:10.1145/641205.641211)

[6]   H. Galanxhi-Janaqi, F. Nah, "U-Commerce: Emerging Trends and Research Issues," Industrial Management and Data Systems, vol. 104, no. 9, pp. 744-755, 2004. (doi:10.1108/02635570410567739)

[7]   K. Siau, Z. Shen, "Mobile Commerce Applications in Supply Chain Management," *Journal of Internet Commerce*, vol. 1, no. 3, pp. 3-14, 2002. (doi:10.1300/J179v01n03_02)

[8]   K. Siau, Z. Shen, "Mobile Communications and Mobile Services," *International Journal of Mobile Communications*, vol. 1, nos. 1/2, pp. 3-14, 2003. (doi:10.1504/IJMC.2003.002457)

[9]   F. Nah, K. Siau, H. Sheng, "The Value of Mobile Applications: A Utility Company Study," *Communications of the ACM*, vol. 48, no. 2, pp. 85-90, 2005. (doi:10.1145/1042091.1042095)

[10] H. Sheng, F. Nah, K. Siau, "Strategic Implications of Mobile Technology: A Case Study Using Value-Focused Thinking," *Journal of Strategic Information Systems*, vol. 14, no. 3, pp. 269-290, 2005. (doi:10.1016/j.jsis.2005.07.004)

[11] K. Siau, F. Nah, "Mobile Technology in Education", *IEEE Transactions on Education*, vol. 49, No.2, 2006 (doi:10.1109/TE.2006.875792)

[12] A.L. Foster, "Can You Hear Me Now?" *The Chronicle of Higher Education*, vol. 52, no. 12, p. A32, 2005.

[13] J. Chen, Kinshuk, "Mobile Technology in Educational Services," *Journal of Educational Multimedia and Hypermedia*, vol. 14, no. 1, pp. 91-109, 2005.

[14] S. Sharma, F. Kitchens, "Web Services Architecture for M-Learning", *The Electronic Journal of e-Learning*, vol. 2, no. 2, 2004

[15] P. Karampiperis, D. Sampson, "Designing Learning Services for Open Learning Systems Utilizing IMS Learning Design", *Proceeding of 4ᵗʰ IASTED Int. Conf. on web-based Education (WBE 2005)*, Swaziland, pp. 165-170, 2005

[16] K. Freed, "A History of Distance Learning", http://www.media-visions.com/ed-distlrn.html, 2004

[17] A.P. Rovai, K.T. Barnum, "On-line course effectiveness: an analysis of student interactions and perceptions of learning", *Journal of Distance Education*, vol. 18, no. 1, 57-73, 2003.

[18] S. Hammad, A.E. Al-Ayyoub, T. Sarie, "Combining existing e-learning components towards an IVLE". *EBEL conference*, 2005.

[19] M.C. Lohman, "Effects of Information Distributions Strategies on Student Performance, and Satisfaction in a Web-Based Course Management System", *International Journal for the Scholarship of Teaching and Learning*, vol. 1, no. 1, pp. 1-17, 2007.

[20] S. Abou El-Seoud, Ashraf M. Ahmad, H. El-Sofany," Mobile Learning Platform Connected to Moodle Using J2ME", *International Journal of Interactive Mobile Technologies (iJIM)*, Vol 3, No 2, pp. 46-54, 2009

[21] C. Lee, "Exploring the Potential of WAP Technology in Online Discussion", *Association for Educational Communications and Technology*, 27th, Chicago, IL, October 19-23, 2004.

[22] S. Myers, D. Talley, "Looking beyond the Whiz-bang technology: using mobile learning technology tools to improve Economic instruction". *The ASSA*, USA, 2007

[23] Sbihli, S. Developing a successful wireless enterprise strategy. *New York: Wiley Computer Publishing*, 2002.

[24] B. Abu-Shawar, J. Al-Sadi, & A. Hourani, "Integrating the Learning Management System with other Online Administrative Systems at AOU", *Proceedings of The International Conference on Algorithmic Mathematics and Computer Science (AMCS'06)*, USA, June 22-25, 2006

[25] B. Sasidhar, B. Kumar. "The effects of mobile devices and wireless technology on E-learning", *Sunway Academic Journal*. vol 2, pp. 45-53, 2005.

[26] G. Rogers, J. Edwards. "An introduction to wireless technology. Upper Saddle River", *NJ: Prentice Hall*, 2003.

AUTHORS

J. A. Al-Sadi is the head department of Information Technology and Computing at Arab Open University, Jordan (e-mail: j_alsadi@aou.edu.jo)

B. A. AbuShawar is an instructor in department of Information Technology and Computing at Arab Open University, Jordan (e-mail: b_shawar@aou.edu.jo)

11

The Role of Podcasts in Students' Learning

I. Nataatmadja and L. E. Dyson

University of Technology Sydney, Australia

Abstract—**Podcasts have been employed extensively in some countries and are now being trialed at a number of universities in Australia. They allow ubiquitous learning whereby students can access a variety of educational material anywhere, anytime on iPods, MP3 players or even desktop computers. There remain many questions about the impact of podcasts on students' learning. One issue is how podcasts can be used to support high quality, experiential learning rather than merely perpetuating the old transmission model of education. In this paper, we explore the reasons why students either use, or fail to use, podcasts provided for their education. We report on the motivation of students enrolled in a large first-year information systems subject. These varied considerably and show that podcasts are a useful adjunct for providing for the diverse range of learning styles of our students. However, we also conclude that further research is needed into the use of podcasts to promote deeper learning in our students and how podcasts can act as a support tool for other forms of m-learning.**

Index Terms—**higher education, m-learning, podcast, ubiquitous learning.**

I. INTRODUCTION

The word "podcasting" comes from combining the word iPod with broadcasting [1]. Currently, the term is no longer limited to broadcasts involving iPods but can also refer to the use of any portable audio player that allows the user to download sound files from the internet [2]. In fact, many "poddies" – users of podcasts – download and listen to these sound files on their computers at home. Typically sound files are compressed to MP3 format to allow for easier download, particularly over dial-up connections or where devices have small capacity. Hence the term MP3 player to describe both iPods and other similar players. Some authors limit the term podcasting to automatic delivery of files to users' devices using RSS (Real Simple Syndication) feeds [3], but often podcasts are accessed by users visiting websites and clicking on links to download podcasts they have chosen individually.

Podcasting is evolving at an amazing rate. In 2004 the number of webpage hits found by the Google search engine containing the term 'podcasts' was 24 [1]. By way of contrast, the number of hits in November 2007 was 124 million.

Podcasts have been used extensively by radio stations for providing access to music and talk shows, and for selling or downloading free music, for example from Apple iTunes. Podcasts have also evolved into vodcasts, used for delivery of television, movies and video [4]. Both podcasts, and to a lesser extent vodcasts, have found a place in university education as part of the new interest in m-learning. Podcasts in the educational setting allow students on-demand access to audio-recordings of lectures or other learning materials at their convenience [3].

At the authors' university a number of faculties are experimenting with podcasts in an effort to support student learning. These trials are attempting to resolve some of the problems which inevitably come with a new mode of educational delivery. This paper reports on one of these experiments. Our research question was whether podcasts would be useful, particularly from the student perspective, and whether different groups of students would find them useful (e.g., international versus local students). We firstly outline findings in the literature about podcasting. Then we describe the implementation of podcasts into a first year information systems subject within the Faculty of Information Technology. The results of our trial are presented and a discussion follows, placing the use of podcasts within the context of current educational theory. Finally, we present an m-learning framework in which to consider podcast implementation.

II. PODCASTING IN EDUCATION

In the United States, podcasts are being used and trialed extensively in universities such as Stanford University [4] and Duke University [5]. A lecturer at Bradford University in the United Kingdom is even using podcasts to completely replace face-to-face lectures [6]. Currently in Australia podcasting initiatives, like other forms of m-learning, are in a state of fragmentation but interest is growing, particularly in higher education [7].

The current generation of students has grown up and lived with digital technology [8]. The fact that most of today's students are very familiar with downloading audio files from the internet and own their own audio players makes the adoption of this method of learning almost automatic.

Podcasts allow anywhere, anytime learning. They permit students to access educational materials at home, while travelling to university or work, or doing any activity they choose. They can play the recordings at any time which is convenient to them rather than be confined to set class times. They have an obvious place in distance education, fulfilling the same role that audiocassettes performed in a previous era. However, through Web delivery, access is much easier, often via e-learning systems already in place at most universities [4].

Podcasts are fairly easy for teachers to generate using audio recorders and commonly available free compression software. For the majority of academics, who are not used to producing sound recordings, podcasts will need preparation to ensure a "compelling listening experience" [1]. This will include editing and post-production. However, once these skills are mastered these difficulties are minimal.

The University of Wisconsin has suggested guidelines for designing educational podcasts [3]:

- Choosing appropriate content
- Setting learning objectives or goals
- Designing the content
- Producing the podcast
- Integrating the podcast into the subject.

We also need to consider the quality of the learning experience, for example, does the podcast provide formal or informal learning, and does it support current educational theories [9].

Another important issue is whether podcasts improve students' academic performance. Research into this is rare in the literature. One of the few studies to look at this issue found no significant difference overall in students' exam marks between those who used podcasts and those who didn't [10]. However, they did find a significant difference in students' ability on essay-type questions in the exams if English was the language they used at home and if they downloaded the podcasts either in the week immediately after the lecture or in the study period immediately prior to the exam.

There are several downsides with podcasting:

- Students must have enough bandwidth to download the podcast, preferably broadband. Otherwise they will be restricted to accessing the podcasts on university computers.
- Discrimination against the hearing impaired.
- Podcasts are not interactive. It is a one-way delivery which precludes student interaction. Since podcasting is an extension of the traditional didactic model of education, will it lead to deep learning?
- Training may be needed for the teacher to improve the quality of their voice, speech patterns and intonation [2].

Podcasting, like other forms of m-learning, still requires much research to evolve a set of proven educational practices. Sustained podcasting will only occur through trial projects and the sharing of outcomes such as those reported in this paper.

III. THE PODCASTING IMPLEMENTATION

A. Teaching and Learning Context

This experiment in podcasting took place in a core subject, Introduction to Information Systems. There were a total of 340 students undertaking the subject, mostly undergraduates with some postgraduates. The majority of students were enrolled in an IT degree, a combined business/IT double degree, or a business degree with an IT major. Though similar subjects had been taught before, the subject was a new one, introduced with a large-scale revision of the undergraduate IT program.

Because this was a first year, first semester subject, scaffolding and support for students' learning was given the highest priority. Many students have difficulty making the transition from the high school learning environment to university. At school they receive a great deal of one-on-one support from teachers, with whom they have a more intimate relationship, whereas at university the environment is strange, student numbers are large and lecturers have too many teaching, research and administrative duties to take a personal interest in every single student. At university students are expected to be adult learners, with self-directed and self-motivated learning the key to success. The problem is exacerbated with core first-year subjects, which tend to have larger numbers of students and where students are compelled to take the subject whether they are interested in the topic or not. Technology has provided large-scale methods of supporting new students. M-learning, and in particular podcasting, has now joined e-learning as a method of giving support, such as making learning materials accessible to students beyond classroom hours.

Another reason for podcasting was to engage with the students' interest in the latest and most up-to-date technology. Mobile technology and wireless networks would form part of the later study of many of the students during their degrees, so encouraging them to use and start thinking about issues of mobile technology was appropriate. The revision of the subject and the undergraduate IT program seemed a good opportunity to introduce mobile devices into the students' learning for the first time.

It should be noted that the podcasts were not meant to replace lectures but to supplement them. Students were encouraged to attend lectures by awarding 5% for lecture attendance based on spot checks through the semester and by making the lectures interesting and engaging, for example by using videos and interactive discussion activities.

B. Podcasting Production and Use

For this subject specially scripted audio-summaries (AS) of the lectures were recorded by the lecturer in charge of the subject. The lecture was delivered twice each week (once in the morning to full-time students, and then a repeat lecture in the evening of the same day for the part-timers. The lectures were mostly 1 hour 20 minutes long. The audio recording was completed usually the day after the lecture had been delivered using an Olympus Digital Voice Recorder DS-4000. This device allowed editing of the recording before it was converted from a proprietary file format into WAV format and then into compressed MP3 format for more convenient download by students. In order to make the recordings as professional as possible, a script was prepared before recording and this took some time to construct: for an average lecture it took ½ day to script, record, convert and upload the podcasts, but some complex lectures could take a whole day where students were struggling with difficult concepts. Following compression the podcast files were uploaded each week to a folder on the Blackboard e-learning system in use by the university.

Summaries of only 9 of the 12 lectures in the subject were recorded. The 3 lectures that were not recorded included a guest lecture, the content of which was not known to the regular lecturer beforehand; a lecture consisting largely of copyright video material; and a lecture aimed at promoting students' self-reflection on their learning over the duration of the subject. None of these was deemed suitable for summarizing and recording.

Recordings varied in length from 11 minutes 51 seconds to 23 minutes 56 seconds, depending on the complexity of the lecture material being delivered. The average length was 18 minutes 36 seconds.

It was decided to record *summaries* instead of the usual practice of podcasting the whole lecture for several reasons:

- *Incorporation of extra material*: the recording of the audio-summaries after the lecture allowed the lecturer to add more material where it was believed this would assist students' understanding. For example, when students had asked questions during or at the end of the lecture, or posted questions on an online discussion board set up specially for the purpose, the lecturer would try to clarify concepts that had not been clearly understood by all the students. In addition, where the length of the lecture was insufficient to cover all important concepts, a discussion of this extra material would be included in the recording. The summaries allowed links to the students' set of lecture notes by referring to numbered slides, and also included references to helpful places in the readings which were included with the lecture slides in a combined book of Notes and Readings which all students purchased as part of their course in lieu of a textbook.
- *Quicker lecture review*: the length of the audio-summaries was on average a quarter of the length of the lectures. This meant that students would be able to review the lecture material much more quickly than listening to an entire recording of the lecture. It was hoped that this would make the podcasting more attractive to the students.
- *Smaller file download*: the shorter length of the summaries also meant a file size on average about a quarter the file size of the original lectures. This would be expected to make it quicker and easier for students to download the podcasts onto their devices, particularly if their iPods or MP3 players had smaller capacity (for example, with iPod "shuffle" models), or if they were downloading to PCs at home over a dial-up connection or with limited download contracts with their provider.
- *Avoidance of copyright issues*: all of the lectures incorporated one, or occasionally more, videos. These were copyright, so recording summaries avoided the problem of editing out the video soundtrack and the inevitable "jumps" in the audio track that would result.

Automatic RSS feeds were not available for use by the lecturer. Instead, students went online and downloaded the podcasts individually when and if they wanted to. They could download them to their PC at home, their laptop or to their iPod or MP3 player. Students were advised of the availability of the podcasts several times in lectures or via emails sent to all students enrolled in the subject.

IV. EVALUATION AND RESULTS

The podcasting trial was evaluated at the end of semester along with other features of the new subject. Out of a total ten questions about the subject, the one on podcasting presented to the student was: "Did you use the audio-summaries of the lectures? If so, what did you find useful about them? If not, why didn't you use them?" All students present in the lab the week when the evaluation was administered were asked to complete the form. However, it was not obligatory and some students may have chosen not to complete it. We received 247 survey forms from the student cohort. Of these, 6 students did not answer the question on podcasting. So 241 responses were received, giving a response rate of 71%.

From the survey forms completed, 87 students (36%) said they used the podcasts while 154 students (64%) said that they did not use them. Answers for the students who used them and the students who said they didn't use them were first separated. Then answers in each group were categorized and tallied up. The results are presented in Figures 1 and 2.

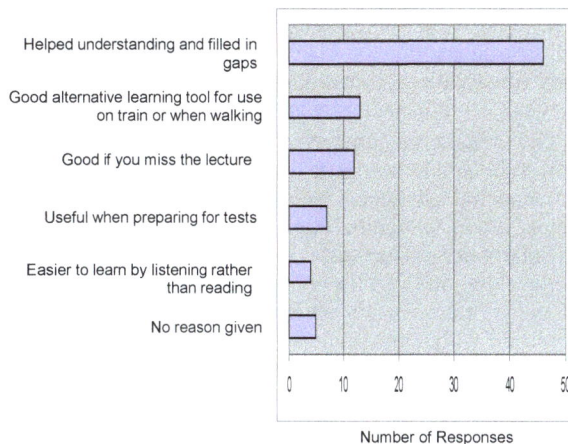

Figure 1.   Reasons why students used the podcasts

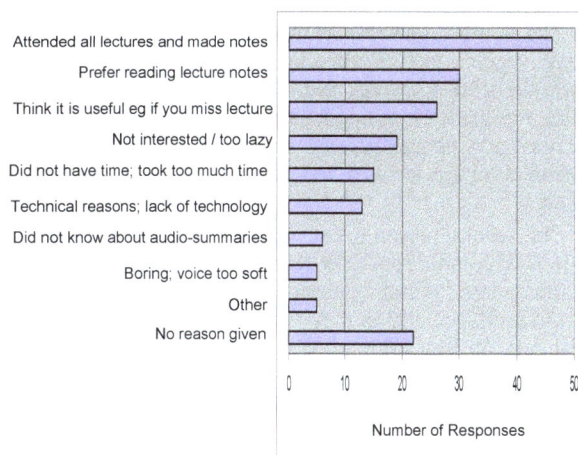

Figure 2.   Reasons why students did not use the podcasts

V. DISCUSSION OF RESULTS

A. *Motivations for Using Podcasts*

Generally, the comments by students who used the podcasts were very encouraging. For example, one student commented, "Yes, more, more, they are so helpful. I am a person who gets the idea of readings easily". Other comments included "They were very useful and I wish she posted all of the lectures in that form, cause that was the main way I used to study", "Yes, before the first lecture, I copied all the AS in my MP4 player", "Yes, it is useful. You can transfer the file into MP3/MP4 and you can always enjoy with it".

Overwhelmingly, the main reason why students used the podcasts was to help their understanding: 53% of responses. A typical comment was "Sometimes I can repeat the part that I don't understand. It helps the lecture notes to become more understandable."

One of the chief uses of podcasting quoted in the literature – anytime, anywhere learning – was supported by our survey. A number of students mentioned that they were very handy for studying on the train, while walking or running, and in fact much easier than studying from the book of lecture notes and readings when travelling.

Other motivations given were to catch up with lectures that had been missed and preparation for tests: "they were useful in picking-out the more important sections of our test, rather than us spending hours trying to remember it all". An interesting finding was that some students found they fitted with their learning style: for their private study they preferred listening rather than reading.

Given the large numbers of international students at our university and in our faculty, it is worth noting that several international students mentioned that podcasts really helped them overcome their language difficulties: "I am an international students, I cannot understand everything in the class. So I can listen the AS after class. It is useful"; "Yes, it help me to review the lecture when I was confused with some point. And it help me to enhance the ability of English listening because I am Chinese."

B. Challenges to Podcast Use

In contrast to the limited range of motivations of those who used the podcasts, there were many reasons why students did not use the podcasts.

The number one reason was that many students (25%) were regular lecture attendees. A couple of comments from the students were: "No, I didn't need them as I didn't miss any lectures, the content was well laid out so for me it was not necessary"; "I didn't use them because I attended all the lectures and understood most to all of the content".

The second reason given was that a number of students (16%) have preference for reading rather than listening to audio tapes: "Aural learning I did not find useful as a medium. Especially as they lacked the … clarification that text has. A transcript of the summary would be ideal." This is the exact opposite of those students who liked to use the podcasts. This shows that students have different learning styles which affect their decision whether to use podcasts or not.

Other reasons given included that it was useful if you missed a lecture, the recordings were boring or they just were not interested. They also mentioned that they were time poor: "I actually downloaded once then I forgot about it because of all the other homework and assignments. Do not have time to listen to them".

There were a few students who tried to access the podcasts but failed for a variety of technical reasons. These included the fact that the file was too big to download, problems with dial-up connections with low bandwidth, taking too long to download, the student's download limit, and the fact that downloading did not work on the students' PC or iPod. One student did not have an internet connection and another did not have an iPod or other device to listen to the podcasts. One student had an unidentified PC issue. Another student had a combined technical and multitasking issue: "No, I am unable to download them. They come up as embedded players on my browser. It was hard to listen to them since I was doing other work at the time I was on my PC."

C. Other Points of Interest

Some of the students who never used the podcasts nevertheless thought they were a very good idea. For example, one student said "however I regret not using them. They are an excellent tool for those who use them." Another confessed that "from deep down I can see that it will help in studying", and another "but I would have liked to. I am sure they would have helped me learn more."

Likewise, some of the students who used the podcasts mentioned problems or areas for improvement. These generally overlapped with the problems raised by the non-users. The main issues listed were that the podcasts were either too long or too boring, that students preferred reading rather than listening, and finally three students experienced technical difficulties. These problems may have led to discontinued use or lower use than might otherwise be expected.

VI. PODCASTS AND THE LEARNING EXPERIENCE

The fact that only one third of students used the podcasts shows the latter should not be regarded as a replacement for traditional lecturer delivery. In addition, there may be many good reasons why on-campus students should be encouraged to attend lectures. These include the ability to ask the lecturer questions, the social value of getting to know their lecturer and meeting their fellow students, and the opportunity for the lecturer to gauge how well the students are coping with the curriculum. Podcasts do not provide these features: they are a one-way, non-interactive transmission mode of learning and a solitary activity.

While podcasting is a useful adjunct to the classroom experience, providing flexibility and catering to different learning styles, we must question whether it results in learning of the highest quality. If m-learning is to transform our educational practices then it needs to foster deep approaches to learning, that is an orientation in the student towards understanding, personal sense-making and active learning. This will achieve better learning outcomes than surface approaches such as memorization of lecture material and the reproduction of the lecturer's knowledge [11] [12] [13].

In Figure 3, we range a number of different m-learning activities against a continuum of shallow to deep learning. Podcasting of lectures is at the shallow end of the spectrum. Ubiquitous access to resource materials also lies here if it consists purely of content delivery and rote learning, although it *can* foster deeper learning if students are encouraged to combine the resources with other sources of knowledge and so use them in the construction of new meanings for themselves. Likewise, interactive classroom activities, usually using clickers and personal responses systems, have been criticized as following a behaviourist educational paradigm if they focus on drill [14], but have been seen by others as improving students' attention and providing student feedback to the lecturer [15], as well as promoting social, collaborative learning [16]. Forms of m-learning which engage students most, allowing them to contextualize their new knowledge and to construct new meanings by incorporating rich learning experiences into their personal knowledge framework, include mobile-supported fieldwork and active simulations [14].

M-Learning	Type of Learning	Depth of Learning
Podcasting	Absorb/revise information from lecturer	Shallow
Anywhere anytime access to diverse resources via PDAs & laptops	Content delivery or reflection on learning materials?	
Interactive classroom using personal response systems	Drill or interactive social learning?	
Communication via mobile phones	Collaborative learning	
M-fieldwork	Contextualized learning	
Multimedia data capture by students for student use; participatory simulations	Constructivist learning	Deep

Figure 3.   M-Learning Framework

From this framework for considering m-learning, we can see that podcasting alone will not revolutionize education in the twenty-first century since it largely perpetuates the traditional didactic, teacher-centred approach. However, it can be used to lay the foundation of knowledge from the lecturer, on which the student can then discover and build their own knowledge, using more engaged m-learning approaches.

VII. CONCLUSION

The study revealed a range of motivations behind the students' decision to use the podcast lecture summaries or not to use them. Despite the fact that only slightly more than a third of the students decided to use the podcasts, the extremely positive comments from many of them have encourage us to continue to provide podcasts for this subject. We, as educators need to take into account the different learning styles of our students and provide a choice in the format of the learning materials we make available to them.

In particular, the favourable comments from international students, who comprise a significant proportion of students in our faculty, suggest that podcasts are a valuable additional learning resource. Our next research project in this area will be to thoroughly investigate the use of m-learning by this student cohort.

While podcasts should be accepted as a normal part of education today – providing, as they do, an additional learning tool for many of our students – at the same time we must remember that podcasts are only a small part of m-learning. If we acknowledge that the focus of learning should be on the student and not on the teacher, podcasts do not fulfill this aim, at least not in their present form of downloadable audio-lectures. More investigation needs to be carried out of how we could improve podcasts to include activities to promote deeper thinking in our students. Research is also needed into how podcasting might support those forms of m-learning which have already been demonstrated to provide high quality collaborative, contextualized and active learning. We must look for ways of how m-learning as a whole can be used to encourage students to adopt deep approaches to study and therefore improve their learning experience and outcomes.

ACKNOWLEDGMENT

The authors thank Andrew Litchfield for presenting the paper on which this article was based at the 3rd International Conference on Mobile and Computer Aided Learning, IMCL2008, Jordan, Amman, and we thank the audience for their comments which were helpful in its revision and improvement.

REFERENCES

[1] G. Campbell, "There's something in the Air: Podcasting in Education," *Educause review*, November/December 2005. Retrieved on November 6, 2007 from http://www.educause.edu/ir/library/pdf/erm0561.pdf

[2] Educause, "7 things you should know about podcasting," *Educause Learning Initiative*, pp.33-46, June 2005. Retrieved on November 6, 2007 from http://www.educause.edu/eli

[3] C. Laing, A. Wootton, and A. Irons, "iPod! ULearn?", *Current Developments in technology-Assisted Education*, pp. 514-518, 2006. Retrieved on November 6, 2007 from http://podcasting.thefutureoflearning.googlepages.com/514-18.pdf

[4] N. Townend, " Podcasting in higher education", *Viewfinder*, Media Online Focus, British Universities Film & Video Council, vol. 61, pp. i-iv, December 2005.

[5] Y. Belanger, *Duke University iPod First Year Experience Final Evaluation Report*, pp. 1-15. Retrieved on November 6, 2007 from http://cit.duke.edu/pdf/ipod_initiative_04_05.pdf

[6] C. Stothart, "Do the iPod shuffle, but don't miss the lecture", *The Times Higher Education Supplement*, May 26, 2006. Retrieved on November 6, 2007 from http://thes.co.uk/current_edition/story.aspx?story_id=2030201

[7] H. Watson and G. White, " mLearning in Education-A Summary", *education.au limited*, pp. i-41, July 2006.

[8] M. Prensky, "Digital Natives, Digital Immigrants, *On the Horizon*, NBC University Press, Vol. 9, No. 5, pp. 1-6, October 2001.

[9] M. Sharples, J. Taylor, and G. Vavoula, " Towards a theory of mobile learning, *Proceedings of mLearn, 2005*.

[10] D. Bond, T. Holland, and P. Wells, "Student utilisation of lecture podcasts and their relationship to student achievement", *Presentation at the UTS Teaching and Learning Forum*, Sydney, November 14-15, 2007.

[11] M. Prosser and K. Trigwell, *Understanding Learning and Teaching: The Experience in Higher Education*, Buckingham: Society for Research into Higher Education and Open University Press, 1999.

[12] F. Marton and S. Booth, *Learning and Awareness*, New Jersey: Lawrence Erlbaum Assoc. Publishers, 1997.

[13] P. Ramsden, *Learning to Teach in Higher Education*, London: Routledge, 1992.

[14] L. Naismith, P. Lonsdale, G. Vavoula & M. Sharples, *Report 11: Literature Review in Mobile Technologies and Learning*, Bristol: Futurelab, 2005.

[15] N. Scheele, A. Wessels, W. Effelsberg, M. Hofer and S. Fries, "Experiences with interactive lectures – considerations from the perspective of educational psychology and computer science", *Proceedings of the 2005 Conference on Computer Support for Collaborative Learning*, pp. 547-556, 2005.

[16] A. Litchfield, L. E. Dyson, E. Lawrence and A. Zmijewska, "Directions for m-learning research to enhance active learning", *Proceedings ASCILITE Singapore 2007*, pp. 587-596, 2007.

AUTHORS

I. Nataatmadja is with the Faculty of Information Technology at the University of Technology, Sydney, Australia (e-mail: indra@it.uts.edu.au).

L. E. Dyson is with the Faculty of Information Technology at the University of Technology, Sydney, Australia (e-mail: Laurel.E.Dyson@uts.edu.au).

12

Simplistic is the Ingredient for Mobile Learning

Issham Ismail, Hanysah Baharum and Rozhan M. Idrus

Universiti Sains Malaysia, Penang, Malaysia

Abstract—This study explored the students' acceptance of Mobile Learning via Short Message Service (SMS-Learning) amongst distance learners in the Universiti Sains Malaysia. This study aims to examine the student's acceptance towards the language use in SMS-Learning content, the cost of communicating and also the navigation of the system. The study employed the qualitative methodology where data were collected through questionnaire that was administered to 105 distance education students from Bachelor of Management, Bachelor of Science, Bachelor of Social Science and Bachelor of Art. The survey responses were tabulated in a 5-point Likert scale and analyzed using the Rasch Measurement Model. The results indicated that the simple language used in SMS-Learning was accepted by the respondents. By using the language precisely, it leads to high usability of SMS-Learning which will allow it to academically assist them in their study.

*Index Terms—***Mobile learning, Rasch model, SMS, text message**

I. INTRODUCTION

The extension of mobile wireless technologies has contributed to the shifting of the educational environment from the traditional setting to an e-learning setting. In traditional education, both teachers and students are physically present together [1]. There are increasing numbers of higher institutions that offer courses using mobile devices as an alternative teaching and learning tools [2]. M-learning or mobile learning is the type of learning characterized by the usage of wireless technology, through the personal control of the learning time and place [3]. SMS or text messaging is the transmission of short text messages to and from a mobile wireless phone, fax machine, and/or IP address.

The rapid evolution of mobile devices and quick development of wireless communication has paved the way for another alternative medium for higher education institution to employ mobile learning as a means to facilitate educational transaction to the learners. Mobile wireless technologies help to improve efficiency and effectiveness in teaching and learning with the advantages of mobility [4]. Mobile learning has been perceived by many educationalists to offer flexibility in learning and present a multitude yet unique educational advantages [5]. Regardless of such interests in mobile wireless technologies in higher education, there is lack of academic research on the use of mobile wireless technologies in the higher education setting [2].

A study done by Malaysian Communications and Multimedia Commission (MCMC) found that in Quarter 2, 2009, the penetration rate for cellular phone in Malaysia is 100.8 %. Penetration rate over 100% occurs because of multiple subscriptions [6]. The mobile phone is a multi-purpose device and not only used for transmitting voice communication but at the same time also provides a number of other functions and services, such as the short messages service. Majority of mobile phone user used SMS as a communication tool for sending and receiving messages [7]. Nevertheless, there are an increasing number of SMS commercial services such as voting, news and sports alert, ringtones/logos and advertisements. The explosive growth of SMS usage can aid marketers in developing appropriate m-commerce services [8].

In this study, lecturers and students send and receive text messages to and from most high-tech mobile wireless phones through SMS. With SMS, messages are produced on the tiny keypad of the phone and users are able to exchange alphanumeric message (up to 160 characters) with other users of digital cellular networks, almost anywhere in the world within second of submission [9][10]. The use of SMS will potentially be increased in the education field as technology improves [11]. In order to have an effective SMS communication, both sender and receiver must understand the message that is delivered to them [12]. Thus, the language used in authoring the message has to be clear and understandable. SMS language is more like speaking than writing and more short-lived than letters. There are also unique formulations in SMS that have a slight foundation in writing and/or spoken language, but seem to be distinctive [12].

In order to evaluate the acceptance of SMS-Learning among students of distance education, satisfaction of the students on SMS-Learning will be considered. Wang developed a comprehensive model and instrument for measuring learners' satisfaction with asynchronous e-learning systems; and he found that satisfaction could be classified into the four following dimensions; content, personalization, learning community and learner interface [13]. Evaluation on learners' satisfaction of the web-based e-learning system was the continuity of the previous research done which indicates that learner interface as being the most important dimension of criteria [14]. In the field of human-computer interaction, users satisfaction is the "subjective sum of interactive experiences" influenced by many affective mechanism in the interaction [15]. The interaction between instructors and learners play an important role in learning activities. Instructor's attitude towards e-Learning has significant effect on e-Learner's satisfaction where learning activities and learner's satisfaction are influenced by instructor's attitudes in handling learning activities [16].

In addition to this study, the usability of the system used is also to be considered. The concept of usability refers to the intention of using a computer system. Nowadays it is usually associated to ease-of-use of a website and is considered a critical factor on the development of electronic commerce [17]. Usability concerns the ease in which the user is capable of learning how to manage the system, the ease of memorizing the basic functions, the grade of efficiency with which the site has been designed, the degree of error avoidance and the general satisfaction of the users in terms of manageability[18]. Thus, website usability is defined as "a quality attribute that assesses how easy user interfaces are to use" [19]. Hence, this study proposes that the satisfaction towards using SMS-Learning and the usability of the system contribute to the students' acceptance of the SMS-Learning project.

II. METHOD

A. Participants

In this study, the distance learners were given the opportunity to register and become a respondent in an SMS-Learning programme. The samples for this study were selected by lecturers at the beginning of the semester. This SMS-Learning programme was conducted for 3 months commencing from February 2009 till the end of April 2009 (10 weeks). One hundred-five undergraduate students consisting of 31 males and 74 females had enrolled from four different programs (Bachelors of Science, Bachelor of Arts, Bachelor of Management and Bachelor of Social Science) in the School of Distance Education (SDE), Universiti Sains Malaysia (USM).

TABLE I.
COURSES OFFERED IN SMS-LEARNING PROGRAMME

Programme	Course	Year of study
Management	Financial Principle	2
	International Business	3
Physics	Mechanics	2
	Optics	2
Economics	Money and Banking	2
	Quantitative Economy	3

The learning materials given to these six groups of students were based on their courses taken. Each group was monitored by their own course manager. The SMS-Learning materials for the courses were prepared by the respective course managers. All of the respondents declared that they are mobile phone owners and were able to participate in SMS-Learning Program.

B. Instruments

This study was conducted using a questionnaire-based survey that consisted of respondents' demographics, respondent's satisfaction towards SMS-Learning program and also the usability of SMS-Learning. The survey utilized a 5-point Likert-type scale that allows students to rate their agreement of each item of the survey. In this study, respondents were asked to rate the items correspond to a Likert-type rating scale where 1=strongly disagree, 2=disagree, 3=neutral, 4=agree, 5=strongly agree.

C. Data Analysis

The student's acceptance towards the language used in SMS-Learning Program, the cost of communicating and the navigation in the system are highlighted in this study. Data was entered for each respondent into WINSTEPS Version 3.68 employing the rating scale Rasch Model [20]. Rasch is mathematically identical to the most Item Response Theory (IRT) model; however it is compara-

tively more viable proposition for practical testing since it can be applied in the context in which persons interacts with items [21].

When applying the Rasch model, data must fit the model, with the assumption of unidimensional domain being measured. In order to evaluate and analyze student's acceptance towards SMS-Learning program, several tables and figures are used in this study. A statistical summary table was produced to describe the separation rate and reliability of the persons and items. Separation is a number of statistically different performance strata that the test can identify in the sample. The reliability rate indicates whether the test discriminates the sample into enough levels for the intended measure [21].

Item and person misfit table was presented and explained. The statistics show how well the data fits the model, with fit implying a meeting of requirement or matching of intentions [21]. Basically, it was an investigation of the match between a group of persons and a set of items, specific to the intent of the measure. The empirical hierarchy of items was illustrated using variable map and connected to the students' level of ability to endorse each item, with each reported in logits. A logit (log-odds unit) is a unit of interval measurement which is well-defined within the context of a single homogenous test [22]. The variable map visually reveals the hierarchy and the order of the items as well as any potential gaps in measure [21].

III. RESULTS AND DISCUSSION

A. Demographic profile of respondents

This section portrays respondent's background such as gender, age, ethnicity, degree program and also type of mobile devices owned. As shown in table II, the number of females responding to the questionnaire slightly outnumbered males with 70% of females and 30 % males ranging in age from 20 to above-50. It was noted that this age group prefer to pursue their studies in distance education because they remain in full time employment. As for ethnic structure, almost half of the respondents were Malay which is 57 %, while 26 % were Chinese, 10 % were Indian and 7 % were from other ethnic group.

About 95% of the respondents were from Management program, 2% for both Science and Social Science program and only 1% from Art program. From a total of 105 respondents in this survey, about 91% of them owned a mobile phone and 3% have PDA/pocket PC/Palmtop. While 6% have both of mobile phone and PDA/pocket PC/Palmtop. The result shows that all of them declared that they have owned mobile phone and able to use SMS-Learning program.

B. Item and Person Misfit Order Table

In order to identify item maps, fit statistic were examined to determine whether the item fits the maps or not. The person and item misfit is helpful when evaluating the acceptance of SMS-Learning program amongst students. The data must fit the model and meeting of requirements or matching a group of persons and items [21].

Table III shows the item statistics in a fit-order table produced by Winsteps. The results demonstrate whether the instrument functions as valid tools for data collection. Outfit mean-square fit statistics (MNSQs) are equivalent to a chi-square statistics; values greater than 2.0 indicate unexplained randomness throughout the data [23]. The item that fall within the infit and outfit limits of 0.6 and 1.5 were accepted in his analysis [24].

In this study analysis, the results demonstrate the outfit mean-square fit statistics (MNSQ $\leq 0.6 \geq 1.5$) for 13 items were acceptable to the model while 9 items fall outside the indicated range, suggesting either are not supporting the underlying construct or items need revamped, likely because respondent are viewing the items differently than that intended by the researcher.

C. Reliability and Separation

In order to have an overall view of the reliability and validity of the instruments and associated responses, the statistical summary tables of the persons and items were produced.

Reliability is the degree to which measures are free from error and therefore yield consistent results. The closer the reliability coefficient (Cronbach's Alpha) to 1.0 the better it is and those values over 0.80 are consider as good [25]. Values in 0.70 are acceptable while less than 0.60 considered as poor. In the reliability analysis, the alpha value that is closer the reliability coefficient to 1.00 is the better. In this study, the Cronbach's Alpha of 0.88 can be considered good.

Person reliability was 0.85, with a separation of 2.37. Item reliability was 0.90, with a separation of 2.99. Given a 0.7 threshold of acceptability, both scales are deemed reliable and usable for the purpose of this study. The person separation of 2.37 means students were roughly separated into 3 groups, later labeled as those who satisfied with the program, those who were fine with the program and those who dissatisfied with the program. For item separation of 2.99, it indicated 13 items a generally separated into 3 groups. Label as items that students satisfied, items that students thought were fine and items that students unsatisfied. The person reliability of 0.85 is good for the sample of 105. The item reliability is 0.90 which is reasonably high considering sample size and small numbers of items. Thus, the survey as a whole appears to have functional reliability.

TABLE II.
DEMOGRAPHIC DATA

Item	Frequency	Per cent
Gender		
Male	31	30
Female	74	70
Age		
20-29 years	44	42
30-39 years	46	44
40-49 years	12	11
50 and above	3	3
Ethnicity		
Malay	60	57
Chinese	11	10
Indian	27	26
Others	7	7
Program		
B. Science	2	2
B. Arts	1	1
B. Social Science	2	2
B. Management	98	95
Mobile Device Ownership		
Mobile phone	96	91
Both	6	6
PDA/Pocket PC/Palmtop	3	3

TABLE III.
FIT STATISTICS FOR STUDENTS' SATISFACTION AND SYSTEM USABILITY

No	Item	Statement	Infit MNSQ	Outfit MNSQ
1	USE140	The language used is simple enough.	0.88	0.81
2	SAT55	The content of the messages are short, brief, useful and powerful.	0.92	0.87
3	SAT54	I prefer more frequent messages from lecturers.	1.23	1.17
4	USE139	You prefer to navigate in 3D (text, picture, mms, audio and video).	1.50	1.48
5	SAT53	I'm satisfied with the time each message delivered to me.	0.82	0.80
6	USE144	You think SMS-Learning is effective to help your study.	0.74	0.73
7	USE136	It is safe to use the system to save your learning content.	0.85	0.71
8	SAT61	The messages send to me can be illustrated in my mind.	1.16	1.17
9	SAT59	The messages send to me promptly.	0.71	0.69
10	USE142	Only one word or term is used to describe any item.	1.00	1.13
11	SAT62	The m-learning is more attractive than the traditional learning method.	1.00	0.90
12	USE138	It is clear to navigate in 2D (merely text messaging).	1.11	1.42
13	SAT58	The cost of communicating in the mobile learning course with the tutor and other students was acceptable.	1.12	1.18
	Mean		1.00	1.01
	S.D.		0.21	0.25

TABLE IV.
INEFFECTUAL ITEMS

No	Item	Statement	Infit MNSQ	Outfit MNSQ
1	SAT56	The fees of the messages charged of RM0.15 are reasonable.	1.74	1.85
2	SAT63	The messages sent to me are disturbing my life.	2.46	3.11
3	SAT60	I wish to receive important news from school through messages.	1.83	1.92
4	SAT57	The fees of messages should be cheaper.	2.32	2.14
5	USE134	The system is easy to use.	0.42	0.42
6	USE135	It is easy to learn by using the system.	0.59	0.58
7	USE137	The system is effective and efficient.	0.57	0.53
8	USE141	Jargon is avoided.	0.59	0.57
9	USE143	Terminology is consistent with general usage.	0.51	0.53

D. Variable Map

Figure 1 presents a map of the items, ranked by level of satisfaction and system usability, and the respondents, ranked by their willingness to endorse with the items. Within the map, items have been labeled by a key word in the statement. The items and person map display a hierarchy of characteristics preferences as rated by participants and indicates that participant's willingness to endorse the items is generally very high. Items that are located at the top of the map have been identified as those that are most difficulties to endorse [26]. Those at the bottom are easier to endorse; thus as you move from bottom to top of the map, items are more difficult to endorse.

Results suggest the easiest item to endorse is item USE140; *The language used is simple enough.* It is an interesting result that students endorse the simple language used in SMS-Learning Program as the most satisfied item. By using the simple language in the program, students easily understood the content and the process of transferring information from instructors to the student were simplified. This result was consistent with item SAT55, *The content of the messages is short, brief, useful and powerful.* It shows that students who are the distance learners are interested to receive a concise message which means the message is brief, short, explicable and very useful in assisting them in revising the subject taken anywhere and anytime. Because of the practicality of the SMS-Learning in term of language used, responses from students shows that they prefer to subscribe to the program as supported by Item 54; *I prefer more frequent messages from lecturers.* They choose to receive more frequent messages instead of once a day so that they are able to keep updating their knowledge and study continuously.

When there were positive responses to the simplicity of the language used in the program and the frequency of receiving the message, there was a disagreement result which shows the most difficult item to endorse by respondents. It was item SAT58, *The cost of communicating in the mobile learning course with the tutor and other students was acceptable.* Responses from the survey explained that the cost of communication will be the barrier in SMS-Learning program. This SMS-Learning program is still in the early stage of development, therefore the cost of communications via SMS slightly outnumbers the cost of learning using electronic learning (e-learning). Learning via e-learning was nearly at zero cost and it will encourage the decreasing in SMS cost.

This study also found students agree with item USE139; You *prefer to navigate in 3D (text, picture, mms, audio, and video).* It explained that most of the respondents own a mobile phone that is able to deliver and receive not just a text message. Respondents show the interest in receiving learning material in form of pictures, sounds and animation in order to enhance their understanding. It was a contrarily result with the item *USE138, It clear to navigate in 2D (merely text messaging).* Respondents had difficulties in endorsing this item. The possible reason behind this result is that respondents are ready to navigate in 3D application and they owned the latest mobile phone which comes with many functions instead of text messaging. This result shows that respondents will easily agree if SMS-Learning program provides various type of message in the development of mobile phones.

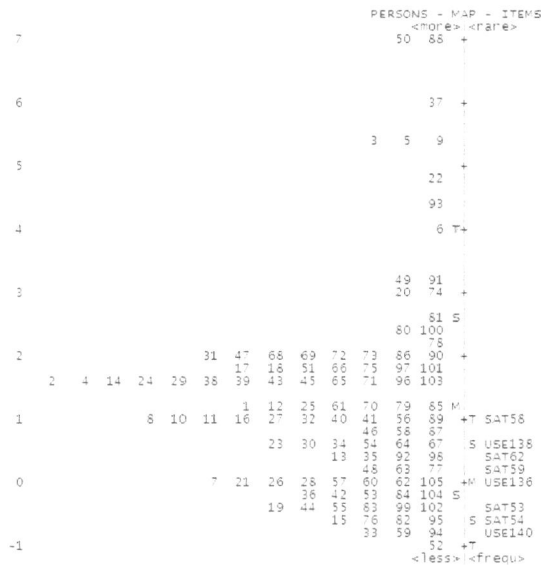

Figure 1. Item/Person map

IV. LIMITATIONS

The first limitation is the number of participants adopting SMS-Learning was extremely small and this precluded much statistical analysis. Larger samples in this study would have been helpful in overcoming the problem. The study participants were enlist from a single university and utilized students from School of Distance Education only. Therefore, the wide-ranging of the study could be limited to the institutions of similar size and courses. At the early stage of study, there was only one-way interaction between instructors and students and it showed a lack of interactive communication between students and instructors. In the future, researchers may implement two way interactions to create an interactive way of learning.

V. CONCLUSION

This study has developed and investigated SMS-Learning program and it showed that the majority of the respondents accepted the SMS-Learning program as a tool of teaching with the language used in this program. It showed that the simplistic is the constituent of mobile learning. The instructors have to ensure that the language used must be short, brief, useful and powerful to help them in their study. Through Rasch results, student's willingness to endorse items and the corresponding items are clearly stated and compared along one scale as they were analyzed and evaluated. The survey instrument is reliable and is able to separate both the sample of students and items. This study provides improvement suggestions to this SMS-Learning program. With the additional capability of mobile phone, SMS-Learning has the potential to become one of the most interactive learning tools in this era. As this occurs, additional studies should take place and more students will have the opportunity to engage in SMS-Learning program which will list more offered subjects.

ACKNOWLEDGMENT

The authors would like to acknowledge Universiti Sains Malaysia for the support under RU grant and USM Fellowship scheme.

REFERENCES

[1] Georgiev, T., Georgieva, E., & Smikarov, A. (2004). M-Learning-A new stage of e-learning. Paper presented at the International Conference on Computer Systems and Technologies.

[2] Kim, S.H., Mims, C. & Holmes, K.P. (2006). An Introduction to Current Trends and Benefits of Mobile Wireless Technology Use in Higher Education. *AACE Journal, 14*(1), 77-100. Chesapeake, VA: AACE.

[3] Sharples, M., Taylor, J., & Vavoula, G. (2005). Towards a theory of mobile learning. *Paper presented at mLearn 2005,* Capetown South Africa. http://www.mlearn.org.za/CD/papers/Sharples-Theory of Mobile.pdf

[4] Maginnis, F., White, R., & Mckenna, C. (2000, November/December). Customers on the move: m-Commerce demands a business object broker approach to EAI. *eAI Journal,* 58-62.

[5] Goodison, T. A. (2001) The implementation of m-learning in UK higher education. *Proceedings of ED-MEDIA 2001. AACE Press, 25-30 June 2001,* Tampere, Finland, pp.613-618

[6] MCMC (2009). Fact & Figures (Statistics & Record). Retrieved September 11, 2009 form http://www.skmm.gov.my/facts_figures/stats/index.asp

[7] Gilligan, R. and Heinzmann, P. 2004. "Exploring how cultural factors could potentially influence ICT use: An Analysis of European SMS and MMS use", Cultural Difference Workgroup COST 269

[8] Harris, P., Rettie, R. and Cheung, C.C. 2005. "Adoption and usage of m-commerce: A cross-cultural comparison of Hong Kong and the United Kingdom", Journal of Electronic Commerce Research, 6(3), 210-224.

[9] Peersman, G and Cvetkovic, S (2000). The Global System for Mobile Communication Short Message Services. *IEEE Personal Communications, June 2000.* The University of Sheffield.

[10] Hard af Segerstad, Y. (2005). Language use in Swedish mobile text messaging. In R. Ling & P. E. Pedersen (Eds.), Mobile Communications: Re-negotiation of the Social Sphere (pp. 313-333). London: Springer-Verlag.

[11] Trifonova, A. (2003). Mobile learning—review of the literature. (Technology Report No. DIT-03-009). University of Trento, Department of Information and Communication Technology.

[12] Ling, R., & Sollund, A. (2002). Final Report for. Youngster Project, EU IST Program. Roessler, P. and Hoeflich, J, pp, 133-157.

[13] Wang, Y. (2003). Assessment of learner satisfaction with asynchronous electronic learning systems. *Information & Management, 41,* 75-86. doi:10.1016/S0378-7206(03)00028-4.

[14] Shee, D. Y., & Wang, Y. (2008). Multi-criteria evaluation of the web-based e-learning system: A methodology based on learner satisfaction and its applications. *Computers & Education, 50,* 894-905. doi:10.1016/j.compedu.2006.09.005.

[15] Lindgaard, G., & Dudek, C. (2003). What is this evasive beast we call user satisfaction. *Interacting with Computers, 15*(3), 429-452. doi:10.1016/S0953-5438(02)00063-2

[16] Sun, P., Tsai, R. J., Finger, G., Chen, Y., & Yeh, D. (2008). What drives a successful e-Learning? An empirical investigation of the critical factors in uencing learner satisfaction. *Computers & Education, 50,* 1183-1202. doi:10.1016/j.compedu.2006.11.007.

[17] Flavia 'n, C., Guinalı 'u,M., & Gurrea, R. (2006). The role played by perceived usability, satisfaction and consumer trust on website loyalty. Information and Management. *The International Journal of Information Systems Applications,* 43(1), 1–14.

[18] Nielsen, J. (1994). *Usability engineering.* San Francisco: Morgan Kaufman.

[19] Nielsen, J. (2005) *Usability 101.* Retrieved August 14, 2009, from http://www.useit.com/alertbox/20030825.html

[20] Linacre, J.M. (2009). A User's Guide to Winsteps Rasch Model Computer Programs. Chicago,IL: MESA Press.

[21] Ren, W., Bradley, K.D., Lumpp J.K. (2008). Applying the Rasch Model to Evaluate an Implementation of the Kentucy Electronics Educations Education Project. *Journal of Science Education and Technology,* 17(6), 618-625. doi:10.1007/s10956-008-9132-4

[22] Winsteps Help (2007) *Logit and probit: what are they?* Retrieved August 17, 2009, from http://www.winsteps.com/winman/whatisalogit.htm

[23] Smith, R. M. (1996). *Polytomous mean-square statistics.* Rasch Measurement Transactions, 6, 516-517.

[24] Fox, C. (1999). An introduction to the partial credit model for developing nursing assessments. *Journal of Nursing Education,* 38(8), 340-346.

[25] Sekaran, U. (2000). Research Methods for Business: A Skill Building Approach. Singapore: John Wiley & Sons Inc

[26] Harris, Jr. (2006). A study of Black University Students' Perception of Marriage. Constructing and Evaluating measures: Applications of the Rasch measurement Model. Symposium presented at the Mid-Western educational Research Associationannual meeting, Columbus, OH, pp.19-24

AUTHORS

Issham Ismail is with the School of Distance Education, Universiti Sains Malaysia, Minden, Pulau Pinang, 11800 Malaysia (e-mail: issham@usm.my)

Rozhan M. Idrus is with the School of Distance Education, Universiti Sains Malaysia, Minden, Pulau Pinang, 11800 Malaysia. He is specialized in Open and Distance Learning Interactive Technologies and e-Learning (e-mail: rozhan@usm.my).

Hanysah Baharum is a student in the School of Distance Education, Universiti Sains Malaysia, Minden, Pulau Pinang, 11800 Malaysia and currently doing her master in Educational Technology in Universiti Sains Malaysia. (e-mail: hanysahbaharum@gmail.com).

13

Using USB Keys to Promote Mobile Learning

M. Rosselle[1], D. Leclet[1] and B. Talon[2]

[1] MIS-UPJV Laboratory, Amiens, France

[2] Université Lille Nord de France, Calais, France

Abstract—M-learning (i.e. mobile learning) is a field of e-learning that provides learners learning environments using mobile technology. In this context, learning can take place anywhere and anytime, in open and distance learning. Depending on the type of technology it may be done through software called nomadic (i.e. prepared to mobility). Among these technologies, there are those composed of digital interfaces and with autonomy of treatment: Smartphone, PDA, calculator and even mp3 key. In this article we propose to take into account storage devices as mobile technologies. Our focus was on the USB key. We present a procedure to test whether a learning environment embarked on a USB key can be described as nomadic or not. This procedure has been tested on a sample of three ILE (Interactive Learning Environment. This approach has allowed us to define criteria of nomadism, criteria which were then included in the design of a synchronous Weblog on USB key.

Index Terms—m-learning, mobile blog, nomadism, USB key.

I. INTRODUCTION

The research described in this article takes place in the framework of m-learning (mobile learning). The definitions of m-learning are manifold [1]. We offer a definition that seems to synthesize the issues conveyed around this concept:

"Learning that occurs when the learner is not at a fixed location and/or benefits of mobile technologies. Thus, we believe that m-learning covers ubiquitous learning, which allows the pervasiveness of computers in the learning environment and pervasive learning, that offers a multimodal and multi-channel access to the educational device" [2].

Thus, issues covered by the m-learning are diverse. How / what to make learn with mobile media? How to adapt software from ILE (Interactive Learning Environment) research domain to mobility? What educational device providing continuity in the learning process can we design? How to adapt content to different equipment and different contexts? What delivery mechanism of content to the learner must be put in place? These are all questions that the m-learning proposes to find answers.

Among the technologies used for mobile learning, we found mainly technologies composed of digital interfaces with a self-treatment (mobile phone, Smartphone, PDA, calculator, game console, MP3 key, etc.) [3][4][5]. They also include technologies such as GPS geolocation or the radio-frequency identification [11][6]. However, to largely cover mobile learning, why not also take into account storage devices like USB key, DVD, or external HD. Thus, is the fact to propose to learners learning environments on such supports in the issues covered by the m-

learning? We believe that the issue of the provision of learning environments on such materials can be considered a scientific lock in mobile learning.

Thus, mobile learning is based on the use of mobile software. This software designed for mobility is called "nomad software." The question here is not to study the portability of software on different machines with different operating systems. This issue is studied in the field of operating systems and virtual machines. The issue that interests us is to "transport" software running on a machine of a certain type (e.g. PC running Windows XP) to an equivalent machine type without installation constraint.

In this context, one of our concerns is to provide continuity in the learning process. Indeed, we found, during the evaluation of a pedagogical device called COOLDA-MAETIC, i.e. evaluating the MAETIC method [7] on a malleable platform called COOLDA [8] that students tended to interrupt their learning process. There was a lack of availability of tools. The USB key seemed to be a way to "ensure" continuity during their learning process. However, it was essential to verify the feasibility and ease of transporting learning environments. Thus we decided to define the criteria of nomadism "feasibility".

II. BACKGROUND

In January 2007, a mobile educational device was developed [9] which incorporated a voice server. Although it has shown its technical feasibility, it has not proved to be entirely satisfactory from the point of view of its usability [7]. In January 2008, we experienced an educational device called COOLDA-MAETIC with first-year students of an Associate degree in computer science and with students of a Bachelor Degree in Digital Imaging at the "Littoral Côte d'Opale" University (ULCO). In this experiment, we found that students tended to interrupt their learning process.

The assumption was that interruptions are due to the fact that students do not have a tool configuration, outside the meeting sessions, similar to that available in the classrooms of the university. However, the mobility of working tools could promote work outside the meetings. To make the tools becoming mobile, we propose to board the working environment on a USB key. This offers the possibility to contribute to the group work offline. The USB key is indeed a light, easy and inexpensive way to transport its environment and to work temporarily offline.

In this part, we propose to justify the choice of using a USB key, then we give some definitions and we present a procedure to test whether a piece of software is nomad on a USB key.

A. The choice of using the USB key

The choice of using the USB key as a support for mobility is based on the fact that it is a widely distributed and inexpensive. Moreover, its use does not require paying a subscription or a package from an operator or to be permanently connected to the Internet. Furthermore, we all have the opportunity to work on different computers and it would be very convenient to carry with us our software and our work environments, as we carry our documents. The value of a program that runs on a USB key is to allow its owner to work on any computer and retrieve his complete environment, his parameters, like his software. Moreover, this work leaves no trace on the host of the USB key.

For the technician who maintains a multimedia room, the advantage is that he does not have to install software specific to a given class. He installs the standard software of the training and provides connectivity with the Internet if necessary. In UMVF (French speaking Virtual Medical University [12]), the study of the "desk of the student" has shown that the work of technicians to ensure a multimedia room is ready to receive a Medicine course was long and tedious. This requires consulting a repository [10] of materials (e.g. sound card) and software (e.g. pdf reader) and configurations (e.g. 640x480 screen). The use of USB key equipped with nomadic software simplifies that work.

For the researcher or the teacher, the advantage is to provide learners adequate software by distributing keys with adapted and configured software. Furthermore, the key is used to store traces collected during the activity with the software.

The idea to bring the software environment on a USB key is technologically possible. Indeed, USB keys already allow carrying a PC on a USB key. The PC can be equipped with a Linux operating system (like [13] or [14]), which operates independently of the host PC's operating system. There are also solutions such as MojoPac [MojoPac], which can carry on a USB key or iPod some virtual work spaces under Windows XP. The data are separated from the host PC, but it is the latter that works in the background. There are finally nomadic software distributions on USB key using the Windows operating system of the host PC [16] [17] [18].

In this article we will only study software running on PCs using the Windows XP operating system. We will now introduce some definitions to clarify what we mean by mobility offered by a USB key.

B. The proposal of Some Definitions

We want to embark on USB key software that is able to operate using only the operating system of the host computer. We call this software: nomad software for USB keys or simply nomad software.

In literature, they are also called portable software. But are these words really equivalent? We offer a definition of these two words: mobile and portable. Furthermore, the literature says that the software running on USB keys is discreet software [18]. We define this term and its relations with the terms nomad and portable.

- Discreet software is software that, when installed or executed, leaves no traces. Under Windows XP, when a software "leaves trace", they are found in the registry or the hard disk in the installation directory or elsewhere (in "c:/documents and setting/software" for example).

- Nomad software is software that runs in a folder and works independently in the latter. So if we move it (e.g. on a USB key), the software is sufficient in itself and only needs the operating system to run. Obviously, if the software writes important data in a folder other than the installation folder or in the registry, and that we want to use the software on another machine, it will not find the right information and not therefore not function properly. So software that comes with a simple zipped file is a better candidate than software that uses an installer because in the latter case, the installer may write to the registry or install files elsewhere that in the installation directory.

- Portable software is software that can run on different machines and on different operating systems. Portable software has a great potential to be nomad because it is less likely to use any subtlety of the operating system of the machine on which it was developed. Free software is good candidates. Indeed, free software developers like to develop for several platforms (Linux, Mac, and Windows). However, a nomad software is not necessarily a portable software and vice versa.

In the context of our work, we talk about nomad software. However, we will not use the portable term that describes a software set, that intersection with the nomad software set is not empty.

C. The Proposal of a Test Procedure

In our study, our wish is to exploit existing software in a context of nomadism. We therefore propose to test the software on the basis of feasibility, to know whether they are nomadic or not. In the study, the Procedure Test (PT) we propose, declares "*this software is nomadic*" or "*this software is not nomadic because....*". This procedure does not make it nomad.

Checking nomadism involves verification of a number of constraints. In the previous part, we highlighted the link between discreet software and nomad software. We will therefore in a first time, verify that the tested software is discreet. If the software is not discreet and stores vital information on the hard disk or in the registry, we can stop. It will not be nomad. Once the discretion is verified, we'll check nomadism. The test of nomadism does not alter the original software, i.e. no line of its source code is affected. Finally, remember that the proposed procedure applies to applications running under Windows XP.

1. *Checking discretion.* We check the discretion of an application during its installation and during its use. For the installation, we scan (with RegShot [19]) the registry first and the tree beginning at "C: \ Windows," then we install the software and we scan the registry and tree above a second time. The comparison of the two scans can identify the keys that have been added, changed or more rarely deleted. When software has no installation program but is unpack from a single file (e.g. zip file), this analysis phase becomes unnecessary.

For the use, we begin to move the installation directory of software on another location on the installation disk. We then scan the registry for the first time and the tree starting at "C:\Windows" and beginning to "C:\Documents and Settings". We could scan the entire installation disk, but it is not profitable because, in general, when software writes elsewhere that in its installation directory, it writes in "Documents and Settings". Then we run the application with as many manipulations as possible, especially changing preferences when possible. The manipulations depend on the tested software. Wherever possible, we do a full type exercise with the software. Then we scan again the registry and trees mentioned. A comparison of two scans gives us the number of keys added or changed, the number of files added or changed and the number of files created. This way, we check the discretion of the application during its use.

The most delicate work begins. The question is now whether the keys and files created during installation or use are crucial or not. This is a case-by-case study.

1. *Checking nomadism.* Once the software is installed on the hard drive, we check the implementation of nomadism in two steps. The first step consists to move the software in another directory than the installation one. This step provides a check of the discretion of the application during its use at the same time we check if the software is running normally and does not use paths and logical units depending on the installation directory. The second step is to copy the installation directory on a USB key. Then we test the software on a computer other than the one on which we made the installation. We can be sure that if the software uses keys from the registry database or files stored on the first computer, it will not find them and therefore it will not be able to run normally.

III. APPLICATION OF THE TEST PROCEDURE ON A SAMPLE OF THREE ILE

In Interactive Learning Environment (ILE) domain, there are many applications, based on Windows, that could be adapted to run on a USB key in an appropriated use context. We conducted an exploratory study to test the mobility of these systems, chosen because they were easily downloadable. Our choice was -Cabri Geometry, Aplusix and Pépite. Our goal was to validate experimentally the feasibility of the test procedure and the "conformity" of our criteria. We have therefore, for each ILE, checked the discretion and the nomadism.

A. Cabri-Geometry

Cabri-Geometry is software to draw geometric figures and manipulate them dynamically. Cabri has already been adapted for a calculator which is a mobile device. But does Cabri already fulfill the conditions to run on a USB key? We tested Cabri II+ version 1.4.2 downloadable at [20]

1) Checking discretion

The table below shows the total number of modifications (keys and files) during the installation phase on a Windows XP and during the use phase once the installation directory moved.

Table 1 shows the number of changes (keys, values etc. on lines) during installation phase of the software (center column) or during the use phase (right column). This table shows that Cabri Geometry is not very discreet. But are the added or updated data sensitive? We will not detail here all changes, but will just comment some of them.

TABLE I.
DISCRETION OF CABRI-GEOMETRY

	installation phase	use phase
Added Keys	146	0
Added values	286	4
Modified values	6	8
Added files	1	1
Updated files	11	21
Added folder	0	2
Total	**450**	**37**

During the installation phase, Cabri-Geometry creates keys to identify MIME types - Multipurpose Internet Mail Extensions - (HKLM\SOFTWARE\Classes\MIME\Database\Content Type\application/Cabri II Plus). It is not blocking as MIME associations can be made later on another computer if necessary. It amends the key installations of software. This is more inconvenient because it is likely to consult later this information (e.g. HKLM\SOFTWARE\Classes\ Installer\Products\CA6F080C32623B2 ... 6D7485\ProductName: "Cabri II Plus 1.4.2).

During the use phase, Cabri creates a folder in "Documents and Settings\All Users\Application Data\Cabrilog. This is typically the kind of creation that can harm the nomadism. We can notice that the total number of changes during the use phase is relatively low. This is encouraging for the future.

2) Checking nomadism

We moved the installation directory of Cabri-Geometry on a USB drive and launched it on a post on which the installation had not taken place. At the launch of the application, we had an error message calling for the product installation. However, once this error message taken into account, the software is still usable in demonstration mode. A message appears on screen stating that the software is not usable in the classroom. But that does not stop to create a figure and handle it

In conclusion, we can say that Cabri-Geometry is not very discreet. However, it operates in degraded mode (demonstration mode) when it is moved on a USB key. We can therefore consider it is nomad. In any case, the authors of this software does not have a big substantive work to provide a completely nomad version.

B. Aplusix II

Aplusix is a program to learn algebra. It allows to edit algebraic expressions and to manipulate them. We used setupAplusixV1_02_fr_Demo.exe version downloadable at [21].

1) Checking discretion

The table below shows the total number of modifications (keys and files) during the installation phase on a Windows XP and during the use phase.

Table II shows that the software is not very discreet especially in the installation phase. During the use phase, it stores settings and information about the exercises. This can make trouble. Moreover, it creates log files and information about the user in "C:\ Windows\" (6 files) and "C:\Documents and Settings\Users\" (2 files).

TABLE II.
DISCRETION OF APLUSIX

	Installation phase	Use phase
Added Keys	503	5
Added values	706	8
Modified values	11	6
Added files	1	0
Updated files	14	8
Added folder	1	0
Total	**1236**	**27**

2) Checking nomadism

We only tested the students' version. The execution of Aplusix in a directory other than the installation directory but on the same machine and the same logical unit has succeeded. We created a new user (an alias) and produced an exercise in developing algebraic expression. The execution of Aplusix on the USB key on another computer, allow us to find the user previously created and to continue the same exercise as above.

In conclusion, we can say that the two aforementioned executions allow us to say that Aplusix is nomad on a USB key even if he is not especially discreet during the installation phase. Files created on the disk and keys created in the registry do not seem to disturb the correct running of the software on a USB key.

C. Pepite

Pepite is software to learn algebra. Pepite has two applications: PepiTest and PepiProfil. PepiTest allows the learner to test their knowledge of algebra. Once the test and data saved, PepiProfil can analyze the results of the learner and generate a cognitive profile that the teacher can adapt for learning. Pepite is available at the following address: [Pepite]

1) Checking discretion

The table below shows the total number of modifications (keys and files) during the installation phase of Pepite on a Windows XP and during the use phase of PepiTest

Table III shows that the software is fairly discreet during the installation phase. The values and modified files blockers do not seem to block nomadism. During the use phase, many more values and files are added, modified or created. In particular, 12 files in "C:\Windows" (including log files) and the file containing data for the user "C:\Documents and Settings\User\ntuser.dat.LOG" are modified. This can make trouble.

TABLE III.
DISCRETION OF PÉPITE

	Installation Phase	Use Phase
Added Keys	0	41
Deleted Values	0	1
Added Values	0	41
Updated Values	4	17
Deleted Files	0	2
Updated Files	5	13
Total	**9**	**115**

2) Checking nomadism

Running pepiTest, the student' version of Pepite, in a directory "C:\" other than the installation directory, was conducted smoothly. Running pepiTest and pepiProfil on the USB key on another computer running on XP went well. However, during the execution of pepiTest, the program tries to install a component of Windows Office Professional. A message then calls the introduction of the Office CD. We just click cancel. The messages are annoying but do not seem to disturb the normal functioning of PepiTest, which remembers exercises done.

In conclusion we can say that Pepite is nomadic for the USB key (even if we tackle this problem inserting the Office CD). In addition, Pepite is pretty discreet during the installation phase and a little less during the use phase.

This exploratory experiment allowed us, on the one hand, to make a first validation of our test procedure and to test its feasibility. On the other hand, it allowed us to identify two criteria: the discretion and nomadism. Indeed, for software to be discreet, it must not use the registry to store information. Similarly, it should not create file or folder outside its installation directory. To be nomadic, it must operate alone. To do this, it does not seek specific functions of the operating system of the host computer. It does not seek specific components of the system as the Java virtual machine or framework.net. Finally, it should not make use of the software installed on your host (such as ACCESS).

It is on these two criteria that we tested nomadism on USB key of a piece of software widely used today by many users: the Weblog.

IV. A SYNCHRONISABLE BLOG NOMAD ON USB KEY

Remember that one of our concerns is to offer students continuity in the learning process. Indeed, when assessing the educational device named COOLDA-MAETIC, we noticed that students working in group to a project, tended to interrupt their learning process between sessions attendance. We propose then to board their working environment on a USB key and to offer them the opportunity to contribute to the group offline. We want them to include ways to edit the logbook of their project (which is carried by holding a Weblog) offline via a USB key

We have designed a nomad synchronisable Weblog: nomad because our goal is that it works on a USB key and synchronisable to provide synchronization of offline blog with the online blog.

As a first step, we conducted a study of existing technologies. We studied software for Weblog editing (blogger) that offer both a download feature (recovery of a blog to a USB key) and an upload feature (update to the blog site changed on the USB key). Offline Weblog technology requiring an offline customer, we also studied two clients: SharpMT [23] and Wbloggar [24].

In a second step, we designed a synchronisable nomad Weblog into two parts: a server and a client. The client is Windows software. It allows the data backup in a SQLite database and the content creation, while being most nomads as possible. The client is developed in C + + to avoid installing the Java virtual machine or the framework .Net on the USB key. The customer can delete, update or see posts or comments. It can also create posts. The blog server-classic is a website with a remote HTTP/XML interface, which returns responses in XML format and allows to download or to send the blog content from a client. It is written in PHP/MySQL.

We then applied the Test Procedure (TP), explained earlier, to verify the discretion and nomadism of the synchronisable Weblog. Since the server part of the Weblog is a website, we have not had to test its nomadism. However, the client Weblog is a program written in C + + which can both run on a PC with Windows XP or on a USB key connected to a host PC with Windows XP operating system. We then tested the nomadic client Weblog.

Finally, we tested the modification and content creation on the key and on the site and the synchronisation of the key and the site. We have now to test the client and server in ecological environment.

V. CONCLUSIONS

The focus of this article was on the choice of USB key as mobile technology. Thus, after introducing the context of our research, justified the choice of USB key, supported by some definitions, we proposed a procedure to test whether a learning environment embarked on a USB key can be described as nomadic or not. This procedure has been tested as part of a study conducted on a sample of three ILE. This approach has allowed us to define feasibility criteria, criteria which were then included in the design of a synchronisable nomad Weblog on USB key: i.e. running a blog offline on the USB key and synchronizing it with the weblog site.

Our goal now is to conduct an experiment of this synchronisable Weblog with an audience of students at the ULCO in the year 2009. The trends of this work are nomadisation of the working environment COOLDA-MAETIC on USB key. The COOLDA-MAETIC working environment is built around a development environment (Eclipse). It has a chat, a CVS, an office suite (open Office). We will add our synchronisable Weblog to this environment. To test our environment, we will apply our testing procedure. We have already seen that Eclipse is nomad on USB key. The office suite is already nomad and available on the Framakey site [18]. We will thus check the nomadisation of the CVS and of the chat to set up the experiment. The results of this experiment will be the subject of future communications.

REFERENCES

[1] K. Masters, "M-learning: how much of what has been diffused? A systematic literature review", ED'MEDIA 2008, 5790-5796

[2] C. Quénu-Joiron, D. Leclet, B. Talon. "Conception de dispositifs pédagogiques intégrant la mobilité". Atelier Apprentissage Mobile. Associé à la conférence EIAH 2007. p17-20. Lausanne, Suisse. 26 juin 2007.

[3] E. Webb, G. Cavanagh "How Mobile is your Podcast?" ED'MEDIA 2008, 3954-3958

[4] C. Salis, M. Ambu, "MOsKA, Mobile Organized Knowledge Access for Science: Astronomy and Renewable Energies Videos for Mobile Phone Delivery", ED'MEDIA 2008, 4388-4393

[5] J. Arreymbi, E. Agbor, M. Dastbaz, "Mobile-Education - A paradigm shift with Technology", ED'MEDIA 2008, 5114-5122

[6] K.-J. Huang, T.-C. Liu, S. Graf, Y.-C. Lin, "Embedding mobile technology to outdoor natural science learning based on the 7E learning cycle", ED'MEDIA 2008, 2082-2086

[7] D. Leclet, and B. Talon, "Assessment of a Method for Designing E-Learning Devices", Proceedings of World Conference on Educational Multimedia, Hypermedia and Telecommunications, ED-MEDIA 2008, AACE/ Springer-Verlag (Ed.), Vienna, Austria, June 30 - July, p 1-8, 2008.

[8] A. Lewandowski, G. Bourguin, "A New Framework for the Support of Software Development Cooperative Activities", Proceedings of the 8th International Conference on Enterprise Information Systems (ICEIS'06), Paphos, Cyprus, 23-27 May, 2006, ISBN: 972-8865-43-0, INSTICC Press, pp. 36-43, 2006.

[9] D. Leclet, E. Leprêtre, Y. Peter, C. Quénu-Joiron, B. Talon, T. Vantroys. "Améliorer un dispositif pédagogique par l'intégration de nouveaux canaux de communication" Environnements Interactifs d'Apprentissage Humain, EIAH 2007. p347-357.

[10] M. Rosselle, P. Gillois, J. Morinet-Lambert et F. Kohler. » Installing the Student's Computer to Access a Virtual Medicine University". International Symposium TICE'2002, Technologies of Information and Communication in Education for Engineers and Industry. INSA, Lyon, France, 11-15 nov. 2002.

[11] radio frequency identification http://en.wikipedia.org/wiki/RFID last visited on 14/11/2008.

[12] Université Médicale Virtuelle Francophone http://www.umvf.prd.fr/ last visited on 14/11/2008.

[13] Linux on USB key www.knopper.net/knoppix/index-en.html last visited on 14/11/2008.

[14] Linux on USB key www.pendrivelinux.com last visited on 14/11/2008.

[15] virtual workspace www.mojopac.com last visited on 14/11/2008.

[16] Windows software on USB key http://portableapps.com last visited on 14/11/2008.

[17] Windows software on USB key www.winpenpack.com last visited on 14/11/2008.

[18] Windows software on USB key www.framakey.org/En/Index last visited on 14/11/2008.

[19] Software to scan registry and file tree sourceforge.net/projects/regshot last visited on 14/11/2008.

[20] Cabri http://www.cabri.com/cabri-2-plus.html last visited on 14/11/2008.

[21] APLUSIX II: Algebra Learning Assistant http://aplusix.imag.fr/en/index.html last visited on 14/11/2008.

[22] Pépite software (in French) http://pepite.univ-lemans.fr/ last visited on 14/11/2008.

[23] blog client http://www.codeplex.com/sharpmt/ last visited on 14/11/2008.

[24] blog client http://www.wbloggar.com/ last visited on 14/11/2008.

AUTHORS

M. Rosselle is a member of the MIS-UPJV Laboratory, 33 rue St Leu, 80039 AMIENS CEDEX 1, France (e-mail: marilyne.rosselle@u-picardie.fr).

D. Leclet is a member of the MIS-UPJV Laboratory, 33 rue St Leu, 80039 AMIENS CEDEX 1, France (e-mail: dominique.leclet@u-picardie.fr).

B. Talon is a member of the LIL-ULCO Laboratory, Maison de la recherche Blaise Pascal, BP 719, 62228 CALAIS CEDEX, France (e-mail: talon@iutcalais.univ-littoral.fr).

14

Using Mobility to Enhance Routing Process in MIS System

K. Oudidi, A. Habbani and M. Elkoutbi
University Mohammed V- Souissi, Rabat, Morocco

Abstract—This paper introduces the original Mobile Intelligent System (MIS) in an embedded Field-Programmable Gate Array (FPGA) architecture. This would allow the construction of autonomous mobile network units which can move in environments that are unknown, inaccessible or hostile for human beings, in order to collect data by various sensors and route it to a distant processing unit.

To have a better performing routing process, we propose a new mobility measure. Each node measures its own mobility in the network, based on its neighbors' information. This measure has no unit and is calculated by quantification in regular time intervals.

Index Terms—Information systems embedded application, intelligent sensors, wireless sensor network, ad hoc networks, OLSR protocol, multipoint relays, node mobility and mobility quantification

I. INTRODUCTION

Sensors have become an essential element in all systems where information resulting from the external environment is to make evaluations and act. To have an exact and complete grasp of the subject requires the deployment of several sensors and, possibly, the combination of all retrieved information to better adjust each parameter's sensor.

A sensor network is composed of a large number of units called nodes. Each node mainly consists of one or several sensors, a processing unit and a communication module, etc. These nodes communicate between each other according to the network topology and the existence or not of an infrastructure (access points) to forward the information to a control unit outside the measure zone. With these features available, we can imagine an adaptive complex system built on several sensors in a wireless communication system. An original system has been designed and realized: MIS (Mobile Intelligent System) project, which allows integrating three main functions: information' acquisition, processing and routing around an embedded architecture such as FPGA (Field-Programmable Gate Array).

Mobility impacts conditions where routing protocols should operate, the context that nodes can use to communicate, and the problems that protocols should solve.

In this paper, we introduce the architecture of the MIS and present a new quantitative measure of mobility reflecting the mobility degree in each MIS. Using OLSR routing protocol, this mobility measure will be exploited by the MIS during the route discovery process to enhance it and adapt it in the presence of high mobility.

This paper is organized as follows: The first section consists of a general introduction. The second one focuses on the functional architecture and the experimental MIS platform and its units. Section 3 shows the importance of mobility in designing ad hoc routing protocols. Section 4 introduces and discusses our network mobility measure. Section 5 presents some experiments of the behaviour of network mobility in different Mobile Ad hoc Networks (MANET) configurations. In Section 6, we present an application of our proposed mobility measure. The last section concludes and presents some future works.

II. MIS PLATFORM

In this section, we present the MIS project and its experimental platform system previously introduced in [1-5].

A. MIS Presentation

MIS (Mobile Intelligent System) is a platform of intelligent wireless sensor prototyping elaborated within the Wireless Sensor Networks (WSN) group of the Laboratory Electronic and Communication (LEC) for topological applications of communicating objects' networks. This platform is based on various sensors (CO, resistive tape recorder, etc.), a routing and treatment unit, a module of wireless radio communication using standard BLUETOOTH or WIFI and a routing and treatment unit based on a microprocessor (IP software).

B. MIS Applications

One of the main applications is to construct autonomous mobile networks capable of moving in environments which are unknown, inaccessible or hostile for human beings or in risky areas (fire, radiation, earthquake, etc.) in order to optimize human assistance. The aim is to provide ground information to establish a strategy of evolution according to the set target. For example, victims can be located during rescue operations thanks to small mobiles capable of infiltrating through rubble or exploring the watery funds. Another equally important application is military exploitation. In this context, the use of sensor networks allows the surveillance of the perimeters, to assist air or ground attacks and to lead espionage operations. To this end, no element should be indispensable for the functioning of the network. Such an ad hoc architecture can maintain the network in activity after the loss of one or several elements and requires a routing module.

C. MIS Architecture

The functional architecture and the experimental platform MIS is built on the development kit ALTERA Cyclone (System One Programmable Chip). It is essentially composed of four units (figure 1): an acquisition unit, a treatment unit, a routing unit and a communication unit.

The detailed architecture of the designed and produced beacon is given below (fig.2-3). It is articulated around the Nios II processor. Several interfaces are used to connect the peripherals to the processor (SPI, UART, Bus Avalon, PIO, etc.).

The system is also composed of different sensors allowing data acquisition and the generation of numerical signals. These signals are treated by target card ALTERA cyclone. After treatment, control signals are routed towards a central station using a routing protocol.

The routing protocol can be implemented on MIS in two different ways, either directly into software or in a hybrid way: the software part of MIS is in C language and material acceleration is implemented using hardware description language VHDL (optimizations to be made to meet the criterion of consumption and execution speed). This implementation has been finalized and made possible by adding an operating system of the μClinux type. The big advantage of μClinux in comparison with other systems is the compatibility of API's programming with the Linux standard systems. It also has all TCP/IP network functions, available on the Linux kernel and supported by the ALTERA card. Furthermore, it does not consume a lot of memory.

In the next subsections, we shall mainly describe our contribution: "*Using Mobility to Enhance the Routing process in the MIS System*'', subject of this paper.

III. IMPORTANCE OF MOBILITY IN PROTOCOL DESIGN

The behavior and characteristics of a network's wireless links is very different when it comes to mobility, which makes the design of communication protocols operating in the presence of mobility more challenging. Mobility also changes the neighborhood in which a given node must share the communication bandwidth available with others. As nodes move, the paths established from sources to destinations can be broken, which leads to the creation of new paths and the reallocation of resources along such paths to meet the application requirements. However, while mobility makes the implementation of several functions and services more challenging, it allows some useful functionality. For example, thanks to mobility, a node can know its location and can therefore exploit location-dependent services. Similarly, mobility introduces a fundamental change in our perception of networking. While the end-to-end connectivity assumption is justified in wired networks, the cost, energy consumption and form factors of computing devices have enabled embedded computing and networking devices that can be used in environments where end-to-end connectivity may at best be intermittent.

IV. THE PROPOSED MOBILITY MEASURE

A node in a wireless network can be found in three states in relation with its neighbour: node moving/ its neighbour static, node static/its neighbour moving, and finally node moving/ its neighbour moving. Consequently, these three possible states result in a change in the link status of the node with its neighbour. Hence, as the node moves in the network, the link status changes over time.

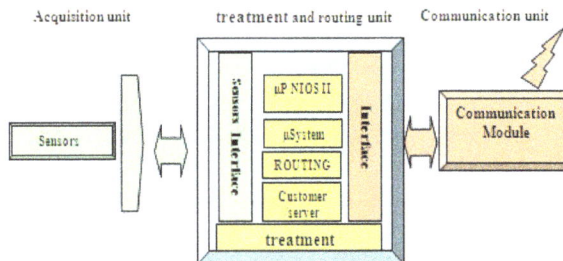

Figure 1. MIS architecture

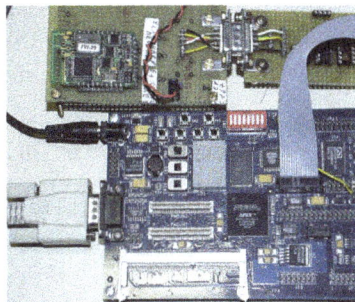

Figure 2. Wireless unit interfaced with the processing unit

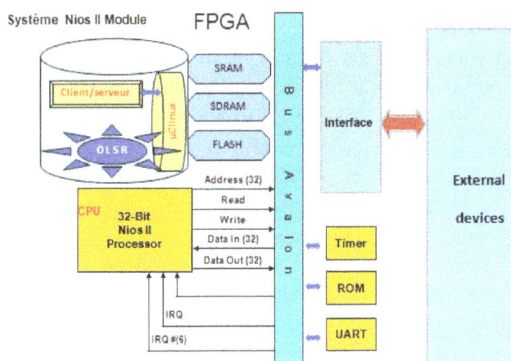

Figure 3. Synoptic of the System SoPC NIOS II for the acquisition and the routing of temperature

Based on this observation, we define *the node (MIS)* and *the network mobility* in an ad hoc environment. *Mobility* is quantified locally. It is independent from the location of a given node. We represent this local and relative quantification by its neighbour's change rate. *The node mobility* at a given time *t* for node *A* in the ad hoc network is defined as the change in its neighbour compared to the previous state at time $t - \Delta t$. Thus, nodes that join or/and leave the neighbour of node *A* will impact the evaluation of its *mobility*. As explained before, we define the *mobility* of node *A* at time *t* by the following formula:

$$M_A^\lambda(t) = \lambda \frac{NodesOut(t)}{Nodes(t-\Delta t)} + (1-\lambda)\frac{NodesIn(t)}{Nodes(t)} \quad (1)$$

Where,

NodesIn(t) : The number of nodes that joined the range of node *A* during the interval $[t - \Delta t ; t]$.

NodesOut(t) : The number of nodes that left the range of Node *A* during the interval $[t - \Delta t ; t]$.

Nodes(t) : The number of nodes in the range of node A at time t.

λ : a 0 to 1 positive value defined in advance to promote incoming/outgoing nodes depending on situations (attacks, rescue operation, etc.).

The choice of the value $(\lambda = \frac{1}{2})$ will ponder equally the NodesIn(t) and NodesOut(t) nodes during Δt and keep the *mobility* node value in the interval [0,1]. In other words, let us take node A that has 11 neighbors at $t - \Delta t$ (Figure 4(a)). During the Δt time interval, its neighbour has changed its position as shown in (Figure. 4(b)): two nodes (red color) have left the range of node A, and three nodes (green color) have joined its neighbour. Consequently, the state (neighbors) of the node will change after Δt (Figure 4(c)). At the end of each time interval, the node will be able to evaluate the change in its neighbour represented by this relative *mobility* which, in this example, is equal to 21%:

$$M_A = 0.5\frac{2}{11} + (1 - 0.5)\frac{3}{12} \approx 21\% \quad \text{(with } \lambda = 1/2).$$

The node *mobility* quantification has no unit, varies between 0 and 1, and does not suppose any mobility model [6] for evaluation. Each node in the MANET can make an autonomous and automatic evaluation of its mobility at regular time intervals. This evaluation can be periodically done while exchanging Hello messages (a characteristic that we find in the proactive protocol family).

Moreover the calculation and recalculation of *node mobility* is fast and does not consume many resources (CPU and memory).

After measuring the node mobility, we can define the *network mobility* measure in regular time intervals as the average of the involved nodes mobility:

$$Mob(t) = \frac{1}{N}\sum_{i=0}^{N-1} M_i(t) \tag{2}$$

Where N is the number of nodes in the network.

In addition, we can define the time average of this network mobility as the average of the simulation period (T):

$$M = \frac{\Delta t}{T}\sum_{k} Mob(t) \tag{3}$$

Where $k \in \{0, \Delta t, 2\Delta t, ..., T\}$. T is the time of simulation.

V. Validating our Moility Measure

In this section, we present the behavior of the *network mobility* relating to some characteristics of the ad hoc network for the default case $(\lambda = \frac{1}{2})$. The default case corresponds to an environment where NodesIn(t) and NodesOut(t) are equally pondered.

As *mobility* is a main constraint with a direct impact on the performance of MANETs (Mobile Ad hoc NETwoks), it is necessary to study its behaviour in different scenarios by changing several properties of the ad hoc network: number of nodes, network dimension, transmission range and speed of nodes.

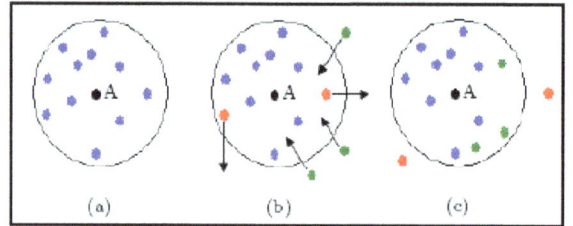

Figure 4. Node mobility quantification in Δt interval

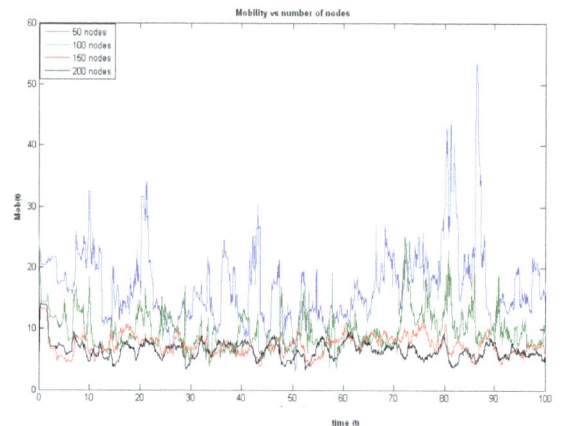

Figure 5. Effect of node numbers on network mobility

We choose the random waypoint mobility model [7] to represent nodes motion in the MANET. This mobility model is the most widely used mobility model, due to its simplicity and capability to synthesize scenarios with varying degrees of mobility. The impact of mobility models [8] [9] is also discussed later in this paper.

In our simulations, we consider a MANET with 1000mX1000m and 100 seconds as a period of simulation. The value of Δt in this mobility quantification work is 0.05 s $(\Delta t = 0.05$ seconds). Moreover, we represent our network mobility M in the following graphs by percentage.

A. The Effects of Node Number

The number of wireless devices is expanding and new multi-user applications are developed to adapt to this increase. This requires the building of larger ad hoc networks with the participation of more and more nodes.

This is why we have decided to simulate the effect of an increase in the number of nodes on *the network mobility* M. In our simulations, we use a square network area of 1000m X 1000m size with different configurations (50, 100, 150 and 200 nodes) and set the transmission range to 100m and the maximum speed to 40m/s (high mobility).

We note from the graphs in Figure 5 that mobility increases (from black to blue) as the number of nodes decreases in the network. More precisely, we can see that the mean and the variance of mobility become important when the node number decreases in the network. This can be explained by the increasing sensitivity of MANET to link state changes when the number of nodes falls down.

B. Impact of Speed

Many simulations concerning the influence of speed on mobility have been made [10]. They show that mobility largely depends on the concerned nodes' speed and direction of movement. In order to confirm this result for our network mobility measure, we consider in our simulations a MANET with 50 nodes. The transmission range is set at 100m. We have taken for simulations the following maximum speed: 0m/s; 20m/s; 40m/s; 60m/s.

Figure 6 shows clearly that an increasing speed automatically implies an increase in the network's mobility. When nodes move at a high speed to a random destination, they have an important change in their neighbours (nodes that join and/or leave their transmission range). In short, the *network mobility M* logically depends on nodes' physical speed.

C. Impact of the Transmission Range

The communication devices available on the market offer a wide range of power levels, which affect the transmission power and connectivity. It is often assumed that the larger the transmission range, the better for data delivery. In this section, we study the impact of the transmission range, on MANET performance. We have found that it affects logically our network mobility's metric. We consider a MANET with 50 nodes in case of high mobility (maximum speed of nodes is 40m/s). In order to show the impact of node's transmission range, we have taken the following values: 50m, 100m, 150m, and 200m.

Figure 7 shows that the network mobility varies inversely with the transmission range.

Otherwise, mobility becomes important if we have nodes with a small transmission range. In the case of a small transmission range, nodes neighbours' that move rapidly in the network have more chance to leave and/or join the transmission range of its neighbour. Consequently, the rate of link state changes and mobility becomes important.

D. Impact of Mobility Models

In the literature, many mobility models have been used to simulate node's motion [6] [11]. In this section we will show the behaviour of our network mobility measure using different mobility models. The simulated mobility models are: mobility waypoint model, Manhattan model and reference point group model.

We notice that the network mobility behaviour is impacted by the mobility model chosen. For the random waypoint mobility and random Gauss Markov models, the average and variance behaviour of the network mobility are stable all over simulation time. The Manhattan mobility model shows an important variance of mobility. The registered variance can be caused by nodes that are found at *column intersections* [6] [10]. In the reference point group mobility model, the variance is equal to 0, but during the simulation time, the mobility changes dramatically. These variations are produced when the groups are close to each other.

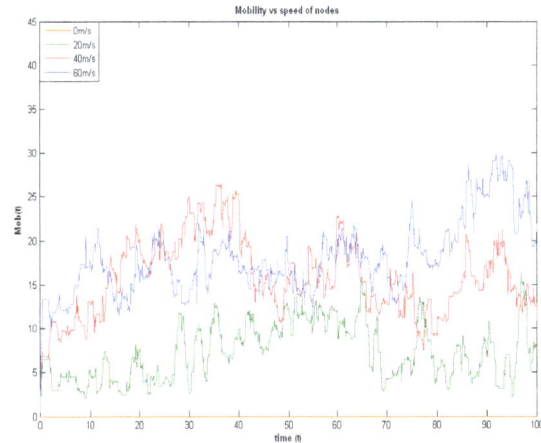

Figure 6.   Effect of speed on the network mobility.

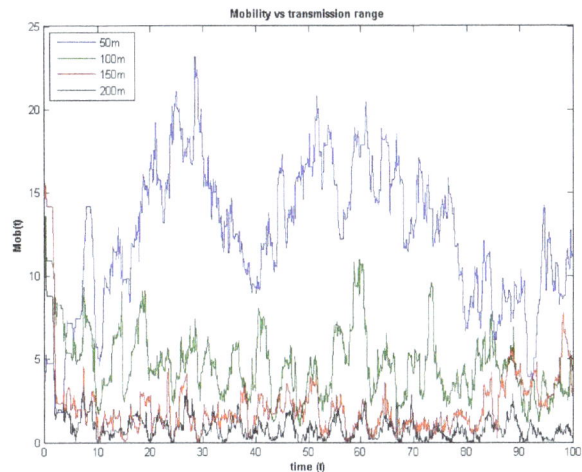

Figure 7.   Effect of transmission range on the network mobility measure

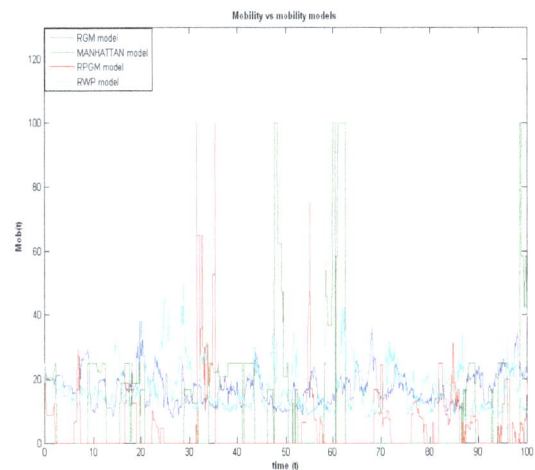

Figure 8.   Behavior of the network mobility using different mobility models

VI. Validating our Mobility Measure

A. Network Scenarios

To generalize the validation of our proposed mobility metric, we have considered two types of network scenarios [8] (entity mobility and group mobility models). In this study, a variety of network scenarios are generated for each type of mobility models as summarized in Tables 1 and 2. For the first type including the RWP and RGM models and to make comparison more challenging, we have considered the worst scenarios by supposing a maximum speed of nodes equal to V_{max} = 40m/s and a minimum speed equal to V_{min} = 0m/s. For RGM model, speed v and direction θ are updated every Δt = 2.5 seconds, where Δv_{max} = 0.5 and $\Delta \theta_{max}$ = 0.125π.

Table 1 shows the first type of network scenarios that simulate a group of nodes that move randomly in a square region. Each scenario has its proper parameters that distinguish it from other scenarios. For the RWP model, parameters are the region dimensions, the number of nodes N and the *pause time*. For the RGM model, parameters are the region dimension and number of nodes N. For example, scenario *S1* related to the RWP model has 30 nodes that are moving with a pause time equal to 0 seconds in an *800mX800m* square region, and scenario *G1* related to the RGM model has 50 nodes moving in *700mX700m* square region.

Table 2 illustrates the second type of network scenarios using the RPGM model where nodes move in a *1000mX1000m* square region. For the trajectory of the logical center of each group, the RWP model is used. Moreover, for the same reason (i.e. to make comparison more challenging), we assume the worst scenario by supposing the maximum speed of the logical centre is equal to V_{max} = 40m/s, the minimum speed is equal to V_{min} =0m/s, and a pause time is equal to 0. The update interval τ =1 is used for the random motion vector. In scenario *P1*, there are 5 groups, where each group contains 10 nodes (total 50 nodes). One of the nodes' reference points is located at the logical centre of each group and the other 9 reference points at the corners of a regular hexagon centred at the logical centre with the length of its side 0.25. The length of the random motion vector has a uniform distribution between 0 and RM_{max} = 0.25. All of the 5 nodes' reference points are located at the logical centre of each group. Scenario *P2* ensures more intra-group motion compared to scenario *P1* by having RM_{max} = 0.5. As our proposed mobility metric depends on parameter λ, we have also studied its impact on the relationship with the rate of link change. The values of λ considered in this work are: λ = 0.00, 0.25, 0.50, 0.75, and 1.00. Finally, to make simulations in the same conditions, the simulation time and communication range for all scenarios are equal to 500 seconds and *100m*, respectively. Moreover, to be more precise, let us note that, henceforth, each measure represents an average of 20 measures.

B. Results and Debate

As Node mobility in ad hoc network depends on the change in link status, validating our mobility metric re-

TABLE I.
Entity Mobility: Nodes Move Randomly

Random WayPoint Mobility Model			
Si	Network dimension	N	*pause time*
S1	800X800	30	0
S2	800X800	40	0
S3	800X800	50	0
S4	600X600	50	0
S5	700X700	50	0
S6	800X800	50	5
S7	800X800	50	10

Random Gauss-Markov Mobility Model		
Gi	Network dimension	N
G1	700X700	50
G2	800X800	50
G3	900X900	50
G4	800X800	20
G5	800X800	30
G6	800X800	40

TABLE II.
Group Mobility: RPGM Model is Used.

Pi	Details
P1	5 groups, 10 nodes/group (total 50 nodes), RM_{max} = 0.25 (small intra-group motion),
P2	5 groups, 10 nodes/group (total 50 nodes), RM_{max} = 0.5 (small intra-group motion),
P3	10 groups, 5 nodes/group (total 50 nodes), RM_{max} = 0.25 (small intra-group motion),
P4	10 groups, 5 nodes/group (total 50 nodes), RM_{max} = 0.5 (small intra-group motion),
P5	10 groups, 4 nodes/group (total 40 nodes), RM_{max} = 0.25 (small intra-group motion),
P6	4 groups, 10 nodes/group (total 40 nodes), RM_{max} = 0.25 (small intra-group motion),

quires a strong linear relationship between this mobility and the link change rate. To this end, we have compared them. The mobility metric in the network being normalized by N (total number of nodes), it is essential to normalize the link change rate measure by $N(N-1)/2$ which represents the maximum number of links in a network with N nodes. Moreover, as the change of the link state occurs in time, it is essential to make this comparison by taking into account the time constraint. Moreover, we suppose at each end of interval Δt that the ad hoc network is steady. In this work, the comparison is made at the end of simulation between the following measures: the time average of ad hoc mobility measure M_λ, and the average normalized link change rate.

In all the scenarios, the calculation of each measure during simulation is evaluated by quantification at the same discrete time intervals. For this study, the step chosen for the quantification of these two measures is equal to Δt = 0.05s. We chose Δt, a relatively small value, so as to have a better estimation of the link change rate. To calculate the average normalized link change rate, we initially define $L(t)$ as the number of link changes that occurred at time interval $[0,t]$. Then, the number of link changes $l(t)$ occurring in Δt time is as follows:

$$l(t) = \frac{\Delta L(t)}{\Delta t} = \sum_k \delta(t - t_k) \qquad (4)$$

where t_k is the time instance of the *k-th* link change. The *time average of the normalized links change rate* is given by:

$$\bar{l} = \frac{\Delta t}{T} \sum_k \frac{l(t)}{\dfrac{N(t)(N(t)-1)}{2}} \qquad (5)$$

where $k \in \{0, \Delta t, \ 2\Delta t, ..., \ E\left(\dfrac{\Delta t}{T}\right)\}$. $E(x)$ is the integer part of x, $N(t)$ is the total nodes at the instant t, and T is the total simulation time. As the total of nodes is supposed to be fixed in this study, Equation (5) can be written as follows:

$$\bar{l} = \frac{2\Delta t}{N(N-1)} \frac{L(T)}{T} \qquad (6)$$

After simulations, the results show that the time average ad hoc network mobility M_λ has a good linear relationship with the average normalized link change rate for the two network scenario types. This good linear relationship is not influenced by the considered values of λ ($\lambda = 0.00$, 0.25, 0.50, 0.75, 1.00), but these values have an impact on the slope of the line approximating the linear relationship. On the other hand, we can consider our metric of mobility measure M_λ as an alternative mobility measure evaluated at discrete times, contrary to the mobility measure proposed in [14] which is based on uninterrupted time. Moreover, our unified mobility metric is more trivial and independent of all pattern motions of nodes.

Figures 9-a and 9-b show the simulation results for the mobility metric M_λ with the first type of mobility models (entity mobility models) with different values of λ. As shown in figure 9, the average normalized link change rate $\bar{l}$ shows a strong linear relationship with M_λ for the entire network scenarios. By considering $\lambda = 0.50$, the first type of scenarios related to the entity mobility model (RWP and RGM models), the good linear relationship is well maintained even if we change the number of nodes N (*S1-S2-S3* for RWP, and *G2-G4-G5-G6* for RGM), the physical dimension of the network (*S4-S5-S6* for RWP, and *G1-G2-G3* for RGM), and the pause time (*S6-S7* for RWP). As shown in figure 10, the same behaviour is detected in the results obtained with the second type of network scenarios relating to the group mobility model (RPGM model), by varying the groups' number (*P1-P3-P6*), total of nodes (*P1-P5*), and the intra-group motion (*P1-P2*, and *P3-P4*).

This shows that our mobility metric M_λ has the same behaviour in terms of link change rate in the ad hoc network.

By construction, our mobility measure approach is based on the link status change undergone in the vicinity of the communication range. We compare our studies to [14], who have found a relationship between the remoteness concept and the link status change.

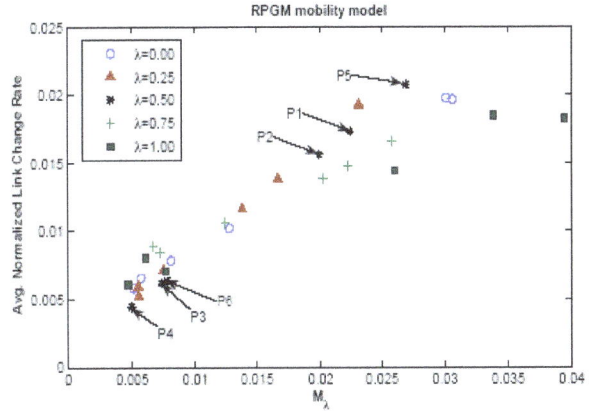

Figure 9.   Normalized link change rate vs. mobility metric in group mobility model

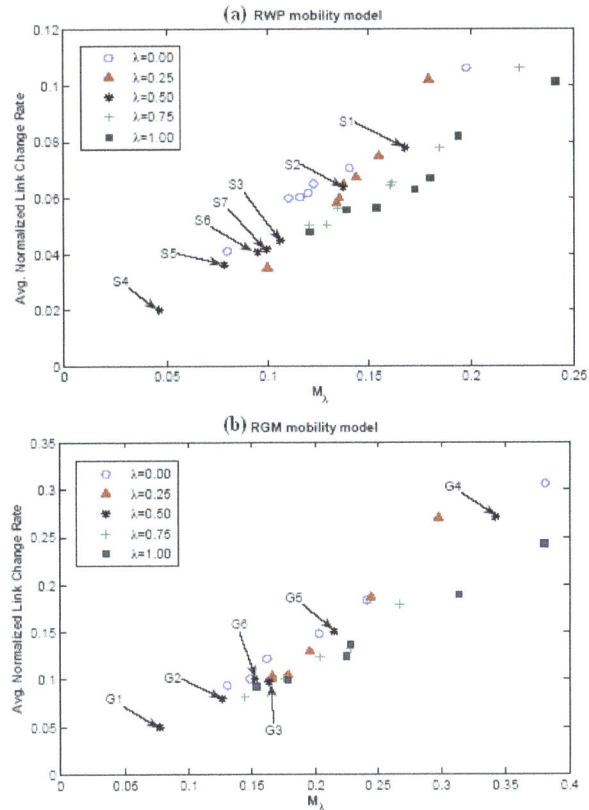

Figure 10. Normalized link change rate vs. mobility metric in entity mobility model

They define a distance function between two nodes as a relation of remoteness between them. Moreover, this distance function should satisfy the following requirements. It increases from 0 to 1 monotonically. The derivative of remoteness is 0 at distance 0. It goes up as the distance increases, reaches its maximum at the communication boundary, then decreases as distance increases further, and approaches 0 as the distance approaches the infinite value. In short, their mobility measure is defined as the average derivative of remoteness over all node pairs. Several remoteness functions fulfilling these requirements can be found. So it is quite clear that the returned measure of mobility depends on the chosen remoteness function. On the other hand, these remoteness functions are complex in a lot of cases and contain many parameters which require hard operations in a network simulator environment where resources are needed (CPU and memory). To measure mobility with a remoteness function, hard operations (such as derivation and integration calculation) are needed; these could saturate resources and make the network down. In a real ad hoc network, we can have two nodes next to each other in the absence of a link (e.g: presence of an obstacle such as a button wall). Therefore, the concept of remoteness cannot reflect exactly the rate of link status change in an ad hoc network.

In the literature, it is difficult to compare the performance results of several protocols simulated in different mobility models, even if they have the same simulation parameters. To this end, we find that comparative studies on protocol performance are based on the same mobility model (often RWP model). This makes it possible to put simulations under the same *mobility condition*; in other words, in the same mobility measure. Then, to meet this condition, it is necessary to know how the changing ad hoc network parameters influence our mobility measure: to know the mobility measure behaviour when the ad hoc network parameters change. The main ad hoc network parameters characterizing it are: number of nodes, network dimension, and node transmission range.

VII. APPLICATION

As an application of this mobility measure, we have exploited the node mobility parameter defined in Section 4 to adapt MANET routing protocols to topology changes. The protocol that we have considered is the OLSR protocol.

A. The Optimized Link Routing Protocol

1) Overview

The OLSR (Optimized Link State Routing) [15] protocol is a proactive table-driven routing protocol designed for mobile ad hoc networks. As a link state routing protocol, OLSR periodically advertises information about links to build the network. However, OLSR optimizes the topology information flooding mechanism by reducing the amount of links that are advertised. It also reduces the number of forwarding messages by limiting them to the set of Multipoint Relays (MPRs). Information topology is sent by a Topology Control (TC) message and exchanged using broadcasted messages into the network. TC messages only originate from nodes acting as MPRs. The latter are selected in such a way that a minimum set of MPRs, located one-hop away from the node doing the selection (called MPR Selector), are sufficient to reach every single neighbour located two-hops away of MPR selector. By applying this selection mechanism, only a reduced number of nodes (depending on the network topology) will be selected as MPRs. Every node in the network recognizes its one-hop and two-hop neighbours by periodically exchanging HELLO messages containing the list of its one-hop neighbours. On the other hand, TC messages will only be advertised between MPRs and their electors. Then, only a partial set of network links (the topology) will be concerned by advertising control messages, also MPRs are the only nodes allowed to forward TC messages and only if messages come from a MPR Selector node. These forwarding constraints considerably decrease the amount of flooding retransmissions.

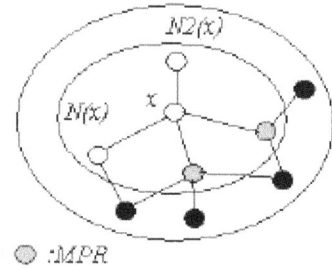

Figure 11. Example of MPR Selection

2) The MPR Selection algorithm

Computing the MPR set with minimal size is a NP-complete problem [11]. The standard MPR selection algorithm currently used in the OLSR implementation is as follows:

For a node x, let $N(x)$ be the neighbourhood of x. $N(x)$ is the set of nodes which are in the range of node x and share a bidirectional link with it.

We denote by $N2(x)$ the two-neighbourhood of node x, i.e, the set of nodes which are neighbours of at least one node of $N(x)$ but that do not belong to $N(x)$ (Figure 11).

The standard algorithm of MPR selection is defined as follows [16]:

Algorithm 1 : standard MPR selection algorithm

1. $U \leftarrow N^2(x)$

2. $MPR(x) \leftarrow \emptyset$

3. while $\exists v : v \in U \wedge \exists! w \in N(x) : v \in N(w)$ do
   (a) $U \leftarrow U - N(w)$
   (b) $MPR(x) \leftarrow MPR(x) \cup \{w\}$

4. while $(U \neq \emptyset)$ do
   (a) choose $w \in N(x)$ such as: $CRITERIA(w) = |N(w) \cap U| = \max(|w' \cap U| : w' \in N(x))$
   (b) $U \leftarrow U - N(w)$
   (c) $MPR(x) \leftarrow MPR(x) \cup \{w\}$

5. return $MPR(x)$

B. The proposed Criteria of MPR selection

1) Link mobility estimation

Some OLSR experiments [15] show that links must be more stable and less mobile to avoid fragile connections which involve data loss and frequent route changes. OLSR protocol constantly maintains the shortest paths to reach all possible destinations in the network. So, it is judicious to estimate the quality of links before adding them in the topological information used to calculate the best routes. The quality of a link can be estimated based on the received signal's power. This information is provided by some wireless cards. If this information is not available, OLSR estimates the link quality based on the number of control messages lost. A link failure can be detected using the timer expiry or by the link layer informing upper layers of the failure with a neighbour node after reaching the maximal number of retries.

With the aim of estimating the quality of links in terms of mobility, we define the mobility of a link $L(A,B)$ between two nodes A and B as the average mobility of the involved nodes (Figure 12), as shown in the following equation:

$$M_{L(A,B)}(t) = \frac{1}{2}\left(M_A(t) + M_B(t)\right) \qquad (7)$$

Figure 12. Mobility of the link L(A,B) is (50%+40%)

Evaluating the link mobility alone is not significant because we can have a normal value of link mobility with a high mobility value of one of the involved nodes. The dependence between the mobility of nodes composing a link (in the network core) at time t can be seen as the probability of the link loss $L(A,B)$, as follows:

$$P_{L(A,B)}(t) = \frac{1}{2}\left|M_A(t) - M_B(t)\right| \qquad (8)$$

Therefore, a reliable link $L(i,j)$ can be seen as a link meeting the two following criteria:

1) link $L(i,j)$ mobility is lower than a threshold link S_L which depends on the characteristics of the wireless network (network density, network mobility, network scalability, network dimension, etc.):

$$M_{L(i,j)}(t) \prec S_L \qquad (9)$$

2) The dependence factor $P_{L(i,j)}(t)$ of nodes (i and j) mobility is near to zero:

$$P_{L(i,j)}(t) \to 0 \qquad (10)$$

The choice of a link meeting these two conditions ensures a low mobility to link, with a strong dependence between the involved nodes.

2) The proposed criteria

In this section, we propose two new criteria for the operation of MPR selection. These criteria are based on estimating the quality of links between one-hop neighbours and two-hop neighbours. The link quality is given by this link's dependence factor. A reliable link is a link with a dependence factor near 0. The new selection of MPRs nodes is a compromise between the number of links towards the nodes at two-hops and these links' reliability is presented by these links loss probability. The selection of a neighbour node as a MPR node can be viewed as a maximization of the selection criterion. The first criterion suggested, 'the simple criteria', is based on nodes' mobility (Equation (11)). The second is based on sum (Equation (12)) and the third on the product (Equation (13)). The principal advantages of these criteria are the facility of calculation and fewer requirements in terms of memory and CPU resources.

$$DIR - CRITERIA(w) = \min_{w \in N(x)} M_w(t) \qquad (11)$$

$$SUM - CRITERION(w) = 1 - \frac{\sum_{i=1}^{N} L(w,i)}{N} \qquad (12)$$

$$PRD - CRITERION(w) = 1 - \prod_{i=1}^{N} L(w,i) \qquad (13)$$

VIII. CONCLUSION AND PERSPECTIVES

In this paper, we have introduced the architecture of an intelligent beacon for ad hoc wireless sensor networks named MIS (Mobile Intelligent System). This beacon may obtain data from the environment and detect possible defaults (great variations). When an alarm is triggered, data is sent on a wireless network such as Bluetooth or Wifi.

In this work, we have also proposed and discussed the theoretical aspect of the proposed relative and lightweight network mobility measure, based on the quantification of link changes at regular time intervals. Simulation results show that the proposed mobility measure have a strong relationship with MANETs environment and MIS parameters. This mobility measure has been implemented in the OLSR protocol (work in progress) and incorporated into the FPGA platform.

In the future, we plan to further this work in two main directions: finding the relation between the network mobility measure and some parameters like link duration and path duration, through variant network scenarios; and positively integrating this new mobility measure in other MANET routing protocols and MIS.

The interest of such a work has a big impact on the applications relating to the networks of wireless mobile sensors, in particular those dedicated to the military sector. The implementation and test are currently underway.

REFERENCES

[1] A. Habbani, J. abbadi, And P. Garda, " versatile platform for AD-HOC, wireless sensor network node in FPGA". International Journal in *Physics* and *Chemistry* PCN (Physical and Chemical News). EL JADIDA, 2007.

[2] A. Habbani, J. abbadi, And Z. Guennoun, "MIS: Mobile Intelligent System" . IJCSNS -International Journal of *Computer Science and Network Security*, Korea, 2007.

[3] Habbani, A., Romain, O And Garda, P. MIS and I2C Bus, IJCSES -International Journal of *Computer Sciences and Engineering Systems*, China, Vol.1, No.2, 2007

[4] T. Camp, J. Boleng, and V. Davies. "A Survey of Mobility Models for Ad Hoc Network Research". In Wireless Communications and Mobile Computing (WCMC): *Special issue on mobile ad hoc networking: research, trends and applications*, volume 2, pages 483–502, 2002.

[5] C. Bettstetter, G. Resta, and P. Santi. "The Node Distribution of the Random Waypoint Mobility Model for Wireless AdHoc Networks". IEEE Transactions on *Mobile Computing*, 2(3):257–269, 2003. (doi:10.1109/TMC.2003.1233531)

[6] K. Oudidi, N.Enneya. and M. Elkoutbi. " Mobilité et routage OLSR", 9th African Conference on Research in Computer Science and *Applied Mathematics* -CARI'08, Rabat, Morocco, 27-30 October 2008.

[7] M. Oudidi, N. Enneya, M.Elkoutbi. "Une Mesure Intelligente de la Mobilité dans les MANETs " . SITA. mai 2008. pp172-174.

[8] F. Bai, N. Sadagopan, A.Helmy,"The important framework for analyzing the Impact of Mobility on Performance" University of Southern California, Los Angeles, CA 90089, USA pp 10-15 2003.

[9] F. Bai and A. Helmy, "A survey of mobility models in wireless ad hoc networks", (Wireless Ad Hoc and Sensor Networks, Chapter 1, Kluwer Academic Publishers, June 2004), pp. 1-29.

[10] Thomas Clausen (ed) and Philippe Jacquet (ed). "Optimized Link State Routing protocol (OLSR)". RFC 3626 Experimental, October 2003.

[11] A. Qayyum, L. Viennot and A. Laouiti. "Multipoint relaying for flooding broadcast messages in mobile wireless networks". In Proceedings of the Hawaii International Conference on *System Sciences* (HICSS'02), Big Island, Hawaii, January 2002.

[12] B. J. KWAK, N. O. SONG, and E. M. Leonard, ''A Standard Measure of Mobility for Evaluating Mobile Ad Hoc Network Performance'', IEICE TRANS. COMMUN., vol. E86-B, no.11, pp 3236-3243, November 2003.

[13] J. Boleng, W. Navidi, and T. Camp, ''Metrics to enable adaptive protocols for mobile ad hoc networks'', In Proceedings of the International Conference on *Wireless Networks* (ICWN02), (pp 293-298, 2002).

[14] T. Clausen (ed) and P. Jacquet (ed). "Optimized Link State Routing protocol (OLSR)". RFC 3626 Experimental, October 2003.

AUTHORS

K. Oudidi is with the School of Computer Science and Systems Analysis (ENSIAS), Rabat, B.P. 713 Morocco (e-mail: k_oudidi@ yahoo.fr).

A. Habbani is with the School of Computer Science and Systems Analysis (ENSIAS), Rabat, B.P. 713 Morocco (e-mail: k_oudidi@ yahoo.fr).

M. Elkoutbi is with the School of Computer Science and Systems Analysis (ENSIAS), Rabat, B.P. 713 Morocco (e-mail: elkoutbi@ ensias.ma).

15

Acceptance on Mobile Learning via SMS: A Rasch Model Analysis

Issham Ismail, Siti Sarah Mohd Johari and Rozhan Md. Idrus

Universiti Sains Malaysia, Penang, Malaysia

Abstract—**This study investigated whether mobile learning via Short Message Service (SMS-learning) is accepted by the students enrolled in the distance learning academic programme in the Universiti Sains Malaysia. This study explored the impact of perceived usefulness, perceived ease of use and usability of the system to their acceptability. The survey was constructed using a questionnaire consisting of statements regarding the participants' demographics, experiences in and perception of using mobile learning via SMS, involving 105 students from management and sciences disciplines. The Rasch Model Analysis was used for measurement correspond to a 5 point Likert. Results indicated that the usability of the system contributed to be effectiveness in assisting the students with their study. Respondents agree that SMS-learning is easy, effective and useful to help them study. However, the results found that there has been a problem in mobile learning that less interaction with lecturers. It implies that the acceptability of students to this mode on communication and interaction is highly endorsed.**

Index Terms—**Mobile, Learning, M-Learning, SMS, Rasch Analysis**

I. Introduction

When the Universiti Sains Malaysia (USM) was established in 1969, it was conferred the unique distinction of offering courses for part-time students, thus pioneering distance learning in the country in 1971. However, few Malaysians took advantage of this mode of learning [1]. In that time, the delivery mechanism in the teaching and learning process also evolved from the use of the basic self-instructional text to audio and video conferencing to the current use of the electronic portal and numerous web 2.0 tools.

One of the most significant changes in the field of education during the information age is the paradigm shift from teacher-centered to learner-centered education [2]. Education is now being transformed by the use of wireless mobile technologies into m-learning, as organization look for flexible methods, unbounded by space and time, to deliver learning materials to reach learners [3]. It sets in place the first building block for the next generation of learning, which is the movement from distance learning and web-based learning to mobile learning (M-learning) [3]. In many ways, electronic and mobile learning will move closer together as the power and sophistication of mobile learning devices increases, however in particular ubiquity and location awareness, will always be certain aspects of mobility that will make mobile learning a unique and special approach to education [4].

Therefore, a pilot test was conducted in determining the appropriate development design of learning contents, the suitable system to use and the proper hardware to employ. The study expected to develop mobile learning system framework offering merely SMS text messages into the existing learning mechanisms in USM as it may comply with the university effort in supporting the "bottom billion".

It is imperative that we ascertain the learner's acceptance toward SMS-learning in order to inculcate its use in their studies. This was corroborated by Tallent-Runnels et al [7] stated that, in order to have a good m-learning system, there should be an effort to identify accurate evaluation measures that are required to continue doing research. The learners' perceptions in SMS-learning may provide some information related to the factors in the use of SMS-learning. Hence, the primary reason of the study was to understand the acceptance of m-learning via Short Message Service (SMS-learning) and to identify the factors that can predict their intention to accept the systems.

A. Related research

A study conducted by Ring [5], highlighted the combination of the Web and WAP to deliver e-Learning in order to determine the effectiveness of a course delivered by wireless phone technology. Results indicated that 93% of students having wireless access reported that the technology made the course more convenient and they could work from anywhere. Students also reported that they were able to access courses while commuting which showed that the wireless technology afforded them freedom to access the course from anywhere and students are able to get an overall feel for the content of the course [5].

The study by Ally and Satuffer [3] attempted to determine how learners perceive the enhancements by using mobile device to distance learning materials or online learning. The results concluded that the majority of students responded that they either agreed or strongly agreed that the use of the mobile device to access the course materials was useful and provided both flexibility and convenience. However, in the study most users would mainly use a desktop for their courses though, with occasional use of the mobile device.

In the research study conducted by Lawrence et al [6], the researchers explored the opinions of the students regarding the use of mobile devices in a university learning environment. Although the respondents identified positive feelings to the use of mobile devices in their university learning, such as the limited use of SMS messages for alerts, the availability of podcasts for lectures, and using the SMS for asking questions anonymously in class, they also identified potential problems. The results showed that students were afraid that the use of mobile devices in the learning environment could weaken interpersonal communication and intrude on their privacy. Using SMS was also found to be insufficient for describing complex tasks.

Ismail and M.Idrus took a similar contribution by conducting a research as an attempt to introduce mobile learning as "Convenience Education" which involved the process of determining the design that appropriate for the system, the suitable learning content and also hardware in order to develop a framework that will contribute to the improvement of the education system in Malaysia [19]. Their pilot study was tested on the second year Physics optics course and has received overwhelming agreement and positive responses which may prove that the mobile phone could make a strong and viable contribution to the educational transaction in a Physics course in distance education.

II. THEORETICAL FOUNDATIONS

A. Mobile learning

Theoretically, it is major to understand what is mean by 'mobile learning' as a way of establishing a common understanding as well as a way of exploring the evolution and direction of mobile learning. 'Mobile learning' is definitely not simply the combination of 'mobile' and 'learning' and according to Traxler [22], it has always implicitly meant 'mobile e-learning' and its history and development have to be understood, thus many wider issues should be addressed in terms of explaining, understanding and conceptualizing it [22]. There has never been a specific definition of 'mobile learning' [22] however, there are many evolving definitions that attempt at identifying and defining mobile learning (see Table I).

In the early approaches at defining mobile learning, the focus is more on the mobility of the technology [23] [24] [27] or on the technology alone [26]. Parsons took a similar contribution, saying mobile learning describes any form of education or training that is delivered using some kind of mobile device [4]. In addition, Parsons also highlighted that in many cases these systems take advantage of location awareness and the ability of wireless devices to support communication between groups members, thus mobility enable individuals to participate in distributed simulations and role play across both space and time [4]. Another view of mobile learning says it is the facilitation of learning and the delivery of educational materials to students using mobile devices via wireless medium [6].

Learning with a mobile device will become an integral part of the general spectrum of technology-supported learning complement with its special characteristics including ubiquity, convenience, localization, and personalization, give it unique qualities that help it stands out from other forms of learning [4]. Whilst Winters [28] grants additional perspectives on what might characterize different types of mobile learning dividing it into four broad categories namely technocentric which the mobile learning is viewed as learning using a mobile device, relationship to e-learning which it is viewed as an extension of e-learning, augmenting formal education and finally learner-centred.

These are some of the points of discernment at defining and identifying mobile learning and according to Traxler [22], irrespective of the exact definition, mobile and wireless technologies, including handheld computers, personal digital assistants, camera phones, smartphones, graphing calculators, personal response systems, games consoles and personal media players, are ubiquitous in most parts of the world and have led to the development of 'mobile learning' as a distinctive but still ill-defined entity.

B. Technology Acceptance Model

Technology Acceptance Model (TAM) was used to address why users accept or reject information technology as it proposes the two main internal beliefs about usefulness and ease of use as the essential elements in determining user's intention towards adopting a new technology. Davis (1989) defined perceived usefulness as "the degree to which a person believes that using a particular system would enhance his or her job performance [8]." This follows from the definition of the word useful specifically "capable of being used advantageously." According to Davis, a system high in perceived usefulness, in turn, is one for which a user believes in the existence of a positive use-performance relationship [8].

TABLE I.
DEFINITION AND TAXONOMY OF 'MOBILE LEARNING'

Studies	Definition/Taxonomy
Quinn [23]	"E-learning through mobile computational devices: Palms, Windows CE machines, even your digital cell phone."
O'Malley et al [24]	"Any sort of learning that happens when the learner is not at a fixed, predetermined location, or learning that happens when the learner takes advantage of learning opportunities offered by mobile technologies"
Naismith et al [25]	Suggest that mobile technologies can relate to six types of learning, or 'categories of activity', namely behaviourist, constructivist, situated, collaborative, informal/lifelong, and support/coordination.
Traxler [26]	"any educational provision where the sole or dominant technologies are handheld or palmtop devices"
Keegan [27]	'The provision of education and training on PDAs/palmtops/handhelds, smartphones and mobile phones.'
Winters [28]	"Current perspectives on mobile learning generally fall into the following four broad categories: • Technocentric. This perspective dominates the literature. • Relationship to e-learning. This perspective characterizes mobile learning as an extension of e-learning. • Augmenting formal education. • Learner-centred."
Jones et al [29]	Makes a contribution based on the motivational or affective aspects of mobile learning as defining characteristics, namely: • control (over goals) • ownership • fun • communication • learning-in-context • continuity between contexts
Traxler [22]	Take learning to individuals, communities and countries that were previously too remote, socially or geographically, for other types of educational initiative.
MoLeNET [31]	"exploitation of ubiquitous handheld hardware, wireless networking and mobile telephony to enhance and extend the reach of teaching and learning"

Whilst perceived ease of use, in contrast, refers to "the degree to which a person believes that using a particular system would be free of effort [8]." This follows from the definition of ease: "freedom from difficulty or great effort." All else being equal, Davis claimed, an application perceived to be easier to use than another is more likely to be accepted by users [8].

Usability is typically described in terms of five characteristics: ease of learning, efficiency of use, memorability, error frequency and severity, and subjective satisfaction [9]. Bevan defined usability as the extent to which a product can be used by specified users to achieve specified goals with effectiveness, efficiency and satisfaction in a specified context of use [10]. Considerations of usability principles for mobile Internet applications suggest that mobile learning solutions warrant a specific approach, Uther suggested it may suited to particular aspects of e-learning courses, such as: quick reminders and alerts; daily tips; glossary information; searching for specific information within a topic and course registration [11]. Thus, this variable was believed to have a significant effect to mobile learner's acceptance.

Technology acceptance, in general, has been widely studied and several models of technology acceptance has been proposed and tested [8] [32]. According to Van Biljon and Kotzé, technology adoption models specify a pathway of technology acceptance from external variables to beliefs, intentions, adoption and actual usage [32]. They had studied mobile phone adoption from a variety of perspectives, including sociology, computer-supported cooperative work and human-computer interaction since there is lacking of model integrating all these factors influencing mobile phone adoption. They had proposed a MOPTAM model representing factors that influence mobile phone adoption and it differs from TAM in the refinement of the external variables, the inclusion of social influence and the adaptation to the mobile context, which includes the addition of facilitating conditions [32].

This study intends to identify the acceptance of extending the learning merely through SMS amongst distance learners in USM and for that reason; this study proposes perceived usefulness, perceived ease of use and usability which may contribute to the students' acceptance of the SMS-learning project.

III. METHOD

A. Participants

The respondents are distance education learners from School of Distance Education (SDE), Universiti Sains Malaysia (USM), consisting of 105 students with a gender distribution of 31 males and 74 females ranging in age from 20 to above-50. Responses included 43.8% second year students, 53.3% third year students and 2.9% are fourth year students. The participants were selected from Management programme, Sciences, Arts and Social Sciences. The ethnic make-up consisted of Malays (57.1%), Chinese (25.7%) and Indian (10.5%) and 6.7% claimed they were indigenous. All of them affirmed that they have or owned mobile phones.

B. SMS M-Learning

Previously, none of the participants has had any experience in this kind of project and they volunteered to participate in the SMS-learning project which was conducted in the second semester of the 2008/2009 academic session. The subjects that were delivered in this study are Financial Principle for second year Management and International Business for third year Management. Besides, there are subjects also delivered to Physics students namely Mechanics and Optics for second year. For Economic students, the subjects that were delivered are Money & Banking for the second year and Quantitative Economy for the third year.

The participants involved in this project had volunteered and agreed to use their own mobile phones and all the expenses of SMS communication were liable on the university research grant. In other words, students received the SMS for free.

This program was conducted for 3 months from February 2009 to April 2009, encompassing the related subject matter in the semester via text message received once a day.

C. Instruments

The survey was conducted based on quantitative research design which is constructed using a questionnaire consisting of statements regarding the participants' demographics, experiences in and perception of using mobile learning via SMS. Questionnaire items appropriate for Rasch analysis included 23 statements that pertain to technology acceptance which stress on perceived ease of use, perceive usefulness and usability. Responses to the items correspond to a 5 point Likert scale, where 1 = strongly disagree, 2 = disagree, 3 = neutral, 4 = agree and 5 = strongly agree.

The questionnaires were sent through email to 170 distance learners involved in the project. The distribution of questionnaires was performed immediately after dissemination of the learning contents through SMS had finished at the end of April 2009. The collection of questionnaires from respondents was conducted one month after the distribution. At the end of the survey, 105 questionnaires were returned.

D. Rasch Unidimensional Measurement Model (RUMM)

Data from the survey was entered into the Rasch model computer program Rasch Unidimensional Measurement Model (RUMM) [31] and this project utilized Winsteps 3.68.2 software. According to the study by Cavanagh and Romanoski which also was using Rasch analysis, RUMM calibrates the score of a respondent against the difficulty respondents demonstrated in verifying particular items by application of the Rasch rating scale model [20]. The model applies a logistic equation in which the probability of choosing a particular category in the scale is an exponential function of the difference between the repondents' ability to agree (agreeableness') and the item's difficulty in permitting agreeable responses ('disagreeableness') [20].

RUMM summary test-of-fit statistics were estimated to test the global fit of data from the 23 items to the Rasch measurement model. The psychometric properties of data from each of the 23 items were also examined by calculating individual item fit statistics. Concurrently, the capacity of the items to elicit logical and consistent responses to the five response categories was examined by calculating the thresholds between the five response categories for each item. A threshold is the minimum level of 'agreeableness' which a student must have in order to go from one Likert scale response category to the next [20]. As mentioned by Cavanagh and Romanoski, when respondents are logical in their choice of response categories, the thresholds should ideally follow in a sequence from lowest to highest; in keeping with the order of the response categories from strongly disagree to strongly agree [20].

Then, the results of the RUMM analysis were scrutinised to see if the measurement capacity of the instrument could be improved by deleting certain items. Consequently, items obtaining data with poor fit to the model were deleted from a subsequent RUMM analysis of data from a modified instrument on the assumption that this version of the instrument would be a more accurate measure.

E. Data Analysis

A stepwise refinement process was undertaken to remove items from the scale that were contributing to large errors of measurement to produce a refined scale that was a more accurate measure whereby at the end, out of 23-item data, 17 items were retained . Therefore, the RUMM verified the fit of the data to the model is acceptable.

In order to justify whether students are actually accepting this new education technology of using SMS-learning or not, the perceived ease of use, perceived usefulness and usability of the system are highlighted. Rasch is mathematically identical to the most Item Response Theory (IRT) model; however, it is a comparatively more viable proposition for practical testing since it can be applied in the context in which persons interacts with items [12]. Data were analyzed using Winsteps software version 3.68.2. Code was written to represent each respondent and Likert-type survey treatment. The Rasch model permits item difficulty for each statement presented in the survey derived by the way suitable participants actually respond to the statements [13].

Throughout the analysis, table and figure are used to describe and analyze the perceived ease of use, perceived usefulness and usability as well as its components. A statistical summary table is produced to describe the separation rate and reliability of the students and the items. Separation is the number of statistically different performance strata that the test can identify in the sample while the reliability rate indicates whether the test discriminates the sample into enough levels for the intended measure [12]. Infit and outfit statistics and variable maps provided the basis for determining how well the items measured each item. The statistics show how well the data fit the model, with fit implying a meeting of requirements or matching of intentions [12]. Variable map or also known as item/person map was constructed to illustrate the empirical hierarchy of the items on the survey, because it was connected to the participant's level of ability to endorse each item [14].The items listed at the lower end of the map represent a higher probability of being endorsed, at the same time opposed to those at the top which represent a lower probability of being endorsed. A similar pattern applied to participants. For example, for those at the lower end of the map represent less willingness to favorably endorse an item. The variable map visually reveals the hierarchy and the order of the items as well as any potential gaps in the measure [12].

IV.   Results and Discussion

Responses from distance education students across the nation were included in this analysis. In order to have an overall view of the reliability and validity of the instrument and the associated responses, the statistical summary tables of the students and the items were produced. The person reliability in this study was 0.94, with a separation of 3.95 while item reliability gained 0.71, with a separation of 1.55. Given a 0.7 threshold of acceptability, both scales for person and item are deemed reliable and usable for the purpose of this study. More so, the survey comes out with 0.97 Cronbach alphas indicated that there was consistency reliability in the instruments. Thus, the survey as a whole appears to have functional reliability.

A.   Fit statistics

Prior to interpreting item maps, fit statistics were examined to determine whether the statements/items "fit" the construct. When applying the Rasch model, data must fit the model, with the assumption of a uni-dimensional domain being measured [12]. Each statement should play a significant part in the way a construct is being investigated.

Table II portrays fit-order statistics presenting items that appear to be influenced by outside factors (outfit) and those that display "off-variable noise" (infit). Rasch measurement summary test-of-fit statistics were estimated for 23-item data. In regard to the issue of uni-dimensionality, Smith [15] suggested that items that produce standardized scores that differ by more than ±2.0 from the actual score are items that are only weakly related to the rest of the items comprising the scale. When addressing infit and outfit, a mean squared value range cutoff is determined by the size of the sample. Specifically, the items of this study was agreed to fall within the acceptable infit and outfit limits of 0.6 to 1.5 ([16] [17] [13]) which is less than 2. At the end, out of 23 items, 17 items were retained.

TABLE II.
Acceptable Fit Statistics for SMS-Learning Acceptance

Item	Statement	Infit MNSQ	Outfit MNSQ
USA17	You think SMS-learning is effective to help your study.	.97	.98
USA15	It is safe to use the system to save your learning materials.	1.27	1.15
PU12	Overall, mobile learning is useful to assist my learning.	.86	.84
PEU4	Learning to get use with mobile learning is easy for me.	.90	.84
USA16	The system is effective and efficient	1.03	.98
PU8	Using mobile learning makes it easier form to facilitate my study.	.85	.87
PU11	Using mobile learning improves my performance of undertaking my study.	.95	.92
PU9	Using mobile learning enables me to focus on my study more quickly.	.91	.95
PU7	Using mobile learning enhances my effectiveness of utilizing learning and education.	.96	.96
PEU5	It is easy for me to become skillful by using mobile devices in my study.	1.06	1.03
PEU1	I find it easy to learn what I want to learn in mobile learning.	.85	1.15

Item	Statement	Infit MNSQ	Outfit MNSQ
PEU6	I find mobile learning to be flexible.	.83	.75
PEU2	Overall, I find it is easy for me to use mobile device in my study.	.82	.77
USA13	The system is easy to use.	1.02	1.00
PU10	Using mobile learning for my study increases my productivity in my job.	.94	1.02
USA14	It is easy to learn by using the system.	.94	.91
PEU3	My interaction with lecturer via mobile learning is clear and understandable.	1.31	1.35

B. Variable map

The variable map is another visual guide to information regarding relative scales. There are gaps presenting in the map, which could call for better scale coverage. In the context of evaluation, it could also indicate that goals are being met or exceeded [12]. A good Likert-type instrument is grounded in items with a varying degree of difficulty to assess a range of attitude held by participants [18]. Moreover, there is a general spread of the items.

Figure 1 presents a map of the items, ranked by level of difficulty to endorse, from the least favorite item (hardest to endorse) to the most favorite item (easiest to endorse). Respondents ranked by their willingness to endorse the items from students who are accepted with the project to those who are least accepted with the project. Within the map, items have been labeled by a key word in the statement. Items that are located at the top of the map have been identified as those that are more difficult to endorse [18]. Furthermore, those at the bottom are easier to endorse. This concludes that, from bottom to the top of the map, items are more difficult to endorse.

Results suggest the easiest item to endorse is item USA17, *You think SMS-learning is effective to help your study*. This indicates the respondents' most preferable quality of SMS-learning is the usability of the technology whereby they claimed this project to be effective to assist their study. It illustrates that the students highly accepted the project provided it is usable to help them in their study even though it may has relative weaknesses. Besides, they feel it is safe to save the learning materials by using the system (USA15). It appears that students feel positive about each item, since the mean of students' acceptance is almost two standard deviations higher than the mean of item. It is reasonable to conclude that students value the outcome more than the process. Moreover, this result is inline and consistent with item PU12, where they claimed "*mobile learning is useful for me to assist my learning*". These outcomes actually showed that students are in need of assistance in their study, and SMS-learning is effective and useful to help them study. Moreover, the interesting part shown in the result is item PEU4 whereby the students felt easy to get use with mobile learning. The possible reason behind this result is that the learning does not require a lot of mental effort as the content was customized for the learners' needs. The results are consistent with Ring [5] which showed that students were able to get an overall feel for the content of the course.

The analysis showed item PEU3 is the most difficult items to endorse, *my interaction with lecturer via mobile learning is clear and understandable*, as well as item

USA14, *It is easy to learn by using the system*. It is possible that students negatively perceive ease of use when they begin to experience the project since it is very new and judgeless. This project of SMS-learning is still in the early stage of development, therefore the interaction between students and lectures are limited by the system. The interaction needed in the learning and teaching process is restricted by the inadequacy of the system which only allows the students to receive the messages. Besides, students might feel quite uneasy to learn by using short and briefed SMS since for years they are "customized" to learn by using thick text books and notes. These results are inline with Lawrence et al [6] since using SMS according to them is insufficient for describing complex tasks.

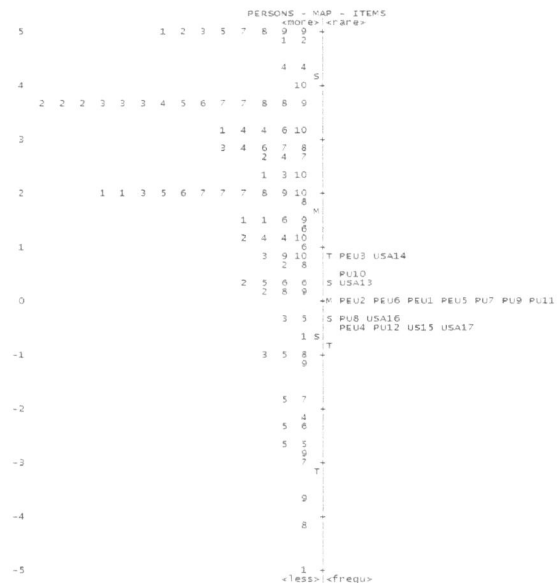

Figure 1.　Variable map of students' acceptance of SMS-learning

V.　Conclusion

This study developed and tested a mobile adoption model of learning through SMS and it was found that the majority of the students accepted the usability of this SMS-learning programme and perceived the whole project as useful. It illustrates that the students highly accepted the SMS-learning is safe, easy, effective and usable to help them in their study even though it may has relative weaknesses. Through the results of the Rasch model for measurement, the students' willingness to endorse items and the corresponding items are clearly stated and compared along one scale as they were analyzed and evaluated. The survey instrument is reliable and was able to separate both the sample of students and items.

This study provided evaluation and improvement suggestions to the use of the SMS in distance education. Unbounded by time and space, the SMS-learning has the potential to embed itself as the new medium of teaching and assisted learning. Moreover, to test the rationale and reasoning for response patterns presented, additional data should be collected utilizing students with various ability and interest. Even so, this calibration frame has allowed for a preliminary evaluation of survey instruments, providing a foundation for future applications. Thus, this study suggested there was no problem to imply SMS-learning as an extension to the existing learning mechanisms provided the system must be usable and useful in order to offer acceptance from its users.

ACKNOWLEDGMENT

The authors would like to thank Universiti Sains Malaysia for the support under RU grant and USM Fellowship scheme.

REFERENCES

[1] Raghavan,S & Kumar, P.R. (2007). The Need for Participation in Open and Distance Education: The Open University Malaysia Experience. *Turkish Online Journal of Distance Education-TOJDE*, 8(4), 102-113

[2] Lee, B-C., Yoon, J-O., Lee, I. (2009). Learners' acceptance of the e-learning in South Korea: Theories and results. *Computers & Education, In Press, Corrected Proof (*doi:10.1016/j.compedu.2009.06.014*)*

[3] Mohamed Ally & Satuffer, K. (2008). Enhancing Mobile Learning Delivery through Exploration of the Learner Experience. *Proceedings of the Fifth IEEE International Conference on Wireless, Mobile, and Ubiquitous Technology in Education, 23 March – 26 March, 2008,* Beijing, China, pp.128-132

[4] Parsons, D. (2007). Mobile Learning, in D. Taniar (Ed.) *Encyclopedia of Mobile Computing and Commerce*, IGI Global, 525-527

[5] Ring, G. (2001). Case study: Combining Web and WAP to deliver e-Learning. *Learning Circuits, ASTD Online Magazine,* http://www.learningcircuits.org/2001/jun2001/ring.html

[6] Lawrence, E., Bachfischer , A., Dyson, L.E., Litchfield , A.(2008).Mobile Learning and Student Perspectives: An mReality Check! *7th International Conference on Mobile Business,7 July – 8 July 2008,* Barcelona, pp. 287-295

[7] Tallent-Runnels, M.K., Thomas, J.A., Lan, W.Y., Cooper, S., Ahem, T.C., Shaw, S.M., & Liu,X. (2006). "Teaching Courses Online: A Review of the Career Decision-making Self-efficacy, Career Salience, Locus of Control and Vocational Indecion". *Journal of Vocational Behavior*, 37(1), 17-31

[8] David, F.D. (1989). Perceived Usefulness, perceived ease of use, and user acceptance of information technology. *MIS Quarterly*, 13(3), 319-340. (doi:10.2307/249008)

[9] Nielsen, J. (1993). *Usability Engineering,* Boston: Academic Press.

[10] Bevan, N. (1997). Quality and usability: A new framework. In: van Veenendaal, E, and McMullan, J (eds) *Achieving software product quality* (pp.1-8), Tutein Nolthenius, Netherlands.

[11] Uther, M. (2002). Mobile Internet usability: What can 'Mobile Learning' learn from the past? *IEEE International Workshop on Wireless and Mobile Technologies in Education (WMTE'02), 29August – 30 August, 2002,* Växjö, Sweden, pp.174Y. Yorozu, M. Hirano, K. Oka, and Y. Tagawa, "Electron spectroscopy studies on magneto-optical media and plastic substrate interface," *IEEE Transl. J. Magn. Japan,* vol. 2, pp. 740–741, August 1987 [Digests 9th Annual Conf. Magnetics Japan, p. 301, 1982].

[12] Ren, W., Bradley, K.D., and Lumpp J.K. (2008). Applying the Rasch Model to Evaluate an Implementation of the Kentucky Electronics Education Project. *Journal of Science Education and Technology,* 17(6), 618-625 (doi:10.1007/s10956-008-9132-4)

[13] Bond, T.G & Fox, C.M (2001). *Applying Rasch model: Fundamental measurement in the human sciences.* Mahwah, NJ: Lawrence Erlbaum Associates

[14] Bradley, K. D., Cunningham, J. D., Haines, R. T., Harris, Jr. W. E., Mueller, C. E., Royal, K. D., Sampson, S. O., Singletary, G. & Weber, J. (2006). Constructing and evaluating measures: Applications of the **Rasch** measurement model. *Symposium presented at the Mid-Western Educational Research Association Annual Meeting,* Columbus, OH, pp.1-54

[15] Smith, R.M. (1996). A comparison of methods for determining dimensionality in Rasch measurement, *Structural Equation Modeling,* 3(1), 25-40. (doi:10.1080/10705519609540027)

[16] Fox, C. (1999). An introduction to the partial credit model for developing nursing assessments. *Journal of Nursing Education,* 38(8), 340-346.

[17] Singletary, G. (2006). An exploratory examination of traditional religious principles and objectivity. Constructing and evaluating measures: Applications of the **Rasch** measurement model. *Symposium presented at the Mid-Western Educational Research Association Annual Meeting,* Columbus, OH, pp.9-13

[18] Harris, Jr. (2006). A study of Black University Students' Perceptions of Marriage. Constructing and evaluating measures: Applications of the **Rasch** measurement model. *Symposium presented at the Mid-Western Educational Research Association annual meeting,* Columbus, OH, pp.19-24

[19] Ismail, I. B., & M.Idrus, R. (2009). Development of SMS Mobile Technology for M-Learning for Distance Learners. *International Journal of Interactive Mobile Technologies (iJIM),* 3(2), 55-57. doi: 10.3991/ijim.v3i2.724.

[20] Cavanagh, R.F.& Romanoski , J.T. (2005). Parent views of involvement in their child's education: A Rasch model analysis. *Paper presented at the 2005 annual conference of the Australian Association for Research in Education, 27 November - 1 December, 2005,* University of Western Sydney, and Parramatta, pp.1-15

[21] Ju, T.L., Sriprapaipong , W., Minh , D.N. (2007). On the Success Factors of Mobile Learning. *5th International Conference on ICT & Higher Education, Knowledge Management, 21 November – 23 November, 2007,* Siam University, Bangkok, Thailand, pp.1-12

[22] Traxler, J. (2009). Learning in a Mobile Age. *International Journal of Mobile and Blended Learning,* 1(March), 1-12. Retrieved from http://wlv.academia.edu/documents/0009/9949/ijmblproof01.pdf.

[23] Quinn, C. (2000) mLearning: Mobile, Wireless, in your Pocket Learning. LineZine, Fall 2000. http://www.linezine.com/2.1/features/cqmmwiyp.htm.

[24] O'Malley, C., Vavoula, G., Glew, J., Taylor, J., Sharples, M. & Lefrere, P. (2003). Guidelines for learning/teaching/tutoring in a mobile environ- ment. Mobilearn project deliverable. Available from http://www.mobilearn.org/download/results/ guidelines.pdf

[25] Naismith, L., Lonsdale, P., Vavoula, G. & Sharples, M., (2004), Literature Review in Mobile Technologies and Learning. Bristol: NESTA FutureLab

[26] Traxler, J. (2005). Mobile learning- it's here but what is it? Interactions 9, 1. Warwick: University of Warwick

[27] Keegan, D. (2005) The Incorporation of Mobile Learning into Mainstream Education and Training. Proceedings of mLearn2005-4th World Conference on mLearning, Cape Town, South Africa, 25-28 October 2005. http://www.mlearn.org.za/CD/papers/keegan1.pdf

[28] Winters, N. (2006). What is mobile learning? .In M. Sharples (Ed), *Big Issues in Mobile Learning* (pp. 1-34). Kaleidoscope Network of Excellence, Mobile Learning Initiative. Retrieved from http://telearn.noe-kaleidoscope.org/warehouse/Sharples-2006.pdf.

[29] Jones, A., Issroff., K., Scanlon, E., Clough, G., & Mcandrew, P. (2006). Using mobile devices for learning in Informal Settings: Is it. *Motivating? Proceedings of IADIS International conference Mobile Learning. July, 14,* -16.

[30] Mobile Learning Network (MoLeNET). (2009). What is mobile learning? Retrieved December 30, 2009, from http://www.molenet.org.uk/

[31] Andrich, D., Sheridan, B., Lyne, A. & Luo, G. (2000). RUMM: A windows-based item analysis program employing Rasch unidimensional measurement models. Perth: Murdoch University.

[32] Van Biljon, J., & Kotzé, P. (2007). Modelling the factors that influence mobile phone adoption. In *Proceedings of the 2007 annual research conference of the South African institute of computer scientists and information technologists on IT research in developing countries* (p. 161). New York, New York, USA: ACM. doi: 10.1145/1292491.1292509.

AUTHORS

Issham Ismail is with the School of Distance Education, Universiti Sains Malaysia, Minden, Pulau Pinang, 11800 Malaysia (e-mail: issham@ usm.my).

Rozhan M. Idrus, is with the School of Distance Education, Universiti Sains Malaysia, Minden, Pulau Pinang, 11800 Malaysia. He is specialized in Open and Distance Learning Interactive Technologies and e-Learning (e-mail: rozhan@usm.my)

Siti Sarah Mohd Johari is a student under the School of Distance Education, Universiti Sains Malaysia, Minden, Pulau Pinang, 11800 Malaysia, and currently furthering her study in Masters of Education Technology (e-mail: sitisarahmjohari@gmail.com).

16

Contextual Mobile Learning
A Step Further to Mastering Professional Appliances

B. T. David, R. Chalon, O. Champalle, G. Masserey, C. Yin

Laboratory LIESP/Ecole Centrale de Lyon, Lyon, France

Abstract—In this paper we describe our approach whose objective is to apply MOCOCO concepts to e-learning. After a short presentation of MOCOCO (Mobility, Cooperation, Contextualization) and IMERA (Mobile Interaction in the Augmented Real Environment) principles we will discuss their use in a project called HMTD (Help Me To Do) whose aim is to use wearable computer for a framework of activities of better use, maintenance and repairing of professional appliances. We will successively describe m-learning scope, contextualization and cooperation advantages as well as learning methods. A case study of configuration of wearable computer and its peripherals, taking into account context, in-situ storage, traceability and regulation in these activities finishes this paper.

Index Terms—M-learning, contextual learning, just-in-time learning, learning by doing, wearable computer, Computer Augmented Environment, cooperative activities.

I. INTRODUCTION

As announced by Weiser in [1], the ubiquitous computing (also known as pervasive computing) seems to become concretizing with the massive propagation of mobile and connected devices (PDAs, TabletPCs, Smartphones, etc) and the use everyday broader of informatics resources as RFID tags [2]. Besides, ubiquitous computing is from 2001 integral part of the Ambient Intelligence (AmI) [3], which merges the "ubiquitous computing" and "social user interfaces" trends to adapt user interfaces to its environment and task context, so to create the proactivity. On the other hand, the Mixed Reality [4] better known as Augmented Reality (AR), for which the founding act can be situated in 1993 [5], is also in full expansion. It attempts to merge the physical and numerical worlds to facilitate the user's task with special devices and particular interaction techniques (i.e. a physical block controls a numerical block). However, the User Interfaces used on these new mobile and connected devices and their uses [6] are similar to the ones of desktop computers and are often inappropriate for mobile users that must realize several tasks simultaneously like talking with other persons, performing technical equipment maintenance, or visiting tourist spot. Besides, whereas these devices can be sensitive to the environment (GPS, RFID tags detection, etc), they rarely made the user benefit of this. Thus we must adapt their behaviour transparently to the user (in a proactive way) as in an Ambient Intelligence (AmI) Environment. AR devices and techniques can be particularly convenient in this respect.

Our objective is to use Ubiquitous Computing and Mixed Reality in learning. To do thus we studied learning situations which could benefit of this approach as well as corresponding learning methods. We found that mastering of domestic or professional appliances is a interesting domain of investigation and we set up HMTD (Help-Me-To-Do) project which objective is to study MOCOCO learning to this domain. In the following sections we will successively describe MOCOCO concepts and IMERA platform and we will discuss M-learning situations and learning methods. Then we will present HMTD foundations privileging industrial situations. We will describe wearable computer configuration process and its use in learning situations.

II. MOCOCO AND ASSOCIATED CONCEPTS

Four main concepts are the fundamentals of our approach:

MOCOCO - acronym expresses main aspects of our approach. Its objective is to indicate that we are creating for different actors an environment allowing MObility, COntextualisation and COoperation during tasks realization. A mobile actor has access to precise and contextualized data and can collaborate with several other mobile or fix actors to solve the problem.

Proactivity - characterizing information propagation to the actors enabled by an Ambient Intelligence Environment and transparent user interface adaptation.

CAE (Computer Augmented Environment) - in the meaning of Augmented Reality and Ubiquitous Computing.

MoUI (Mobile User Interfaces) - denoting the user interfaces for the wearable computers as those of PDAs, Smartphones, mobile phones and other devices appropriate for mobile users working in a collaborative way with an elaborate contextualization (access to contextual and/or personal precise data) in a CAE.

III. PLATFORM PRESENTATION

For our studies we defined IMERA platform (French acronym: Mobile Interactions for Computer Augmented Environment). This platform is composed of a main workplace and three auxiliary distant workspaces. The main working area is a CAE (Computer Augmented Environment) where different actors are moving about. For us, this CAE is a more or less large area covered by a

WiFi network, able to receive RFID tags, either freely set or integrated to real objects located in this space; RFID technology is our first support for the Ambient

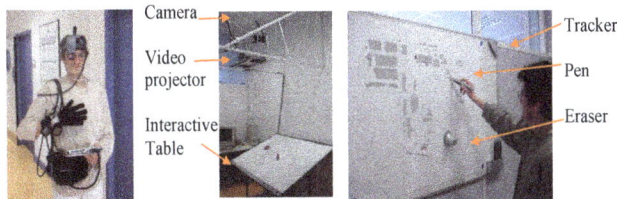

Fig.1: User with wearable computer   Fig.2: The Table Gate   Fig.3: The Tool Tribe

Intelligence Environment. Some RFID fixed readers can also be introduced on this area.

The actors (Fig.1) are moving freely in this area with their wearable computers (PDAs, TabletPCs, etc), each of them equipped with a WiFi card and an RFID reader. These wearable computers are thus connected to the network and are able to access contextual data through RFID technology and communication with mobile and fix actors. The WiFi network allows actors to be both connected between them and with centralized systems (database servers, etc) so they can communicate and access large amount of data. Independently of this working area, several separate distant management and observation workplaces complete this platform.

For our experimentations, we have at our disposal three other workplaces in our lab. A first workspace is intended to be central workplace for observation and management of the collaborative activities involving a coordination, i.e. to supervise the actions made by the actors moving on the platform main workplace. For this purpose, a TableGate equipment (fig.2) [7] is used. It's an interactive pressure sensitive flat-mounted table supporting Mixed Reality thanks to a video projector and a camera. This device is able to recognize the physical objects placed and shifted on it and can also act as a touch pad.

The second workspace, located in another room of our lab, is mainly observation oriented but can be used as second supervision place. It is based on a Tool Tribe device (www.tool-tribe.com), an interactive whiteboard (fig.3) hanged to the wall and completed by a video projector to display numerical data. For example, the video-projector can display the position of the actors on the platform in real-time, a paper map of the platform can cover the panel for that, but a numerical map is also usable. The interactions with the panel are done by physical pens that the system tracks. Some pens are physically writing, whereas others are just used as pointers, so we can select a position, an actor or others objects moreover than to physically write and erase drawings on the panel as on a whiteboard.

Main difference between the TableGate and the Tool Tribe is that on the TableGate, the user can manipulate indifferently physical and numerical objects. In this way, the TableGate allows realizing Mixed Reality tasks, and either Augmented

Reality (tasks in the Real World) or Augmented Virtuality (tasks in the Numerical World). On the contrary, the Tool Tribe doesn't allow interacting with real objects; it is used in the same way as a touch screen, to manipulate only virtual numerical objects.

The last workspace located in another lab room is devoted to the observation and the evaluation of the platform experimentations. It holds a trace server which acts as a UI message loop hook; filtering and storing all the UI generated messages sent through the different networks Ethernet, WiFi, …) either normally (collaborative applications) or dedicated to this purpose (single user applications).

IV. PLATFORM ADAPTATION PROCESS

IMERA platform is used in several collaborative situations (educational, industrial, cultural and sporting events). Its main working area takes place on the corresponding space while distant workplaces can be located anywhere as soon as a WiFi network is accessible. For each situation, it is important to identify the actors and their tasks with the data to be collected and manipulated for that. We determine in this way the technologies to exhibit on the main working area and the most appropriate equipments for each actor. Firstly scenarios are expressed and formalized in a structured way following the method proposed in [8] to describe as precisely as possible all collaborative aspects. Secondly a synthesis leading to the Collaborative Application Behavior (CAB) model is elaborated. Then, we are able to extract the roles of each actor in analyzing the model from the actors' point of view; jointly to the required environment, artefacts, etc. This process helps in the choice of wearable computers and peripherals needed to realize the tasks.

V. M-LEARNING SITUATIONS AND METHODS

Mobile learning (M-learning) is a new approach of learning using wireless device in e-learning. M-learning is the result of two faced evolution: the development of mobile technologies, including the network and wireless devices, and the evolution of learning theory. There are many definitions for Mobile learning; a significant one is the following: M-learning (Mobile learning) is any sort of learning that happens when the learner is not at a fixed, predetermined location, or learning that happens when the learner takes advantage of the learning opportunities offered by mobile technologies [9]. Without discuss deeply different taxonomies [10] of M-learning we can in this paper only separate M-learning in two categories in relation with the context. Either learning activity is totally independent of location of the actor and the context in which he is evolving taking into account only the opportunity to use mobile device(s) to learn (in public transportation, waiting the bus, …) or at the opposite, learning activity is in relation with the location (physical, geographical or logical) of the actor and the context in which he is evolving. We are naturally mainly concerned by this second category of M-learning.

In this way we can also separate learning methods which are for the first category mobility context independent i.e. related only to the learned subject and corresponding learning methods. At the opposite, naturally, learning methods used in situated mobile learning activities are in relation with the context. Global characteristics are "just-in time learning", "learning by doing" and "learning & doing" which can take various forms:

Problem-based learning: [11] defines that the PBL (Problem-based learning) as oriented to development of problem solving skills as well as the necessity of helping learners to acquire necessary knowledge and skills. Problem Based Learning assists learners to solve problems by the process of solving ill-structured problems on which confronted adults or practicing professionals are daily confronted. Generally, PBL is an example of a collaborative, case-centered and learner-oriented method of learning. Mobilearn project [12] studied deeply this approach and proposed several adaptation of PBL to different application domains [9].

Case-based Learning: [13] use concrete situations, examples, problems or scenarios as a starting point for learning by analogy and abstraction via reflection. However, a new research field relevant to Case-based learning is case-based reasoning (CBR) which aim is machine learning. CBR goal is to utilize the specific knowledge previously experienced, creating concrete problem situations (cases) [14]. A new problem is solved by finding a similar past case, and reusing it in the new problem situation. Case-based reasoning might be the new area for case-based learning of human beings, especially for mobile learning. Because it firstly make the machine learns from the human and then human learns from this structured knowledge, which should be the main approach of all cognitive sciences.

Scenario-based Learning (Situated Learning): Scenario- based learning is learning that occurs in a context, situation, or social framework. It is based on the concept of situated cognition, which is the idea that knowledge cannot be known and fully understood independent of its context. Two main principles of this kind of learning are that (1) knowledge needs to be presented in the authentic context, i.e., settings and applications that would normally involve that knowledge; (2) learning requires social interaction and collaboration [15].

VI.   HMTD-INDUSTRIAL MAINTENANCE SCENARIO

HMTD project objective is to allow to the users to master better domestic, public professional appliances. Main idea is to propose to the user in a precise situation (use, maintenance, diagnostic or repairing) to learn appropriately about the appliance to understand functioning principles and command or other actions [16].

One of industrial scenarios supported by the IMERA platform is the following (fig.4). An engineer in charge of maintenance of industrial machines is called on a factory where such machine is out of order. Once in the factory, he equips himself with See-Through goggles connected to

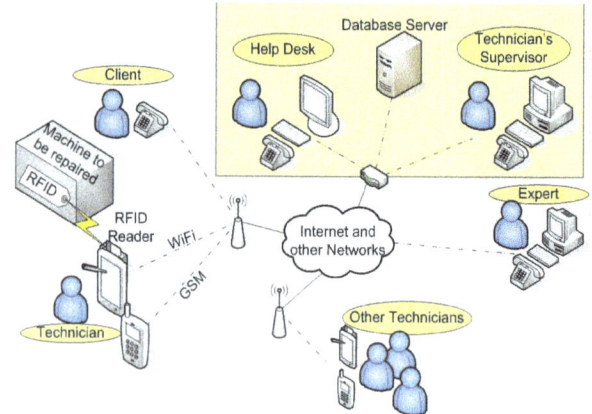

Fig.4: IMERA Platform industrial scenario

a WiFi PDA including a RFID reader. By reading machine RFID tags, he gets all its features and its reparation history stored on an internet database server, through an available WiFi access point connected to the internet. He proceeds to a first analysis and try to formulate a diagnostic. At each moment, he can stop his activity and choose to learn about it, i.e. to receive more complete and precise information either about functioning principles or about actions (commands) which he is asked (guided) to execute. If he has complementary questions or if he failed in making the actions alone, he can contact his supervisor or another expert (appliance manufacturer). He can contact him by chat or contextualized email in which machine references are automatically included to avoid typing error and to provide exhaustive information. Then they are trying to produce the diagnostic together.

Accurate product plans and guides are at the disposal of the engineer through the internet connection to help him on the recognition of the different pieces. He can visualize them on his see-through goggles whereas he is looking at the machine. Simple vocal commands are enabling him to browse the guides. These commands are captured by his PDA microphone and are processed either on a server, being transferred through WiFi and internet or directly on the PDA, depending on the complexity of the command, and the capabilities of the PDA. If diagnostic is still not successful, he can contact a machine manufacturer expert to help him realize the diagnostic.

As soon as the diagnostic is established, and the malfunctioning pieces determined, he highlights them via his wearable computer on a plan of the machine displayed on his augmented goggles. Afterwards, the availability and delay for future reparation is computed. Later, when the parts are delivered, the reparation process is described on his wearable computer with eventually the visualization on his see-through goggles of an assembly plan or others relevant data. As soon as the machine is repaired, he updates the machine reparation history and replaced parts, on the server.

A.   Choice of wearable computer and its peripherals

For different actors of a particular scenario as for different scenarios, it's important to find the most appropriate wearable computers and peripherals (fig.5). Various solutions are possible (light and small hand free equipment, heavier but with better visualization capacities or better interaction performance …). These choices are

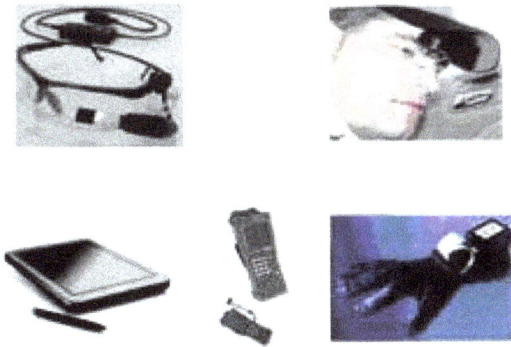

Fig.5: Goggles with integrated screen, See-Through goggles, TabletPC, RFID reader and Data glove

established after a study of all actors' tasks, matching requirements concerning graphics information complexity (textual, graphic schemas or precise blueprints …), interaction complexity (writing, observation, manipulation) and working conditions (seating, standing, hands availability …).

A precise selection process based on a selection space allows comparing different interaction ways and system implementations, with their typical supporting devices organized onto axes and classified for each axis by one of their most relevant characteristics [17]. This process results in different configurations proposals and helps to determine the most convenient ones. The criteria are those of the designer, e.g. the devices number minimization, the interaction continuity maximization (in and between the tasks) and the adequacy with working conditions.

Main possible choices are done through the following axes: gesture interaction of the hand, arm and/or head; vocal interaction with or without feedback; eye interaction, also called lazy interaction; writing and input capabilities via a physical or virtual keyboard or a touch screen; display capabilities as screen integrated in glasses or see-through screen in goggles or the screen of a mobile device (PDA, TabletPC, etc); data contextualization; localization of users and objects; communication support as WiFi, and Bluetooth. Contextualization is done by RFID tags reading. The readers are mobile or grounded and the users wear tags to be identified and/or read tags to force their mobile user interfaces become "contextualized" (updated) with the context described in the tag content. This axis doesn't explicit the data storage, that is often a database server. Localization is geographical (using GPS), logical (using RFID), or combination of both techniques for a better accuracy. Others devices and peripherals aren't dismissed, these axes are the basis for the definition of our configurations, but they aren't exhaustive and a configuration defined through them can be modified with other relevant peripherals. Besides, it isn't mandatory to use each axis, since they are not useful for each task.

B. Examples of meaningful configurations

We are describing here three configurations with their purposes.

– **Hand free highly mobile actor**: Purpose: Eyes continuity and at least one hand free. Equipment: Goggles with integrated screen, Control through a data glove, Voice command with vocal feedback, Backpack computer.

– **Hand free Mixed Reality mobile actor**: Purpose: Integration of numerical data in the real world for tasks generally in the real world. Equipment: See-through goggles, Control through a data glove, Voice command with vocal feedback, Backpack computer

– **Head free mobile actor:** Purpose: Sizeable data support and handheld device with interactions by pointing and writing. Equipment: TabletPC (WiFi) with RFID reader.

VII. ARCHITECTURAL CONSIDERATIONS

Without repeating here all aspects of architectural and process considerations, which can be found in [8], we limit our explanation to the management of learning information. In relation with the nature of task on charge by the user, the learning units are either functional understanding of the appliance oriented or oriented to command, maintenance, and diagnostic or repairing. Connection with the real appliance is either one way (from appliance to HMTD system) to collect concrete information (at least identification, or more complete information about main variables and parameters) or two ways allowing to HMTD system to send commands to the appliance to coach its behavior.

By creation of Mix Reality environment it is possible to communicate deeply between appliance and HMTD system. These learning units are of different nature, textual, graphic, simulation, historical collection of data in relation of the kind of operation which is learned (from functioning to repairing). All these units are expressed in XML to allow adaptation. They contain also metadata description respecting LOM. We are also studying SCORM use to be able as easily as possible to adapt these learning units to different platforms which selection process was described previously.

VIII. TESTS AND EVALUATIONS

The aim of the evaluation and test of different configurations of the mobile devices (in the AR and ubiquitous environment) are their utility, utilisability and acceptability. For that, we gather several kinds of traces. Among them, the messages that generate the UI are sent through the network and stored on a trace server; these are either user oriented for single user applications or user and group oriented (messages exchanged inside a group) for collaborative applications.

The tests themselves take place in the following manner. Firstly, the subjects' profiles are determined by asking them to fill a pre-test questionnaire. During the experimentations, the subjects are filmed to supplement the UI logs; they are asked to verbalize their actions and difficulties while two observers follow them and take notes of these problems and attitudes in an observation grid. As soon as the test is finished, each subject fills up a post-test form with a set of multiple choice questions and some open questions. Finally, crossed analyses of these different data allow extracting results of these evaluations.

IX. CONCLUSIONS

We have presented an approach for contextual learning which is for us not only contextual, but also and mainly mobile and collaborative. In this way the choice of appropriate wearable computer characteristics and interaction devices is fundamental in relation with the nature of activity and the context (physical, geographical or logical location, nature of tasks to provide and user preferences). Choice of learning units, their learning methods and associated learning materials is also very important.

An experimentation platform for these studies of new interaction techniques and devices and several configurations deployed for collaborative work with mobile actors in a Computer Augmented Environment is now available at Ecole Centrale. This platform is also an Ambient Intelligence Environment by the integration of new communicating objects grounded or mobile, active or passive, and most recent sensors and effectors are considered, including position and orientation sensors or more original captors as presence detectors. The platform supports the appraisal of concrete scenarios issued from industrial maintenance situations (machines on-site repairs, etc), for the discovery and validation of new interaction ways or devices uses. We are open to other applications to validate our approach and other scenarios are currently studied, mainly in the industrial field, especially with our partner Assetium, and some ECL students for their end-of-year works.

REFERENCES

[1] Weiser M., *The Computer for the Twenty-First Century, Scientific American*, 1991, pp. 94-10.

[2] Srivasta L., *Ubiquitous Network Societies: The Case of Radio Frequency Identification, Background paper of ITU Workshop on Ubiquitous Network Societies*, Geneva, Web: http://www.itu.int/ubiquitous/, 2005.

[3] The Ambience Project, *ITEA project on Ambient Intelligence.* Website: http://www.hitech-projects.com/euprojects/ambience/, 2004.

[4] Renevier P., Nigay L., Salembier P., Pasqualetti L. *Systèmes mixtes mobiles et collaboratifs, Colloque sur la mobilité*, LORIA, Nancy, France, 2002.

[5] Wellner P., Mackay W. and Gold R., *Computer Augmented Environments: Back to the Real World. Special Issue of Communications of the ACM*, vol. 36, 1993.

[6] Plouznikoff N., Robert J.-M., *Caractéristiques, enjeux et défis de l'informatique portée, In Proceedings of IHM'04*, 2004, pp. 125-132.

[7] Chalon R., *Réalité Mixte et Travail Collaboratif : IRVO, un modèle de l'Interaction Homme-Machine*, Thèse de doctorat, Ecole Centrale de Lyonm, 2004.

[8] Delotte O., David B., *From Scenarios to Tasks Model for Capillary Systems. Proceedings of HCI International 2005*, Las Vegas, USA, July 25-27 2005.

[9] O'Maley C. et al. *Guidelines for Learning/Teaching/Tutoring in a Mobile Environment*, Mobilearn project (www.mobilearn.org), on-line report, 2005.

[10] Meyer C., Chalon R., David B., *Caractérisation de situations de M-Learning, Proceedings of TICE 2006*, France, 2006.

[11] Stepien W.J., *Problem-based Learning: As Authentic as it Gets*, Stepien, W.; & Gallagher, *Educational Leadership*, 1993, v50 n7 p25-28.

[12] Mobilearn, http://www.mobilearn.org

[13] Kolodner J.L., Owensby J.N., and Guzdial M., *Case-Based Learning Aids.* In Jonassen, D.H. (Ed.) *Handbook of Research for Educational Communications and Technology*, 2nd Ed. Mahwah, NJ: Lawrence Erlbaum Associates, 2004, pp. 829 – 861.

[14] Aamodt A., *Case-based reasoning: Foundational issues, methodological variations, and system approaches.* Agnar Aamodt, Enric Plaza, AICom, *Artificial Intelligence Communications*, IOS Press, Vol. 7: 1, 1994, pp. 39-59

[15] Lave J., Cognition in Practice: Mind, mathematics, and culture in everyday life. Cambridge University Press, 1988.

[16] David B., Masserey G., Champalle O., Chalon R., Olivier Delotte O., *A wearable computer based maintenance, diagnosis and repairing activities in Computer Augmented Environment, Proceedings of EAM06* : European Annual Conference on Human Decision-Making and Manual Control, Valenciennes, September 27-29, 2006.

[17] Masserey G., Champalle O., David B., Chalon R., *Démarche d'aide au choix de dispositifs pour l'ordinateur porté, Proceedings of ERGO'IA 2006*, Biarritz, France, 2006.

AUTHORS

B. T. David, R. Chalon, O. Champalle, G. Masserey and **C. Yin** are with the laboratory LIESP, Ecole Centrale de Lyon, 36 Avenue Guy de Collongue, 69134 Ecully Cedex, France. (e-mail: Bertrand.David@ec-lyon.fr, Rene.Chalon@ec-lyon.fr, Olivier.Champalle@ec-lyon.fr, Guillaume.Masserey@ec-lyon.fr, Chuantao.Yin@ec-lyon.fr).

17

Transformable Menu Component for Mobile Device Applications: Working with both Adaptive and Adaptable User Interfaces

V. Glavinic[1], S. Ljubic[2] and M. Kukec[3]
[1] University of Zagreb, Zagreb, Croatia
[2] University of Rijeka, Rijeka, Croatia
[3] College of Applied Sciences, Varazdin, Croatia

Abstract—Using a learning system in a mobile environment is not effective if barriers are not overcome in the interaction with targeted users. For that purpose all mobile services, including m-learning ones, demand special attention being paid to interaction with the user. While mobile device applications are becoming more powerful, their development process must utilize the concepts of universal access and universal usability. This paper describes the model of both adaptable and adaptive mobile user interface, through the introduction of a transformable menu component capable to be personalized to each individual user with respect to her/his preferences and interaction style. We discuss the use of customization and adaptation techniques, with the aim to both enhance mobile HCI and to increase user satisfaction, particularly when working with graphically rich m-learning applications.

Index Terms—Adaptive user interfaces, adaptable user interfaces, personalization, mobile human-computer interaction (mobile HCI), mobile learning, universal usability, universal access

I. Introduction

Contemporary software systems are recommended to focus on three universal usability challenges: technology variety, user diversity, and gaps in user knowledge. While technology variety requires supporting a broad range of hardware, software, and network access methods, user diversity involves accommodating users with different skills, knowledge, age, gender, disabilities, and so forth [1]. Regarding mobile services, usability must look at the mobile user and surmise what interfaces are appreciated and anticipated by the user [2].

Fast developing mobile computing devices (mobile phones, pocket PCs, Tablet PCs, and personal digital assistants - PDAs) and mobile technology altogether provide end users with new powerful mobile device applications (MDAs). Moreover, improved capabilities of mobile devices allow these applications to run graphical user interfaces with increased complexity. Popup menus, toolbars, icons and dialog boxes are becoming common graphical user interface (GUI) elements within MDAs. The addition of such new features can be advantageous to the mobile device user (regarding enhancements in interaction speed and efficiency), but on the other hand, sophisticated interfaces can foster user diversity. In fact, most users only use a small fraction of the available

functions while wading through many unused ones, and different users tend to use different functions even when they are performing similar tasks [3]. Just like in desktop application environment, the MDA development process should emphasize both the user interface and the interaction aspects, the overall goal being to personalize the MDA interface to each individual user and provide easy access to the functions that she/he would use, thus addressing the concepts of universal access [4].

User interface personalization can be implemented through adaptable and adaptive mechanisms. Generally, the difference between adaptable and adaptive systems is presented in [5]: an *adaptable* system is one in which the system performance automation resides in the hands of the user; while conversely in an *adaptive* one the flexibility in automation behavior is controlled by the system. According to the aforementioned, adaptable user interfaces have user-driven customization ability, while adaptive user interfaces have a built-in self-adaptation one. Authors in [6] denote customization and self-adaptation as the only scalable approaches to personalization.

Although adaptation paradigms are device-independent, little work has been done in the area of MDA user interface personalization, as opposed to desktop PC applications. In this respect we discuss in this paper the possibility of implementing a both adaptive and adaptable GUI in MDAs, where personalization is primarily presented and visualized through the model of transformable and movable menu component. We believe that using both interaction techniques could provide optimal time division between working on related tasks and organizing the application interface. Since communication between the user and the system interface will have a direct influence on the user experience of the service provided [7], by so doing we expect bringing user satisfaction at a higher level.

The paper is structured as follows: Section II shortly explains the motivation to develop adaptable and adaptive menu component, Section III describes the personalization capabilities regarding to customization and self-adaptation of the proposed menu model, Section IV shows some layout snapshots of initial menu implementations within different development environments and respective mobile device emulators, while Section V presents a brief conclusion, together with the outline of our future work plan.

II.   MOTIVATION

The idea to build a "smart" menu component for an MDA comes as the continuation of our previous research on virtual mobile laboratories [8]. In this work, we have already introduced the model of a system that would enable distance-based laboratory training using wireless interconnected mobile devices (m-Lab). Such a system, which offers much more interactive learning possibilities than commonly used ones (which are for content retrieval only), could completely utilize all the improved capabilities of new mobile devices (e.g. display screens, processor power and memory capacity), by introducing applications with rich GUI elements (Fig. 1). Within the complex process of mobile virtual laboratory implementation, we are now primarily focused on the development of appropriate MDAs, where we expect to fully show the advantages of mobile HCI. In that respect, our motivation objective is to create a user interface model with the ability to (i) implement and run within different graphical MDAs, (ii) provide easy access to numerous application options, and (iii) personalize with respect to each individual user, according to her/his preferences and interaction style, as well as her/his knowledge level.

In present MDAs, finding and selecting desired application option could often be a difficult and time consuming problem. The main reason for this lies in the fact that the user is forced to continuously navigate through built-in multilevel menus, while at the same time she/he is not provided with the information on the menu level position. In such configurations, separate menu levels are usually displayed on the full screen, while the main application content is temporarily hidden. Additionally, this kind of "abrupt interaction" can often result with distraction in user's concentration within the given task.

To improve mobile user interaction with menu navigation, a new model of menu component for graphical MDAs is required. The proposed transformable menu component would have the same form as PC desktop applications menus, as well as suitable size and shape with respect to limitations of mobile device displays. The above mentioned motivation goals (i.e. HCI issues) would be encompassed by using both adaptable and adaptive procedures, thus making the menu component both personalized and easy to use.

III.   THE CONCEPT

Even though mobile devices are nowadays experiencing rapid improvement, their displays are still not at a level where MDAs could compete with some standard desktop applications. Relatively small screens, being the main restrictive factor, implicate application design with optimal usage of viewing area. In such an environment it must not be the goal to present all available information, but rather some smaller set of substantial data.

A.   Adaptable Menu Component: Customization

When working with graphical applications (such as those supported in a mobile virtual laboratory), every part of the display area could be significant and/or contain some useful information. Since menu components also occupy a portion of visible space, the user must have customization control over the menu visibility property. In this case, when strong attention must be directed to the application's graphical context (e.g. learning schematics or analyzing graphs), the user can hide the temporarily not required navigation system without trouble (Fig. 2).

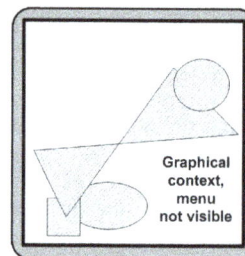

Figure 2.   Mobile device screen layout. Running state of MDA is assumed, showing some graphical context, while menu component is hidden

Additionally, menu docking position can also be a considerable issue when interacting with an MDA. Usually a menu component is located at the top of the application user interface. If the predominant part of the essential content is also located at the top of the mobile device display, menu navigation can be less efficient, due to focus elimination and possible repetitive menu actions. This can be avoided by offering the user more customization control, providing her/him the ability to easily change the menu docking position and menu orientation among four supported locations/orientations: horizontal upper, horizontal bottom, vertical left, and vertical right (Fig. 3).

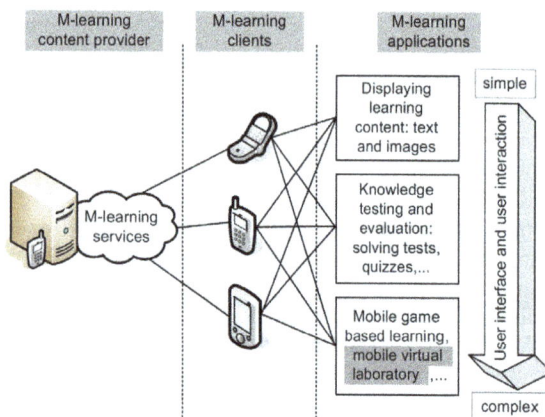

Figure 1.   Improved capabilities of new mobile devices combined with new m-learning services introduce MDAs with complex GUI

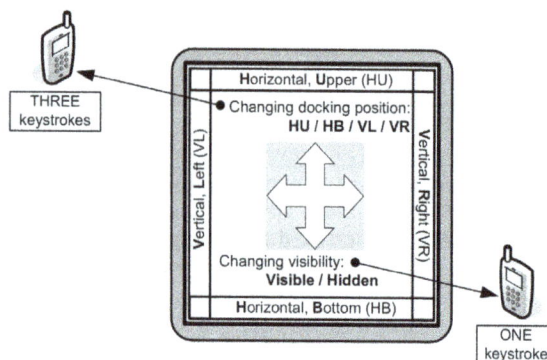

Figure 3.   Changing menu visibility and its docking position/orientation must be accessible with just few keystrokes

From the user standpoint adaptable user interface procedures must be designed as simple as possible. The time spent for UI customization must represent the cost of a marginal set of actions within the overall working time in an MDA. If this is not fulfilled and the customization process appears to be time consuming, then either its purpose is rather questionable or the respective adaptable technique is poorly designed. Taking this into consideration, the control of the visibility property in the proposed menu component is enabled through exactly one keystroke. The respective mobile device key is reserved for changing the menu state from visible to hidden and vice-versa. Additionally, changing the docking position is enabled by a three keystrokes sequence: (i) pressing the reserved button to enter the moveable state, (ii) pressing the navigation key (up/down/left/right) to dock the menu to the desired location, and (iii) pressing the same reserved button, this time to leave the movable state.

Examples of different menu docking positions in the active menu state are presented by two sketches in Fig. 4. Numbers from 1 to 9 represent menu headers, which can be text-based or icon-based at runtime. Clearly, icons or abbreviations of header full names represent more suitable options for vertical menu orientation. Therefore, the best approach is to always define shorter header names. Upon selecting the menu header, a set of popup items becomes visible on the screen. As our intention is to prevent hiding all the underlying graphical content, the upper limit of visible popup items must be defined. This has to be a device-dependent factor, depending on the mobile device screen size, which is computed at runtime (a small integer is assumed, being 5 in the figures below). The remaining popup items become hidden but easily accessible via simple navigation. Since the menu header can contain either representative icons or text abbreviations, it is useful to provide full-text information about selected option, thus enabling the user to quickly learn icon or abbreviation meaning.

It is highly probable that different users will use different menu interaction patterns, even when working on very similar tasks. In that respect, a set of frequently used menu items will vary from one person to another. User interaction diversity combined with the fact that navigation to hidden items can be very unattractive (especially in cyclic actions) is the reason for menu

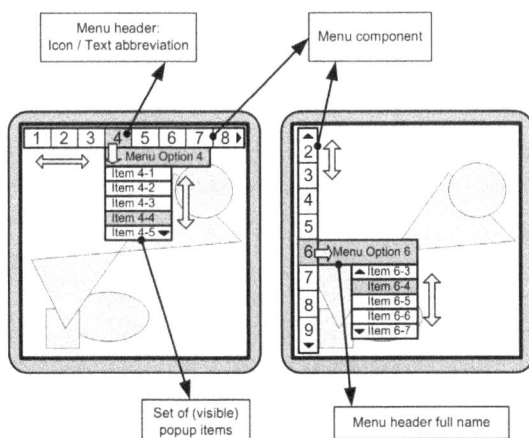

Figure 5.   Customization process sequence: (1) Using navigation keys to find required menu item, (2) Using built-in menu option for selecting required item, (3) Using navigation keys to move selected menu item to desired position, (4) Using built-in menu option for accepting changes

component personalization. Rearranging menu items within a popup set could be a very useful customization potential, especially for experienced users working with a familiar set of tasks. It is highly probable that the most frequently used items will be manually replaced to the beginning of the popup set, while non-used items will be moved to the hidden subset. Such a menu component personalization procedure as described above must be presented to the user as a trouble-free process, designed for simple usage. The best way to accomplish these requirements is to take advantage of the existing navigation model for items positioning. A detailed description of the customization process sequence is given in Fig. 5. The user can change item positions using familiar navigation patterns, while the customization mark

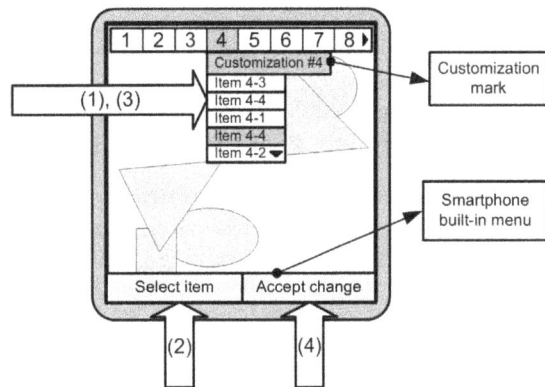

Figure 4.   Different menu docking positions and appropriate navigation routes in the active state

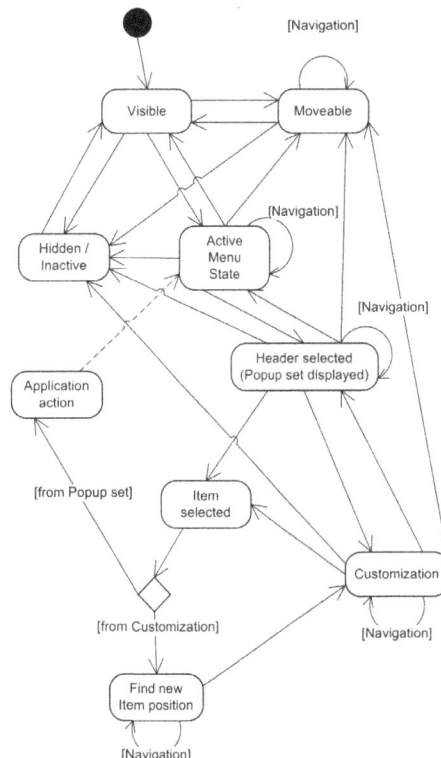

Figure 6.   UML state diagram of proposed menu component. All label-free transition arrows represent exactly one keystroke action

(at the place where the header full name usually stands) is used to notify her/him that the customization mode is temporarily active. Combining the mobile device built-in menu, which is usually presented with only two options enabled, with our menu component, could be a supplementary advantage in the interaction process, as shown in Fig. 5.

Following the concepts of universal access and universal usability, personalized menu components must provide easy access to all of the existing menu functions. Within MDA, this is primarily done by simplifying menu state transitions. In our case, all state transitions are available through exactly one keystroke on the mobile device input (see Fig. 6), thus providing a platform for optimal interaction efficiency.

B. Adaptive Menu Component: System-driven Personalization

When speaking of MDA user interface automatic adaptation, we must distinguish (i) adaptation to the used terminal characteristics and (ii) adaptation to the user preferences, interaction style and knowledge level. The former feature employs available technologies to enable the development of applications with a unified user interface for different types of handheld devices, most often by using XML-based languages for UI description and specification ([9], [10]). We are however focusing to the latter feature by analyzing the possibilities for implementing automated personalization of the menu with respect to user interaction patterns.

While the user interacts with the MDA, her/his menu actions (navigation and selections) must be actively monitored. This involves installing proper adaptation algorithms which should operate "in the background" and, based on data thus collected, change the momentary menu configuration (Fig. 7). The most suitable and well-known adaptation algorithms are based on the user's most frequently and recently used items ([3], [11]). While recency-based algorithms are used to dynamically determine and promote the most recently used items to preferable menu position, frequency-based ones are used

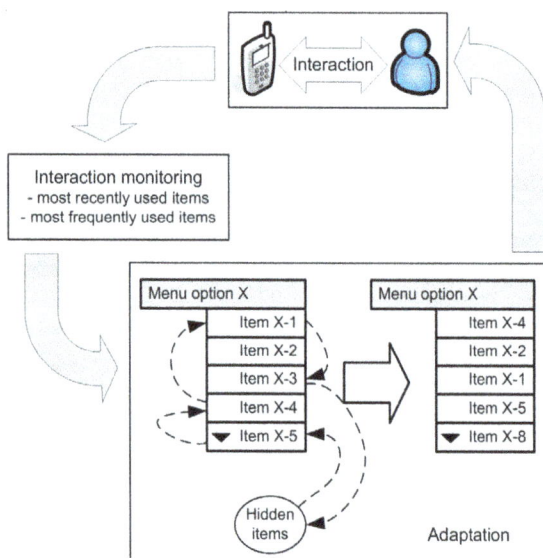

Figure 7.   Basic principle for automated menu personalization, using recency-based and/or frequency-based adaptation algorithms

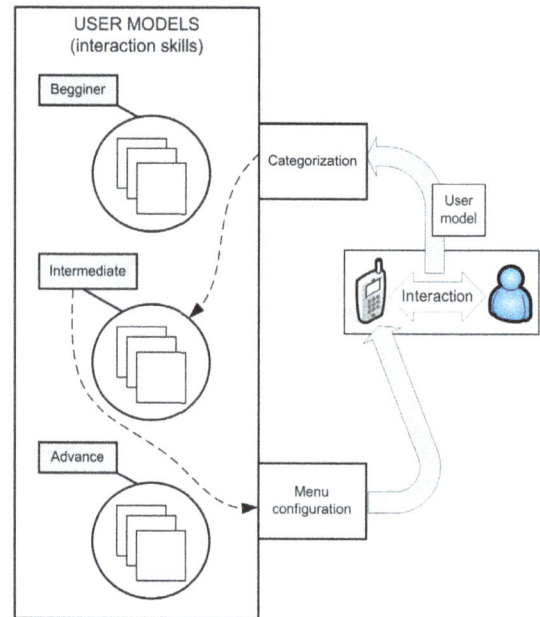

Figure 8.   User models in automated menu personalization

to promote the most frequently used ones. Because of the user diversity, adaptation algorithms can generate completely different menu configurations for different people.

The use of the aforesaid background algorithms allows modeling a user behavior profile, which describes the preferred way of user interaction. This profile can also be used in anticipating future user objectives in specific tasks of her/his work process. Such objective anticipation can extend menu usefulness by introducing completely new menu items for multiple action shortcuts. Along with behavior profile modeling, it is also possible to construct user models containing (and updating) information which would help the system to determine the interaction skill category (e.g. beginner, intermediate, advanced), see Fig. 8. Based on this information it would be possible for the system to automatically and quickly adjust the menu configuration. In such a model, the *sparse data* problem (which is related to the *cold start* problem of personalization) can emerge, in situations where there is insufficient information to build user model and categorize users [12]. Using adaptable procedures, i.e. customization, can be used to avoid that kind of problem, since constructing the user model is then explicitly done by performing customization actions. This is yet another benefit of using both adaptable and adaptive techniques (customization and adaptation altogether). User models, initially created by users themselves, can then be altered again, through profile learning algorithms [13].

IV.   INITIAL IMPLEMETATIONS

In this section we briefly describe some initial implementations of the proposed menu component within different developing environments and mobile device emulators.

Fig. 9 shows the layout of an MDA with a desktop-like menu component being implemented. It can be seen that the menu contains text-based headers (which are for the moment being represented by characters and not icons).

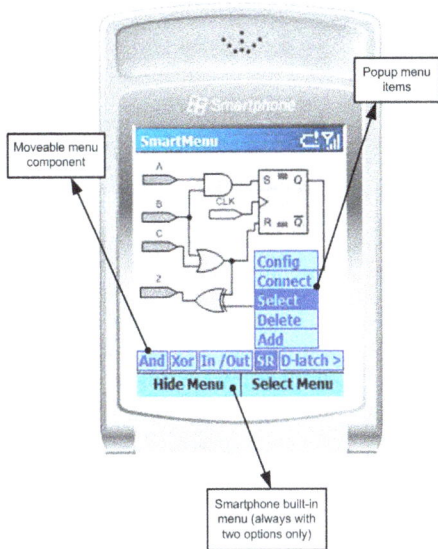

Figure 9. MDA layout with active text-based moveable menu component. Smartphone emulator display size is 176 pixels in

The application is running on the *Smartphone 2003 SE Emulator* (*Microsoft Device Emulator*, a part of *Visual Studio 2005* packet). The smartphone emulator built-in menu options are used to switch the menu visibility state, and to trigger menu navigation (active menu state). The menu component is applicable even on such limited displays (176 pixels in width, and 180 pixels in height). In comparison, many of present smartphones have screen resolutions of 240x320 pixels, while PDA devices have displays with sizes up to 480 pixels in width and 640 pixels in height.

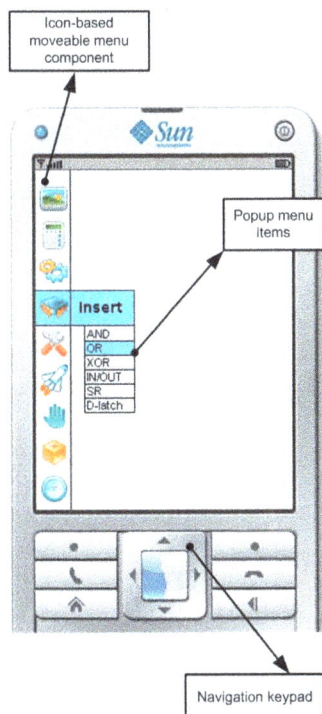

Figure 10. MDA layout with active icon-based moveable menu component. Smartphone emulator display size is 240 pixels in width, and 320 pixels in height

Conversely, Fig. 10 shows the menu component within an MDA which is running on the *Sun Java Wireless Toolkit 2.5* emulator. This example shows the advantage of icon usage for menu headers when running MDAs with vertical menu orientation.

V. CONCLUSION AND FUTURE WORK

M-learning systems will certainly become an additional advantage in the comprehensive process of lifetime learning. Contemporary development of mobile technologies and the associated information-communication infrastructure do make possible the introduction of new and high quality m-learning services and interactive learning contents. At the same time, there already exists a powerful platform for bringing the respective mobile device applications at a higher level. Although present m-learning applications don't take full advantage of this platform (mobile games, satellite navigation, and media streaming are still the only representative applications which completely utilize mobile technology upgrowth), we can suppose that they will soon become very powerful and graphically rich.

When developing these applications special emphasis must be given on the user interface design and the quality of user interaction. The concepts of universal usability, universal access and user experience represent state-of-the-art design guidelines for present day mobile device applications. As opposed to desktop applications, little work is already done for bringing personalized user interfaces to mobile device displays.

This paper proposes a model of such an MDA user interface, where personalization is presented and visualized through a transformable and movable menu component with adaptable and adaptive features. We claim that the usage of these techniques altogether can be a representative model for efficiency enhancement in mobile HCI, and additionally that MDAs integrating the proposed menu component can definitely be considered a significant benefit in the field of m-learning. Developing a proposed full scale menu component (by using the advantages of the object-oriented paradigm) will result with both API extension and component reuse possibility in all likewise applications.

We are continuing our work on the implementation of the presented model by primarily focusing on the development of menu adaptive algorithms. The menu based MDA prototype will be tested in various working environments among the student population. We intend to carry out both usability testing and efficiency measurement (with respect to interaction speed) in controlled experiments, initially run with mobile device emulators. We believe that the obtained results will confirm our expectations, in particular about increasing user satisfaction.

REFERENCES

[1] B. Shneiderman, "Universal Usability: Pushing Human-Computer Interaction Research to Empower Every Citizen," *ACM Communications*, vol. 43, pp. 85-91, May 2000.

[2] S. Greene and J. Finnegan, "Usability of Mobile Devices and Intelligently Adapting to a User's Needs," in Proceedings of the 1st International Symposium on Information and Communication Technologies (ISICT'03), *ACM International Conference Proceeding Series*, vol.49, pp.175-180, 2003.

[3]   L. Findlater L and J. McGrenere, "A Comparison of Static, Adaptive, and Adaptable Menus," in Proceedings of the ACM SIGCHI Conference on Human Factors in Computing Systems (CHI'04), 2004, pp. 89-96.

[4]   C. Stephanidis, "Editorial," *Int'l J. Universal Access in the Information Society*, vol.1, pp.1-3, June 2001.

[5]   R. Oppermann, Ed., *Adaptive User Support: Ergonomic Design of Manually and Automatically Adaptable Software*. Mahwah, NJ, USA: Lawrence Erlbaum Associates, Inc., 1994.

[6]   D. S. Weld, C. Anderson, P. Domingos, et al., "Automatically Personalizing User Interfaces," in Proceedings of the 18th International Joint Conference on Artificial Intelligence (IJCAI'03), Morgan Kaufmann, 2003, pp. 1613-1619.

[7]   B. von Niman, A. Rodriguez-Ascaso, S. Brown, and T. Sund, "User Experience Design Guidelines for Telecare (E-Health) Services," *Interactions*, vol.14, pp.36-40, October 2007.

[8]   V. Glavinic, M. Kukec, and S. Ljubic, "Mobile Virtual Laboratory: Learning Digital Design," in Proceedings of the 29th International Conference on Information Technology Interfaces (ITI'07), 2007, pp.325-332.

[9]   W. Mueller; R. Schaefer, and S. Bleul, "Interactive Multimodal User Interfaces for Mobile Devices," in Proceedings of the 29th Annual Hawaii International Conference on System Sciences (HICSS'04), 2004, pp.10.

[10] M. Bisignano, G. Di Modica, and O.Tomarchio, "Dynamic User Interface Adaptation for Mobile Computing Devices," in Proceedings of the 2005 Symposium on Applications and the Internet Workshops (SAINT-W'05), 2005, pp.158-161.

[11] K. Z. Gajos, M. Czerwinski, D. S. Tan, and D. S. Weld, "Exploring the Design Space for Adaptive Graphical User Interfaces," in Proceedings of the 8th International Working Conference on Advanced Visual Interfaces (AVI'06), 2006, pp. 201-208.

[12] D. Albrecht and I. Zukerman, "Introduction to the special issue on statistical and probabilistic methods for user modeling," *User Modeling and User-Adapted Interaction*, vol. 17, pp. 1-4, March 2007.

[13] B. Mrohs and S. Steglich, "Architectures of Future Services and Applications for the Mobile User," in Proceedings of the 2005 Symposium on Applications and the Internet Workshops (SAINT-W'05), 2005, pp. 128-131.

AUTHORS

V. Glavinic is with the Department of Electronics, Microelectronics, Computer and Intelligent Systems (ZEMRIS) at the Faculty of Electrical Engineering and Computing (FER), University of Zagreb, Unska 3, HR-10000 Zagreb, Croatia (e-mail: vlado.glavinic@fer.hr).

S. Ljubic is with the Department of Automation, Electronics and Computing (ZAER) at the Faculty of Engineering (RITEH), University of Rijeka, Vukovarska 58, HR-51000 Rijeka, Croatia (e-mail: sandi.ljubic@riteh.hr).

M. Kukec is with the College of Applied Sciences, Hallerova aleja 5, HR-42000 Varazdin, Croatia (e-mail: mihael.kukec@velv.hr).

18

Managing Social Activity and Participation in Large Classes with Mobile Phone Technology

A. Thatcher and G. Mooney

University of the Witwatersrand, Johannesburg, South Africa

Abstract—Within the context of a developing country, such as South Africa, access to technology is severely limited. However, most South Africans have relatively good access to mobile phone technology in relation to other portable and mobile technology. In this initiative, students were encouraged to use mobile phone text messaging to send questions to the lecturer during classes or between classes. A total of 86 text messages were sent to the lecturer during a 7-week, second year psychology course. At the end of the course 136 responses to questionnaire distributed in class was obtained. This data was analysed using activity theory as a framework for the discussion. The results indicated that students had strongly favorable perceptions of this initiative and respondents had spontaneously suggested other uses of mobile phone technology to enhance the learning experience. Activity theory provided a useful framework for evaluating the use of mobile phone text messages to enhance student participation and learning.

Index Terms—large class teaching, mobile phone messaging, activity theory; m-learning.

I. INTRODUCTION

It has been argued [1] that mobile and portable technologies will be the most influential technologies for teaching and learning in the next decade. Traxler and Kukulska-Hulme [2] highlight some of the important considerations for adopting mobile technologies for teaching and learning. These include the issues of cost of the technology (including implementation costs), the appropriateness of the technology for the target students, the relative novelty of the technology (where novelty might be considered to be a limiting factor), and the requisite level of efficient interactivity. These factors were important in choosing an appropriate technology for this study.

It might be argued that one of the appropriate uses of technology within educational systems is for building information technology skills for future workforces. There has been extensive research at both contact and distance tertiary institutions concerning the role of technology in teaching and learning. These studies have largely focused on computer and Web-based technologies, although there are also studies that have looked at using PDA/mobile phone hybrids [3] [4] and a small number of studies that have examined mobile phone text messages [5] [6]. In fact, Attewell [1] found that using a mobile device for teaching and learning encouraged students in their sample to use other technological devices.

There are a number of large-scale projects that are extending the boundaries of mobile and portable teaching and learning. These include the m-learning project [5], the MOBIlearn project [6], HandLeR [7], skoool™ [8], and the WiTEC project [9]. In addition, a recent South African initiative, MobilED [10], used mobile phone technology to allow learners to access online content using text messages. Only a limited number of studies have investigated the role of short text messaging for educational purposes. Some projects have involved using text messages as a method of English language instruction [11]. In another project, Garner et al [3] used simple alert messaging to inform a (relatively) small class of psychology students about course information (e.g. test dates), emergency information (e.g. cancellation of a lecture), and information prompts (e.g. essay collections). Students' perceptions of this use of short text messages were highly positive. Smyth [6] also found that short text messages were useful in providing study and examination tips to learners. Learners (especially border-line poor performers) responded positively to short text messages. Perhaps the studies that most closely parallel this study are those that use classroom response systems to facilitate classroom discussions. One of these initiatives was the MAPLE project [8]. The MAPLE concept involves the lecturer posing questions and the students responding immediately using mobile devices designed specifically for this purpose. Roschelle [12] refers to this as a classroom response system. A classroom response system, such as ClassTalk [12], is far more structured than the initiative that we attempted in this study. Nevertheless, there are some useful outcomes that parallel what we were attempting to achieve. According to Eboueya et al [8] MAPLE "actively promotes equal opportunities ... giving them an opportunity to contribute anonymously and on an equal footing" (p. 156). However, the results of the MAPLE project had not been published at the time this paper was written. Other studies on classroom response systems have found positive improvements in students' academic results and classroom engagement [13]. In short, Roschelle [12] has noted that lecturer-controlled communication and short communications have successfully dominated other forms of mobile and portable teaching and learning.

A. M-Learning

The origins of m-learning are sometimes traced back as far as the work of Kay and Goldberg [14] in 1977 working on early designs of notebooks. M-learning essentially refers to learning taking place through the use of mobile devices and wireless technology [15]. A number of different terms have been used to characterise learning with a mobile device, including wireless learning, ubiquitous learning, network learning, and mobile learning (usually abbreviated as m-learning) [16] [17]. M-learning is characterised by learning that is synchronous, spontaneous, and where communication responses are potentially immediate [17]. Furthermore, m-learning can also be geographically dispersed with reduced travelling time, and offers students greater flexibility in when and where they learn. M-learning is associated with a number of technologies including laptops, palmtops, hiptops, PDAs, iPods, MP3 players, hand-held gaming tools, and mobile phones. The advantages of m-learning include mobility and portability (i.e. having access to learning material and initiatives in multiple locations), increased literacy and numeracy skills, removal of formality from the education transfer process, collaborative information sharing, and increased self confidence and self-efficacy, [3] [18]. However, mobile technology (in particular access and the use of technology) does not occur in a vacuum. Instead, m-learning takes place within a particular social and cultural context. Barriers to m-learning, especially in developing countries include the high costs of technology and telecommunications services, prior exposure and experience with technology, and the fact that not all pedagogic practices lend themselves to effective m-learning [18]. M-learning theorists argue that contemporary views on adult learning (i.e. personalized, learner-centred, situational, collaborative, and lifelong) match the qualities of emerging mobile technology extremely well (i.e. personal, user-centred, mobile, networked, and durable) [5] [7].

B. Mobile Phone Use in South Africa

Before considering what types of technology might be feasibly applied in teaching and learning initiatives, it is worthwhile to consider the availability of information and communication technology within the South African environment (where this study was carried out). South Africa's Apartheid history (formally terminating in 1994) meant that the vast majority of the population was systematically denied access to resources (especially educational resources, but also financial and property resources). Parties within the higher education system in South Africa are therefore acutely aware of how access to technology for education purposes might further entrench existing disparities and increase the digital divide [19]. However, some authors have argued for a social inclusion policy whereby the adoption of technology [20], particularly wireless technology [6] is, in fact, an important remedy to the digital divide.

Compared to other forms of information and communication technology, mobile phones have the highest penetration rate in South Africa. According to the Vodafone Policy Report [21] South Africa had a mobile phone penetration rate of 36% for mobile phones compared to an 11% fixed line penetration rate in 2005 (note that these numbers may be significantly higher for the mobile phone users two years later given the rate of growth in mobile phone provision and access). It has been estimated that up to 70% of mobile phone subscribers do not have a fixed line at home. The Internet penetration rate is similar to that of fixed lines at approximately 10% of the population [22]. The OECD [23] provides similar statistics for 2005 with 4.7 million fixed line subscribers (10%), 33 million mobile phone subscribers (72%), 5.1 million Internet subscribers (11%). However, the majority of mobile phone users do not have access to the Internet

from their phones. According to the OECD [23] there are only 60 000 broadband (e.g. 3G or GPRS) subscribers (or < 1%). Even access to computers (let alone mobile computing) is low in comparison to developed countries with the OECD report [23] estimating that there were only 3.7 million personal computers (8%). No figures were provided for mobile computing in the OECD report although anecdotal evidence puts mobile Internet access (via mobile phone technology) at 11% in South Africa. The OECD report [23] notes that while the growth in the number of fixed lines, Internet users, and personal computers has remained fairly static in South Africa as a proportion of the population from 2000 to 2005, the number of mobile phone users has continued to grow.

According to Samuel, Shah and Hadingham [24] South African mobile users have been characterised as young (40% of mobile phone users in their sample were younger than 25 years old), have at least secondary education as their highest qualification (64% have finished a secondary education or higher), but yet earned less than $170 per month (95% of users). These qualities are strongly reminiscent of university student characteristics. It has been argued that it would be extremely useful to consider the use of mobile phones as a learning tool in developing countries [18]. Given the relative ubiquity of mobile phone use, especially amongst the youth in South Africa (and the relative paucity of access to other forms of mobile technology), this study focuses on the use of "low-tech" mobile technology, in the form of mobile phone short text messaging, as a tool in teaching and learning in a university class.

C. Issues in Large Classes Teaching

Large class sizes forms a second reason why one might want to use technology to facilitate teaching and learning activities. South African classes are, on average, substantially larger than university classes internationally [25] [26] [27] [28] [29] [30]. Interaction in large classes is thought to be limited [31] or assumed not to be possible [32]. When interaction does occur, particular gender and racial groups dominate classroom discussions. Accordingly, males [33] [34] [35], particularly White males [28] [36] have been found to be dominant in classroom interactions. The primary explanation for the dominance of White males is the notion that the majority of university lecturers are White males and behaviourist explanations of modelling and identification are thought to be influential [36] [37].

Problems resulting from large class teaching include that students lack advice on how to improve, students lack opportunity for discussion, and the lecturer is unable to cope with the diversity of students [38]. In addition, lecturers receive lower ratings for classroom interaction in large classes [27] and experience more difficulties in classroom management [30] [39]. This rather pessimistic view represents a particular challenge to the present study where large, diverse classes (of approximately 227 students) were present.

Diversity in students, as a challenge to pedagogic practice [40], is conceptualised in different ways. Diversity has been conceptualised in terms of gender and racial differences between students. On this topic, gender differences in class participation have been found [41] [42]. Race has also been a factor that has been investigated in studies of diversity, with Black students experiencing a sense of alienation in university classes [43]. However, the use of racial categories has been a contested debate in which racial categories have been criticised as ignoring individual differences [44]. Gravett and Henning [45] understand diversity in terms of the way students view the lecturer and believe that these notions often encompass strict notions of authority. They also, perhaps more importantly, uphold that there is necessarily student diversity in participation in class. While technology has the ability to ameliorate some of these differences, one must recognise that there is also diversity in access and use of technology [19].

D. Vygotskian Theory and the Relationship Between Task and Tool

This study is based in a Vygotskian approach to activity theory in order to examine the impact of the technology in this educational setting. It is therefore worthwhile to review the Vygotskian approach. Vygotsky's ontological argument regarding the importance of tools was based on Marx's technological determinist theory of the development of society. Both Marx and Vygotsky used the dialectical historical materialist method (epistemology) in order to investigate society (Marx) and the mind/ cognition (Vygotsky). The importance of historical development to this analytic method cannot be over-stated. Vygotskian theory, and the activity theory approach adopted by Russian theorists, suggests that tools and artefacts shape the way in which we interact with the world around us and that this idea should be viewed within a particular social/cultural and historical framework. From an historical perspective, a tool or artefact is a result of its historical development. In turn, from a social/cultural perspective how we use tools/artefacts is based, in part, on the social context as well as the cultural meaning for how the tool/artefact should be used. The historical development of a tool/artefact and its current use within a particular cultural milieu determines behaviour and mental functioning. It is the tools, or the things that we have created to control our environment, that demonstrate the creativity of humans and serve as the frameworks and patterns of our internal processing, or cognitive functioning. Any new form of technology will structure the ways in which individuals think and interact with world. Individuals are attached to many ideological practices or the relations of power in the production of knowledge [46] [47] [48] [49] [50] [51]. In other words, the type of information that one is exposed to, and, thus, the ways in which one's thinking will develop, depends on many factors, including, but not limited to, those that relate to the language that one speaks, ability to use a computer or mobile phone, and economic resources. Language and technology are the tools, or the things that we use to accomplish tasks. It is the use of tools that represents the higher forms of thinking that characterise human development.

These tools are "cultural" or "ideological" in the sense that they are used by distinct groups of people and are related to the exercise of power in society, or between groups of people. Ideology, in this sense, is a system of ideas and ideals forming the basis of a political or economic theory and is, more generally, the set of beliefs or ways of doing things that are characteristic of particular social groups. Students who use mobile phones have access to far more information and ideas than those who

are not and are able to communicate with others in novel ways.

Vygotsky conceptualised the ways in which individuals in societies interact with one another as cultural development. The most central part of "culture" was the role of language, both as a mechanism for interaction between individuals and as framework for the structure and content of consciousness. Communication between individuals, or social interaction, is "based on rational understanding, on the intentional transmission of experience and thought, (which) requires some *system of means*" [47, p. 48]. The importance of historical factors in Vygotsky's theory cannot be over-emphasised. Semiotic systems have evolved in societies over time from grunts and gestures to the multiple ways in which we communicate today, including mobile phones. Signs and tools are two facets of the same phenomenon. "Signs" indicate the inward movement of objects in the social plane (external) to the individual plane (internal) and are psychological in nature. While "tools" are technical in nature and indicate labour operations, or an outward movement from the individual as he/she engages with the environment [52]. This exposition of Vygotsky's work utilises the concept of the tool, following the Russian interpretation, because tools accord the individual an active role and demonstrate that the individual has actually appropriated the way of thinking. Signs only indicate ways of thinking that are externally present, or exist in the world, and do not adequately indicate that the sign has been incorporated into the individual's consciousness.

Tools alter the characteristics and course of mental processes. These instruments of learning re-create and re-organise the entire structure of our thinking and behaviour [48]. It was in the concept of the tool functioning to re-organize our thinking that Vygotsky separated himself from the circular forms of logic proposed by the Behaviourist school of thought. Vygotsky [46] represented this re-organisation of thinking and interacting in the world in the following manner (see Figure 1). The dotted line between the task and the response to the task represented the explanation provided by the Behaviourists – a simple stimulus-response bond. For Vygotsky, this dotted line represented an individual's automatic response, encompassing the ways of thinking that the individual had already acquired. Vygotsky was interested in determining how the individual learnt new ways of thinking (as depicted by the solid lines). These new ways of thinking incorporated new cultural tools, which altered the way in which tasks were understood and how problem-solving occurred. What is of central importance to Vygotsky's ideas is that the new cultural tool fundamentally alters the process of responding to tasks. Vygotsky was attempting to describe how the use of cultural tools becomes automatic in an individual's functioning, or how we automatically use the tools of thinking. The tool used by the individual could not be separated from the response to the task using that tool because the tool would represent a different way of thinking. If two individuals used two different tools in order to solve the same problem, then their responses to the same task would be qualitatively different [46].

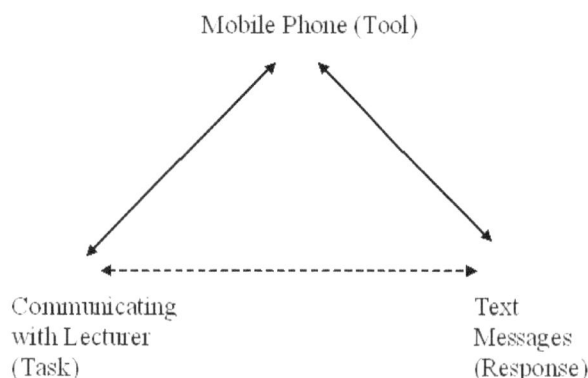

Figure 1.   Vygotsky's original organisation of behaviour, adapted from Vygotsky [46, p. 420].

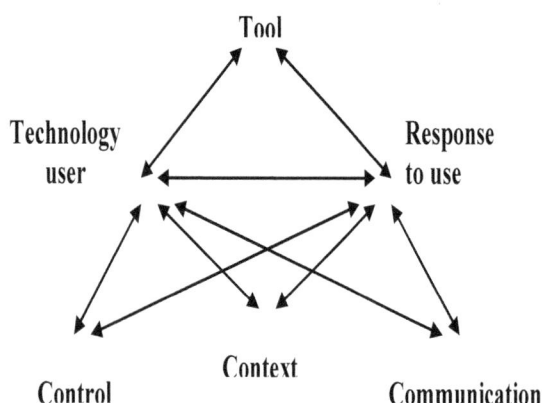

Figure 2.   A framework for understanding mobile learning adapted from Sharples et al. [5]

Vygotsky [46, p.424) provided a "complete cycle of cultural development of any one psychological function". There were five phases in the cycle of tool acquisition as the tool, which is externally present is internalized and, consequently, restructures cognition. Thus, students may have internalised the mobile phone as a tool for the performance of a wide range of tasks, assuming that students in a university currently make use of a mobile phone. In addition, being a mobile phone user has already structured the way in which these students think and interact with others. However, the tool (mobile phone) is being utilised in a novel way (due to University policies preventing mobile phone use in class), or to perform a new task, namely to communicate with the lecturer within a large class environment. Even if the student is a mobile phone user and is one who normally participates in the large class interaction, the task (i.e. communicating with the lecturer) is performed in a new way.

Vygotsky's ideas were developed by Leontiev [54] into a general activity theory and more recently have been applied to understand information systems [55] and human computer interaction problems [56]. Activity theory understands human activity and behaviour as part of the socio-technical environment. Sharples et al [5] have already proposed a model for understanding m-learning adapted from Engeström's [57] activity model. This

model is represented in Figure 2. This figure is essentially the same as Figure 1 except that a third level is introduced to acknowledge the contextual factors, renamed (from the original of rules, community, and division of labour) by Sharples et al [5] as control, context, and communication. Control refers to the degree of influence (e.g. delivery mode and access to learning materials) that learners have on the learning process as well as the social rules and conventions that govern acceptable interaction with the learning technology. Context refers, in this instance, to the learning context (i.e. large class lectures in a developing country context) as well as the communities of people (e.g. students, colleagues, and lecturer) involved in the interaction. Finally, communication refers to how students might adapt their communication and learning interaction with the introduction of a particular technological system [5].

E.   Study Rationale and Research Aims

This study reports on a modest attempt to use short text messaging for in-class interaction within a developing country context. The primary mode of teaching delivery for the course where this initiative was used involves full-time, contact lectures, supported by an online teaching system, WebCT. Therefore, a blended learning approach [58] was adopted, where traditional classroom delivery methods were integrated ('blended') with other education delivery methods (in this study the focus of the investigation was on mobile phone text messaging). There were two primary reasons for introducing mobile phone text messaging into the classroom. First, the class size was large (more than 200 students were registered for this course). This study intended to investigate whether using mobile phone text messages to ask questions in class might have an impact on the number and type of questions as well as who was asking questions during classes, bearing in mind that there have been observed biases in who asks questions in large classes at this University [29]. Second, given the nature of the social context, a developing country, it was necessary to be sensitive to the equitable availability of information and communication technology. Mobile phone technology was chosen due its widespread availability and use amongst the target population. This study therefore aimed to investigate the role of this technology within the framework of Vygotskian theory. Unlike the majority of neo-Vygotskian interpretations, we have examined both semiotic mediation (text messaging) and social activity (large class interaction). In particular we were interested in how the control, context, and communication might influence the users (i.e. the students) in their interactions with the lecturer and with the learning material.

II. METHOD

A.   Procedure

During a large, second year cognitive psychology course (taught over seven weeks), students were encouraged to send a short text message to the lecturer during class or between classes. Students were encouraged through two primary means: (a) on the bottom of each PowerPoint slide presented, a short message was displayed encouraging students to send a message (to the number displayed) if they wished to ask a question; (b) at various points during class the lecturer would ask a question and invite students to respond verbally in class or through sending a text message. The lecturer explained that these messages would only be answered during class and would not be answered via return message or through sending or receiving a voice call. The mobile phone was kept on silent mode (so as not to cause an auditory disturbance while lecturing) but using the battery vibration function to alert the lecturer to incoming messages. The lecturer ensured that all messages were read out and answered during class. After each class, the short text messages were captured electronically with the lecturer making a short note explaining the context of the message. In the last week of the course a questionnaire was distributed to all students present in class and students were invited to complete the questionnaire in class. A total of 136 usable questionnaires were returned (a response rate of 59% of the total class enrolment).

B. Sample

The sample was taken from a second year psychology course of 229 students. From the 136 students who responded to the questionnaire, 98 were female and 37 were male, 75 students had English as their home language with 60 students indicating a range of different home languages including Setswana (N=14), Sesotho (N=10), isiZulu (N=8), Sepedi (N=6), iXhosa (N=6), or other African (N=12) or European (N=4) languages. The majority of respondents had received their secondary education in English (N=129) although a small minority had received their secondary education in an African language (N=6). Most respondents were completing this course with the intention of completing Psychology as a major (N=103), although some respondents were completing a Social Work (N=22), Human Movement Science (N=8), or a Speech and Hearing Therapy (N=6) major. The average age of the respondents to the questionnaire was 21 years and 6 months. The details of respondents' use of mobile phones are provided in the results section. One must bear in mind that the students who responded to the questionnaire may not necessarily be representative of all students taking the course.

C. Analysis of the Text Messages

The short text messages were analysed by two raters working independently. A 100% inter-rater agreement was achieved. The primary method of analysis was thematic content analysis. For the purposes of categorizing the content questions Bloom et al's [55] cognitive abilities and skills taxonomy of comprehension, application, analysis, and synthesis was used. Other categories that spontaneously emerged during the thematic content analysis were: assignment questions, interaction management, social issues, direct responses to the lecturer's questions, and a single response about the use of a mobile phone for learning.

D. Questionnaire

The questionnaire consisted of the following sections: biographical information (e.g. age, gender, home language, and subject-area), mobile phone usage data (e.g. length of time using a mobile phone, frequency of use, and type of use), the use of a mobile phone for the target course (e.g. frequency of text messaging, facilitators of text messaging, and inhibitors of text messaging), and an open ended section where respondents could provide any

additional information of using a mobile phone during lectures. Finally, there was one close-ended question that asked respondents to indicate whether "Having a cellphone [mobile phone] number to SMS [text message] for this course was an excellent idea". This question was scored on a Likert-type scale from "Strongly disagree" to "Strongly Agree" with a higher score indicating strong agreement with the statement.

III. RESULTS

A. General Mobile Phone Usage

Only 2 (1.5%) of the 136 respondents indicated that they had never used a mobile phone. The majority of respondents had used a mobile phone for longer than 5 years (N=102; 75%) and usually sent a few text messages per day (N=57; 42%). While the largest proportion of respondents used a mobile phone for sending and receiving text messages, and sending and receiving voice calls, it was also evident that there was a large degree of sophistication in their use of the mobile phone. Mobile phones were also commonly used as a camera, as a diary, for playing games, for listening to music and the radio, for accessing the Internet and as a cheap instant messaging service. Certainly, these results suggest that (apart from a very small minority) the use of a mobile phone is not unusual for this sample of students. More details on general mobile phone use are given in Table I and Table II.

B. Use of a Mobile Phone During this Course

Due to the fact that it was the intention of the researchers for the text messages to be anonymous it was not possible to directly gather any biographical information from the 86 short text messages received from 52 unique mobile phone numbers (please note that this does not necessarily imply 52 unique students) during the course. However, in the questionnaires we asked respondents to indicate whether they had sent a text message to the lecturer. Forty five respondents indicated that they had sent at least one short text message to the lecturer. The self-reported number of times that respondents indicated that they sent a text is contrasted with the actual recorded number of text messages (based on the unique mobile phone number) in Table III. While

TABLE I.
GENERAL MOBILE PHONE USE

Length of time using a mobile phone	N	%
Longer than 1 year but less than 2 years	1	< 1
Longer than 2 years but less than 5 years	31	23
Longer than 5 years	102	75
Frequency of sending a text message		
A few times a week	24	18
A few times a day	57	42
More than 5 per day	15	11
More than 10 per day	12	9
So many per day I can't remember	19	14

TABLE III.
USES OF A MOBILE PHONE

Mobile phone uses	Have used N	Use Regularly N
Send and receive text messages	131	125
Send and receive voice calls	127	99
Camera facility	106	45
Diary facility	91	42
Games	87	27
Send and receive images	83	23
Link to the Internet	76	34
Listen to music/MP3s	70	39
MXit[a]	63	38
Listen to the radio	49	24

[a] Instant messaging service for mobile phones sent via CSD, GPRS or 3G

TABLE II.
NUMBER OF TEXT MESSAGES SENT TO LECTURER

Number of text messages	Self-reported	Actual
1	17	35
2	14	8
3	3	6
4	3	1
5	5	1
6	1	0
7	1	0
9	0	1

the figures do not match up exactly they show a similar trend. The majority of students sent only one (or two text messages) and a minority of students in this sub-sample sent text messages on multiple occasions.

In this group there were proportionally (in comparison to the group that indicated that they didn't send any text messages) more males than females (33% males in this group compared to 24% in the group that didn't send a text message), more respondents whose first language was English (62% compared to 52%), they had used a mobile phone for longer (87% of the respondents in this group had used a mobile phone for longer than 5 years compared to 71% in the group that didn't send a text message), and used short text messaging more frequently (54% of this group sent at least 5 text messages per day compared to 25% for the sub-sample that did not send a text message).

C. Facilitators and Inhibitors of Sending Text Messages to the Lecturer

The most common cited reasons that encouraged respondents to send a text message were issues related to not having to speak in a large class (N=37), providing

anonymity for asking questions (N=32), and the ability to catch the lecturer's attention (N=24). Surprisingly few respondents indicated that sending a text message was a cheap means of communication (N=6). This can easily be explained when compared to the relatively negligible financial and time costs associated with asking the lecturer directly in person or by sending an email question to the lecturer from an on-campus location (which is free).

Respondents only indicated two reasons that facilitated sending a text message: the mobile phone number on the PowerPoint slides (N=28) and reminders from the lecturer during class (N=4). These results are supported by the most frequently cited reasons for not using text messages which included: easier to speak to the lecturer in person (N=41), the relatively high cost of sending a text message (N=28), not actually having any questions to ask (N=21), easier to email the lecturer (N=16), it takes too long to type out a text message (N=13), or the questions have already been asked by other class members (N=11).

Overall, the majority of respondents strongly agreed with the statement that using text messaging during this course was an excellent idea (96 students strongly agreed with the statement, a further 31 students agreed with the statement, and only 3 students disagreed or strongly disagreed with the statement). There was therefore overwhelming support for this initiative.

D. Qualitative Responses to Sending Text Messages to the Lecturer

The qualitative responses largely provide support for the quantitative results. Most respondents indicated that they found the use of text messages in class useful because it enabled shy students to ask questions and provided some form of anonymity. Respondents indicated that this was especially so in large classes. The following student statements capture the essence of these responses: "Some people don't feel comfortable asking questions in a big class and by using the cellphone [mobile phone] to SMS [text message] the lecturer can get those peoples' questions answered without having to speak in front of the class" or "They would work better for courses with big classes as some people do not have the confidence to speak out in class but can make important contributions or ask good questions by means of an SMS [test message]". Some respondents felt that sending text messages had increased the quantity and quality of discussions in class: "The quality of questions being asked is highly improved, having to send them via SMS [text message] seems to make people think before sending, and detect those people who simply wish to be heard without actually adding anything to the class" and "A lot of people ask a lot of questions now because of the SMSs [text messages] made." In general, respondents felt that using a mobile phone for text messaging in class was a useful teaching and learning innovation: "I think it is an extremely innovative and effective way of allowing everyone in the class to participate, without feelings and inhibition", "Cellphones [mobile phones] are resourceful in class as it allows you to contact the lecturer during the times that you are not present in [sic] varsity", and "It saves time and allows other students in class to also be included in what is being discussed (the question) and makes them to gain knowledge about the things they might have been clueless about."

Respondents also provided a number of suggestions for additional uses of a mobile phone, including reminder notifications about test or essay dates and providing other class interaction information: "Notifications about essay/test dates can be sent via SMS [text message] to students as a reminder, as well as other important notices such as cancellation of a specific lecture." Other suggestions included using the mobile phone as a data capturing device to voice record the lecture or to take pictures of particular PowerPoint slides.

Not all the comments made by respondents were positive. One respondent felt that using a mobile phone to send text messages discriminated against economically disadvantaged students: "Some students don't have the financial resources to engage in this venture, it may make them feel out of place and inferior, it is indirectly discriminatory towards them." Other students felt that using a mobile phone in class is useful only if the phone is switched to the vibration mode otherwise the mobile phone has the potential to disturb lectures rather than facilitate discussion. One student felt that typing out a text message distracted them from listening to the lecturer or the class discussions and one student felt that text messages actually decreased the level of debate in class.

E. Content Analysis of the Text Messages

The single largest use of the text messages was for specific requests about the assignments (N=26); a test and an essay. Typical questions including information about the structure and content areas to be covered in the test, information about the content areas and relevant readings required for the essay, and information about how to deal with specific referencing in the essay. The second largest use of text messages was for interaction management (N=21). Typical interaction management issues included requests to take a break, for other students to stop talking while the lecturer is speaking, for assistance with accessing the course website, and information on whether scheduled lectures would take place. Only a very small number of text messages (N=6) were related to social issues. Social issues included requests to make social announcements (an engagement of one student; a birthday of another student) and questions about whether a certain television programme or sport event had been seen. Surprisingly, only three text messages were direct responses to questions asked by the lecturer in class. It is only possible to speculate on the reasons based on the responses to the questionnaires, but it is possible that students found it easier to respond verbally in class rather than to spend time typing out a text message (and possibly missing out on the class discussion).

An analysis of the text messages based on Bloom et al's [55] taxonomy demonstrated that the majority of text

TABLE IV.
TEXT MESSAGES BASED ON BLOOM ET AL'S [55]
TAXONOMY

Bloom's taxonomy	N
Comprehension	16
Application	9
Analysis	4
Synthesis	1
Evaluation	0

messages, were at a fairly low order; either at the level of comprehension (N=16) or application (N=9). Questions at the level of comprehension were usually requests to repeat content material that had already been covered in the lecture or for clarification and elaboration of explanations. Questions at the level of application were usually applications of theory into different practical domains, usually to check whether their assumptions of the practical implications were correct. There were only a small number of text messages at the higher order. In questions at the level of analysis students not only theory into a new domain but demonstrated that they could critically question the application. At the level of synthesis (N=1) the student the student demonstrated that they could evaluate the new application on the basis of related knowledge (incidentally, this was based on content that we had not yet covered in class, so the student had evidently read ahead in the syllabus). Details of the frequencies of text messages based on Bloom et al's [55] taxonomy are provided in Table IV.

IV. DISCUSSION

The results are discussed using Sharples et al's [5] activity analysis (Figure 2) model as a framework.

A. Technology User/s

The target tool for this study was the mobile phone. When considering the technology users, one must consider those students who used the technology (a mobile phone) for the intended purposes of this study (specifically to send text messages to the lecturer). The demographics of the text message senders were inferred from the responses to the questionnaire. The technology user group was slightly biased towards English-speaking male students (when compared to the general demographics of the class). This said, the majority of respondents who sent short text messages were female (this is unsurprising since the majority of students in the class were female). Respondents in the group that indicated that they had sent a text message were also slightly more experienced with a mobile phone and used text messaging slightly more frequently.

One of the intentions of adopting this initiative was to look at a possible means of realigning student demographics with active student participation in class (approximately half the class was English-speaking and almost three quarters of the class was female). Previous studies have indicated that class participation is dominated by White males [28] [36] and that this is partially a result of social modelling of the lecturer [36] [37] (in this study the lecturer concerned was also a White male). Based on the demographics of respondents who indicated that they had sent a text message, the results of this study therefore do not fully support the claims of the MAPLE project [8] that m-learning initiatives can promote more equitable participation. In fact, one respondent claimed that the initiative used in this study was inadvertently discriminatory.

However, basing equality only on racial or gender criteria alone may disguise some of the other individual differences between users [44]. In particular, the quantitative and qualitative responses demonstrated that students felt that sending text messages to the lecturer allowed students who were shy or who wished to remain anonymous to contribute to discussions. In fact, these responses were specifically targeted at large class interaction situations. These results therefore demonstrate that fixations on race and gender, without considering other differences between students (e.g. shyness when speaking in front of large groups), may be misleading.

B.   Responses to Use

The students' response to the initiative was not overwhelming but was encouraging (86 text messages from 52 unique mobile phone numbers over a 7 week period). Even if we accept the argument that some students are too shy to speak in class, it is possible that these comments and questions might have been addressed by other students during lectures. While the actual number of text messages sent was modest, the vast majority of respondents to the questionnaire felt that the initiative was an excellent idea. When we examined why students did not send a text message the most common responses were that students either preferred to speak to the lecturer in person or preferred to send an email. Within the context of large class teaching this initiative effectively divides the class into smaller units to facilitate greater class participation [59] with some students preferring to send a text message, other students preferring to speak in class, and other students preferring to send an email. Overall, it is likely that participation generally increased. It is evident that different sections of the class preferred different communication technologies to actively participate in class discussions.

It is not surprising that the majority of questions raised through text messages were at a fairly low level of Bloom et al's [60] taxonomy. There are three reasons for this. First, it is likely that second year students might emphasise lower order cognitive abilities and skills. Second, the reminder on each PowerPoint slide invited students to send a text message if they wanted to ask a question. This probably inadvertently discouraged students from sending comments and other critical input that might have been at a higher level. Third, mobile phone text messages are usually limited in length (both in terms of physical mobile phone capacity and network capacity, and in terms of the time it takes to input a text message). Length limitations might have made it difficult for students to formulate comprehensive and critical responses.

C.   Control

For this initiative the lecturer specified a number of controls. First, the lecturer specified which technologies were acceptable for in-class communication (i.e. direct questions by raising a hand and by text message). Second, the lecturer decided on the rules for how these communication technologies should be used (i.e. for text messages, these could be sent at any time to the number provided but responses to text messages would only be addressed during a lecture). Third, the lecturer maintained control over whether a text message was read out and addressed during a lecture. One other point noticed by the lecturer was mobile phones did not typically ring during classes (two times across the whole course compared to two or three times each week previously). It would appear that foregrounding the technology heightened students' awareness of the pedagogic uses and abuses of the technology.

There is also evidence from these results that the initiative provided students with some perceived control over their learning. Providing students with multiple means to contribute to discussions and to ask questions was perceived positively by respondents to the questionnaire. Students were also able to send a question to the lecturer when they thought of the question and not only when they were in class (although only 16 text messages were sent outside of lecture times for this course). Additionally, the most common theme from the thematic content analysis of the text messages was interaction management. The interaction management text messages were used by the students to control social aspects in the classroom. For example, one aspect of the interaction management involved students requesting breaks. Combined with the fact that many of the comprehension level questions were requesting the lecturer to repeat material that had already been covered in class, the text message system therefore provided students with some control over the pace at which content was delivered. Another aspect of the interaction management involved informing the lecturer about people who were talking during class and disturbing the lectures. This was essentially a "naming and shaming" exercise mediated through the authority vested in the lecturer. Further, some students used the text messages to obtain information about whether an upcoming scheduled lecture period would be used. These questions provide an element of control for students since they enable forward planning and scheduling. Finally, it would appear that for a number of students the mere presence of an additional communication medium was sufficient to engender perceptions of control. The attitudes towards the introduction of mobile phone text messaging in class were highly positive with students indicating that they felt the initiative was valuable despite never having used text messaging in this class.

D.   Context

The obvious observation about the context being a large class teaching situation within a developing country with relatively low access to many forms of technology (compared to European, North America, and Far East Asia) has already been made earlier in this paper [21] [22] [23]. Within this context, access to and use of mobile phones is relatively high. The majority of student respondents in this study indicated that they used a mobile phone (98.5%). This is probably far higher than the general South African population with an estimated 36% penetration rate [21]. The respondents in general (i.e. not just those who sent a text message) came from a variety of different educational and language backgrounds. Given this diversity, and the historical socio-economic differences, this study only attempted a modest use of mobile technology appropriate to the students own past experiences and access to technology.

Obviously there are many different types of mobile phones with various different key and peripheral features. In this study, the issue of various different brands and functionalities of mobile phones used by the students was not directly assessed in the questionnaire. However, we are able to infer the use of different mobile phone applications from an analysis of the different ways in which mobile phones were used by this sample of students. It was also clear that students were very familiar with this technology. Most respondents had used a mobile phone for longer than 5 years, sent text messages regularly and were familiar with a wide range of mobile phone applications. In fact, it was quite surprising to note the relatively sophisticated mobile phone applications that respondents were utilizing (e.g. 34 respondents who used their mobile phone to link to the Internet on a regularly basis).

According to Sharples et al [5], the lecturer in this study also forms part of the context. The lecturer was highly experienced with integrating technology into teaching and learning including receiving the University's Vice-Chancellor's award for teaching and learning and has published on this topic [61]. The lecturer's attitude towards this intervention was obviously critically important. The lecturer was the primary driving force behind the initiative ensuring that each PowerPoint slide had a standard statement inviting questions in the form of text messages, verbal invitations during lectures, and by purposefully ensuring that text messages were read out to the whole class. Two of these behaviours were mentioned by the questionnaire respondents as important in facilitating text messages.

E. Communication

The communication aspect in activity theory refers to the ways in which the technology is adapting communication strategies (within the classroom) [5]. For a small group in the class, the text messages initiative enabled the formation of a community of learning that also involved sending text messages on social issues. In this way, other students in the class were included in the interactions between student and lecturer. Many respondents to the questionnaire noted that reading out the text messages and discussing the answers in class enabled students who were interested in finding answers to similar questions were also able to feel included and could learn from the questions being asked. It is perhaps too early to determine whether this initiative has fundamentally changed the classroom interaction, but there were signs that students were starting to think about other ways in which the technology could be used for pedagogic practices. For instance, some students noted that mobile phones could also be used as recording devices (recording audio and/or video from lectures) and students also suggested that text messages should be used as an information service, a use that has shown to be successful in other studies [2] [6]. Clearly this initiative has encouraged students to start thinking how this technology might enhance their learning activities and not just its use as a social communication and entertainment device.

F. Study Limitations

Educators scarcely need to be warned that allowing students to bring their own mobile technology into an existing formal classroom situation has the potential to be disruptive rather than facilitative to the education process [11]. Indeed, the most common negative response from respondents to the questionnaire was the potentially disruptive role that ringing mobile phones might cause in a classroom situation. In addition, there was also the fear that the mobile phone would be used for non-educative purposes (although only a small proportion of the text messages in this study were used for this purpose).

Some authors have argued that it is the user that must be considered mobile, not the technology [6] [58]. In the context of this study it might therefore be argued that the mobile technology is being adopted for a non-mobile use. In this case, this study is therefore not referring to m-learning at all. However, within an activity theory approach one must bear in mind that one is interested in seeing how the technology might be adapted for multiple uses. In this study there is evidence that classroom interaction is changing and that students are beginning to see new possibilities for how this technology can enhance their learning (e.g. the mobile phone for message alerts and as data capturing devices). Further, students were able (even encouraged) to send text messages to the lecturer outside of normal lecture times. While only a small proportion of the text messages were sent outside of lecture times for this course (N=16, i.e. 19%), these text messages were sent in addition to those questions asked via email or face-to-face.

In this study we only collected data on the text messages, but not on the number and type of face-to-face questions. It is difficult to evaluate whether this initiative had a significant effect on changing the traditional social activity and class participation for this course.

G. Recommendations and Directions for Future Research

Students, in their responses to the questionnaires, made a number of suggestions for future use of mobile phone technology in the classroom. One recommendation for a future initiative would be to use text message quizzes [3]. This involves students being given a paper-copy of a quiz and responding to questions via text messages to a text message quiz engine that provides the student with an automatic reply. Using text messages as an alerting and information service is also worth pursuing in addition to the purposes used for this study. Our future investigations should also collect data on the face-to-face classroom discussions. Of course this would be far more resource intensive than collecting the mobile phone text messages as it would involve a research assistant to collect data during each class. The presence of a research assistant in the class might be disruptive to the normal classroom interactions, particularly if students are made aware of the data collection (as would be required by ethical research procedures).

V. CONCLUSIONS

As Roschelle [12] has noted, small-scale mobile pedagogical interventions are often as effective as expensive interventions. This paper has reported on an example of a small-scale mobile phone text messaging intervention to assess the impact of classroom participation and pedagogic interaction in large class teaching. The initiative was analysed using activity theory as a framework for the discussion. The results from the questionnaire suggest that students were overwhelmingly in favor of the initiative despite the fact that only a relatively small number of students participated by sending a text message (i.e. at most 52 students from a class of 227 students). This is borne out in a quantitative and a qualitative assessment of questionnaire responses. Despite the limitations of activity theory for designing learning technology found in the critique by Taylor et al [62] we found activity theory to be a useful evaluative tool for the mobile phone text message learning initiative. While Taylor et al's [62] critique is certainly valid; it largely outlines some of the limitations of a Western interpretation of activity theory. The results are sufficiently positive to strongly recommend that other courses attempt similar initiatives.

REFERENCES

[1] J. Attewell, *Mobile technologies and learning. A technology update and m-learning project summary*. London: Learning and Skills Development Agency, 2005.

[2] J. Traxler, and A. Kukulska-Hulme, "Evaluating mobile learning: reflections on current practice," *Proceedings of mLearn 2005, 4th World Conference on m-learning*

[3] I. Garner, J. Francis, and K. Wales, "An Evaluation of the Implementation of a Short Messaging System (SMS) to support undergraduate students," In *Proceedings of the European workshop on Mobile and Contextual Learning Birmingham*, UK, pp. 15-18, 2002.

[4] Z. Ezziane, "Information technology literacy: implications on teaching and learning," *Educ. Techn. & Soc.*, Vol. 10, pp. 175-191, 2007.

[5] M. Sharples, J. Taylor, and G. Vavoula, "Towards a theory of mobile learning," in *Proceedings of mLearn 2005, 4th World Conference on m-learning..*

[6] G. Smyth, "Wireless technologies bridging the digital divide in education," in *Proceedings of mLearn 2005, 4th World Conference on m-learning.*

[7] M. Sharples, "The design of personal mobile technologies for lifelong learning," *Comp. and Educ.*, Vol 34, pp. 177-193, 2000.

[8] M. Eboueya, D. Lillis, J. Jo, G. Cranitch, and P. Martin, "Mobile participative learning environments for the 21st century classroom: the MAPLE project," in *Proceedings of the 2nd EUI-Net conference on European Models of Synergy between Teaching and Research in Higher Education*, pp. 155-158, May 3-6, 2006.

[9] T.C. Liu, H.Y. Wang, J.K. Liang, T.W. Chan, and J.C. Yang, "Wireless and mobile technologies to enhance teaching and learning," *J. Comp. Assist. Learn.*, Vol. 19, pp. 371-382, 2003.

[10] M. Ford, and T. Leinonen, "MobilED – A mobile tools and services platform for formal and informal learning," in *Proceedings of mLearn 2006, 5th World Conference on m-learning.*

[11] R. Godwin-Jones, "Emerging technologies. Messaging, gaming, peer-to-peer sharing: language, learning strategies and tools for the millennial generation," *Lang., Learn.& Techn.*, Vol. 9, pp. 17-22, 2005.

[12] J. Roschelle, "Keynote paper: Unlocking the learning value of wireless mobile devices," *J. Comp. Assist. Learn.*, Vol. 19, pp. 260-272, 2003.

[13] R.J. Dufresne, W.J. Gerace, W.J. Leonard, J.P. Mestre, and L. Wenk "Classtalk: a classroom communication system for active learning," *J. Comp. High. Educ.*, Vol. 7, pp. 3-47, 1996.

[14] A. Kay, and A. Goldberg, "Personal dynamic media," *IEEE Comp.*, Vol. 10, pp. 31-41, 1977.

[15] H.U. Hoppe, R. Joiner, M. Milrad, and M. Sharples, "Guest editorial: wireless and mobile technologies in education," *J. Comp. Assis. Learn.*, Vol. 19, pp. 255-259, 2003.

[16] B. Alexander, "Going nomadic: mobile learning in higher education," *Educause Review*, pp-29-35, September/October, 2004

[17] Y. Laouris, and N. Eteokleous, "We need an educationally relevant definition of mobile learning," in *Proceedings of mLearn 2005, 4th World Conference on m-learning.*

[18] A. Barker, G. Krull, and B. Mallinson, "A proposed theoretical model for m-learning adoption in developing countries," in *Proceedings of mLearn 2005, 4th World Conference on m-learning.*

[19] M. Warschauer, "Demystifying the digital divide," in *Sci. Am.*, 289, pp. 42-47, 2003.

[20] E.O. Mashile, and F.J. Pretorius, "Challenges of online education in a developing country," *South Afric. J. High. Educ.*, Vol. 17, pp. 132–139, 2003.

[21] D. Coyle, "Africa: the impact of mobile phones," in *The Vodafone Policy Paper Series*, No. 2, pp. 3-9, March 2005.

[22] Internet World Stats. *Internet usage statistics for Africa*. Retrieved from the WWW, 20 March, 2007: http://www.internetworldstats.com/.

[23] OECD, *Information and Communications Technologies, OECD Outlook 2007*. OECD, 2007.

[24] J. Samuel, N. Shah, and W. Hadingham, "Mobile communications in South Africa, Tanzania and Egypt: results from community and business surveys," in *The Vodafone Policy Paper Series*, No. 2, pp. 44-52, March 2005.

[25] N.J. Allers, and N.J. Vreken, "Active learning in physiology practical work," *South Afric. J. High. Educ.*, Vol. 19, pp. 853-863, 2005.

[26] C. Nel, and C. Dreyer, "Factors predicting English second-language students' use of Web-based information systems: implications for student support," *South Afric. J. High. Educ.*, Vol. 19, pp. 129-143, 2005

[27] D. Kember, and A. Wong, "Implications for evaluation from a study of students' perceptions of good and poor teaching," *High. Educ.*, Vol. 40, pp. 69-97, 2000.

[28] D.S. Pollard, "Gender, achievement, and African-American students' perceptions of their school experience," *Educ. Psychol.*, Vol. 28, pp. 341-356, 1993.

[29] P. Stein, and H. Janks, "Collaborative teaching and learning with large classes: a case study from the University of the Witwatersrand," *Persp. in Educ.*, Vol. 17, pp. 99-116, 2006.

[30] E. Walker, and A.E. Wright, "Medical education begins in first year: problem-based, community-oriented teaching in a pre-clinical curriculum," *Acad. Devel.*, Vol. 2, pp. 17-29, 1996.

[31] C. Thomen, and J. Barnes, "Assessing students' performance in first-year university management tutorials," *South Afric. J. High. Educ.*, Vol. 19, pp. 956-968, 2005.

[32] D. Jaques, D. *Learning in Groups (2nd Ed.)*, Essex: Kogan Page, 1991.

[33] J.B. Kahle, L.H. Parker, L.J. Rennie, and D. Riley, "Gender differences in science education: building a model. *Educ. Psychol.*, Vol. 28, pp. 379-404, 1993.

[34] D.J.S Mpofu, M. Das, T. Stewart, E. Dunn, and H. Schmidt, "Perceptions of group dynamics in problem-based learning sessions: A time to reflect on group issues," *Medic. Teach*, Vol. 20, pp. 421- 429, 1998.

[35] J. Scott-Jones, "The complexities of gender and other status variables in studies of schooling," *Hum. Devel.*, Vol. 45, pp. 54-60, 2002.

[36] P.J. den Brok, J. Levy, R. Rodriguez, and T. Wubbels, "Perceptions of Asian-American and Hispanic-American teachers and their students on teacher interpersonal communication style," *Teach. & Teach. Educ.*, Vol. 18, pp. 447-467, 2002.

[37] R.S. Feldman, "Nonverbal behaviour, race, and the classroom teacher," *Theory into Pract.*, Vol. 24, pp. 44-49, 1985.

[38] I. McGill, and L. Beaty, *Action learning. A practitioners' guide*. London: Kogan Page, 1992.

[39] R.D. Pea, "Learning scientific concepts through material and social activities: conversational analysis meets conceptual change," *Educat. Psychol.*, Vol. 28 pp. 265-277, 1993.

[40] A.L. de Boer, T. Steyn, T., and P.H. du Toit, "A whole brain approach to learning in higher education," *South Afric. J. High. Educ.*, Vol. 15, pp. 185-193, 2001.

[41] R.L. Dukes, and G. Victoria, "The effects of gender, status, and effective teaching on the evaluation of college instruction," *Teach. Soc.*, Vol. 17,pp. 447-457, 1989.

[42] S. Saunders, "Market segmentation using quality perceptions: an investigation into a higher education institution," *South Afric. J. High. Educ.*, Vol. 19, pp. 144-154, 2005.

[43] C.M. Loo, and G. Rolison, "Alienation of ethnic minority students at a predominantly White university," *J. High. Educ.*, Vol. 57, pp. 58-78, 1986.

[44] B. Rogoff, and C. Angelillo, "Investigating the coordinated functioning of multifaceted cultural practices in human development," *Hum. Develop.*, Vol. 45, pp. 211-225, 2002.

[45] S. Gravett, and E. Henning, "Teaching as dialogic mediation: a learning-centred view of higher education, " *South Afric. J. High. Educ.*, Vol. 12, pp. 60-68, 1998.

[46] L.S. Vygotsky, L.S. "The Problem of the Cultural Development of the Child," *J. Gen. Psych.*, Vol. 6, pp. 26-39, 1929.

[47] L.S. Vygotsky, *The Collected Works of L.S. Vygotsky. Vol. 1, Problems of General Psychology.* New York: Plenum Press, 1987.

[48] L.S. Vygotsky. *The Collected Works of L.S. Vygotsky. Vol. 2, The Fundamentals of Defectology (Abnormal Psychology and Learning Disabilities..* New York: Plenum Press, 1993.

[49] L.S. Vygotsky, *The Collected Works of L.S. Vygotsky. Vol.3, Problems of the Theory and History of Psychology.* NewYork: Plenum Press, 1997a.

[50] L.S. Vygotsky. *The Collected Works of L.S. Vygotsky. Vol. 4, The History of the Development of Higher Mental Functions.* New York: Plenum Press, 1997b.

[51] L.S. Vygotsky, *The Collected Works of L.S. Vygotsky. Vol. 5, Child Psychology.* New York: Plenum Press, 1998.

[52] L.S. Vygotsky, *The Collected Works of L.S. Vygotsky. Vol. 6, Scientific Legacy.* New York: Plenum Press, 1999.

[53] H. Daniels, *Vygotsky and Pedagogy.* London: Routledge, 2001.

[54] A.N. Leontiev, "Principles of child mental development and the problem of intellectual backwardness," In B. Simon and J. Simon (Eds.), *Educational Psychology in the U.S.S.R.* pp. 68-82, London: Routledge and Kegan Paul., 1963.

[55] B.A. Nardi, (Ed.), Context and consciousness: activity theory and human-computer interaction. MIT Pres: Cambridge, MA, 1996.

[56] L. Bannon, and S. Bødker, "Beyond the Interface: Encountering Artifacts in Use," in J.M. Carroll (Ed.) *Designing Interaction: Psychology at the Human-Computer Interface*, pp. 227-253, New York: Cambridge University Press, 1991.

[57] Y. Engeström, *Learning by expanding: An activity-theoretical approach to developmental research.* Helsinki: Orienta-Konsultit, 1987.

[58] D.R. Garrison, and H. Kanuka, "Blended learning: uncovering its transformative potential in higher education," *Internet and Higher Education, 7*, pp. 95-105, 2004.

[59] R.M. Felder, and R. Brent, "Navigating the bumpy road to student–centered instruction," *Coll. Teach.*, Vol. 44, pp. 43-47, 1996.

[60] B.S. Bloom, M.D. Engelhart, E.J. Furst, W.H. Hill, and D.R. Krathwohl, *Taxonomy of educational objectives. The classification of educational goals. Handbook 1, cognitive domain.* New York: David McKay Company Inc., 1957.

[61] A. Thatcher, "Using the World-Wide Web to support classroom lectures in a psychology course," *South Afric. J. Psych.*, Vol. 37, pp. 345-347, 2007.

[62] J. Taylor, M. Sharples, C. O'Malley, G. Vavoula, and J. Waycott, "Towards a task model for mobile learning: a dialectical approach," *Int. J. Learn. Techn.*, Vol. 2, pp. 138-158, 2006.

AUTHORS

A. Thatcher is with the Department of Psychology, University of the Witwatersrand, Johannesburg, South Africa (e-mail: Andrew.Thatcher@wits.ac.za).

G. Mooney is with the Department of Psychology, University of the Witwatersrand, Johannesburg, South Africa (e-mail: Gillian.Haiden-Mooney@wits.ac.za).

19

Mobile Learning in Context – Context-aware Hypermedia in the Wild

Frank Allan Hansen and Niels Olof Bouvin
Aarhus University, Århus, Denmark

Abstract—Modern project-based education requires students to be able to work with digital materials both in and out of the classroom. Field trips are often an integral part of such projects and greatly benefit students' learning by allowing them to engage with real-world environments firsthand. However, the infrastructure for accessing context sensitive information and supporting in-situ authoring by students while in the field is often lacking. In this paper we present the HyCon framework for mobile, context-aware, and multi-platform hypermedia that aims at supporting several aspects of fieldtrips and project-based education. The framework and its applications utilize information about the users' digital and physical context in order to support digital-physical linking of multimedia materials, and it supports browsing, searching, and creation of new materials which can be linked to points in time and space for later inspection. The HyCon framework is based on a conceptual model for handling context in its myriad forms, as well as a extensible software architecture, and we describe in this paper both. We also describe a number of mobile applications developed with the framework and targeted especially towards mobile learning. Based on a number of field tests with students and teachers from local schools, we evaluate the mobile applications, discuss the experiences gained during the development and evaluations, the requirements for mobile learning tools and the implications for leaning.

Index Terms—Nomadic Learning, Mobile learning, Context-awareness, Context-aware Hypermedia, Geospatial Hypermedia, Physical linking, Tagging.

I. INTRODUCTION

It is becoming didactically desirable as well as technically possible to move learning outside of the classroom and take advantage of the rich sources of information available beyond books and computer screens. It is e.g., possible to gain basic knowledge of what constitutes working at a construction site by reading a book, but the book cannot convey how work is coordinated, how noisy the environment is, how safety is ensured through the action of the workers etc. Taking a field trip to a construction site is a much richer source of information if we wish to proper grasp working conditions. Schools, however, are ill-equipped in supporting this kind of project-based education. In particular, the process of *collecting*, *producing* and *presenting* information from heterogeneous sources and working with this information in and out of the classroom; most schools have dedicated computer rooms and Internet connection, but cannot move the digital information outside the computer rooms.

However, the development in pervasive and mobile technology makes it possible to combine digital information with the physical environment and enables the students to carry a digital context around with them. We see great potential in using knowledge about location and purpose to frame the information made available at a given point in time and space and to let teachers and students play an active part in building the digital context made available on site and sharing it with others.

In this paper we focus on the technical challenges and techniques required to support mobile information work that make sense in a project-based education scenario, where information needs to be understood in context. Based on an empirical study and extensive work with teachers and students from several elementary schools in the Århus area in Denmark, we identify and describe a number of requirements for project based education taking place in- and outside the classroom. We focus on three aspects of project-based education that support learning outside of the classroom through context-aware hypermedia and contextualization of information:

Browsing with your feet is the activity of searching and browsing in context and taking advantage of knowledge about time, place, and purpose when investigating information in the field. Searching and browsing on-site can be dramatically improved by using mobile and context-aware technology. When students are out in the field, they will often need information about the place, they are moving through or are standing in front of. By using knowledge about the user's location, search and browsing can be focused to the context and drastically improve the number and relevance of the search results.

Annotating the world refers to the activity of producing information and writing "digital graffiti" i.e., leaving information in the world tied to the physical place of creation. Annotations are meant to support the students' project activities outside the classroom by enabling them to produce material on-site tied to the location. The students can leave traces of their ongoing activities in the environment that can be revisited later or "bumped into" by others as they travel through the same zone.

Overview at a glance is the ability to quickly provide an overview of the available digital material based on contextual meta-data. Knowing what you have done and where you have been are important aspects of most project presentations. The collection and preservation of the contextual information can be used to create an overview of the process, and allow the children to retrospectively trace their journey, project material, and knowledge building both spatially, temporally, and conceptually.

Peter is a biology teacher in Århus. He has been teaching his 7ᵗʰ graders about animals and their habitats and as part of the project, the class is now going to an area near the school to investigate a number of different biotopes. The area includes a small wood, a waterhole, and a patch of wild plants. Peter inspects a map of the area in the hypermedia application on his laptop and links the Web pages he has used in his classes to points on the map. He then writes assignments and questions related to each of the three biotopes and create a guided tour through the area, which the students should follow. The next day, the children begin the field trip equipped with Smartphones connected to RFID readers and GPS receivers that supply sensor data to the applications on the phones. As they arrive at the wood, the mobile hypermedia application on the phones presents them with a map of the area with Peter's links and assignment overlaid on top of the map. The students begin answering the question by taking pictures of plants and animals and writing short notes. At the waterhole they collect insects and place them in jars marked with RFID tags. Once back in the classroom, the students can browse the material on Web pages generated by the hypermedia services. They create multimedia presentations with facts about the three habitats illustrated with the notes, pictures, and videos produced in the field and with the material Peter provided. While they present their work to each other and to Peter, they bring out the jars with insects and plant samples and places them on a RFID reader in the classroom. The pictures they took at the waterhole appear on a Smartboard in the classroom and they explain where the different insects were found and how they live. Both the children and Peter are satisfied with the project and Peter decides that he will use the material, the children created and linked to locations in the three biotopes, in next years assignments to give the coming students an understanding of the change of the animals and their habitats over time.

Figure 1.   Scenario: Context-aware Hypermedia for Mobile Learning

To see how these activities can be supported by context-aware hypermedia, consider the scenario described in Figure 1. This scenario illustrates several important requirements. First of all, the scenario applies hypermedia to mobile computing and suggests hypermedia systems for Smartphones to support access to information in the field. Through a set of simple sensors the system gets information about the physical environment (in this case information about the location from the GPS and identity of the nearby objects, the children work with from the RFID reader) and uses this context information to find and present structures and documents that are potential interesting for the user in the current situation. In the scenario, the students also create new material with their Smartphones, and use (implicitly) the gathered context information to link documents to physical places and to the plants and insects they collected. Hence, the classical hypermedia support for browsing and structuring has been applied to physical phenomena and augmented with context-aware capabilities. However, the scenario also hints at requirements for the infrastructure needed to run this kind of system. Information is produced and accessed from a number of different devices (laptops, Smartphones, Smartboards, etc.) in a number of different places (Peter's office, the classroom, and in the field). This is a typical example of a ubiquitous computer environment where a number of heterogeneous devices with diverse capabilities participate across users' activities and utilize the same data and infrastructure. Both the hypermedia infrastructure and the

components that utilize it must be flexible enough to integrate into this technology web where devices come and go and new devices, sensors, services, and data types are introduced over time.

We present in this paper a conceptual framework for context-aware hypermedia that support mobile learning. We discuss the design and evaluation of an implementation of the conceptual model, namely the *HyperContext* (HyCon) framework and some of its applications, which explore different aspects of our general model, and discuss the challenges in developing these kinds of systems both technical and conceptual. The HyCon applications have been evaluated in an educational setting with sixth to seventh grade classes. The framework and its applications have previously been described in [4][5][22][24].

The rest of the paper is structured as follows: Section II discusses the notion of context and context-aware computing and presents our conceptual model for context, which is used as the foundation for our implementations. Section III describes our data model for context-aware hypermedia and the HyCon framework architecture. Section IV discusses the requirements for mobile learning, learning activities and the applications we have built to support these activities. Section V describes the technical implementation of the prototype applications. Section VI discusses evaluations and lessons learned and Section VII presents a number of related efforts. Finally, Section VIII concludes the paper.

II.   CONCEPTUAL FRAMEWORK

The notion of computers responding according to their users' implicit stated context is an intriguing and challenging one. In this section we describe our conceptual model for context-aware hypermedia and in Section III we describe two different implementations of this conceptual model. However, before presenting the model we discuss a number of previous efforts to define and model context.

A.   Context-aware Computing

Context-aware computing refers to software systems that can adapt their behavior, interface and structures according to a user's context. Information about the context can be made available to the system either explicitly by the user (e.g., by entering a login name) or implicitly through associated sensory systems. Adding context-aware capabilities to systems is desirable for several reasons: firstly, interaction with applications can be greatly improved if an application adapts to a user's situation and only provides information relevant to that situation. For example, filtering information about a city based on a user's location can make it much easier for users to gain an overview, especially given the limitations of browser user interfaces on small, mobile devices [24]. Secondly, knowledge about the physical environment surrounding a user can be used to create new powerful applications. Hypermedia, for instance, has traditionally been concerned with linking digital resources stored on computers, but the ability to identify physical resources through tagging or sensor input enables hypermedia systems to also support linking to locations and physical objects.

B.   Previous Definitions of Context

The idea of utilizing location in computing systems was advocated by Mark Weiser in his seminal 1991 paper on ubiquitous computing [44]. Some of the early research on location- and context-aware computing was performed on the Olivetti Active Badge systems in 1992 [41]. The term "context-aware computing" was however not defined until 1994, where Schilit et al. listed three important aspects of context: where you are, who you are with, and what resources are nearby [37][38]. Schilit et al. defined context as a location and its dynamic collection of nearby people, hosts, and devices. A number of similar definitions exist that all try to define context by enumerating examples of context elements [6][14][36]. However, as discussed by Dey [15], these types of definitions may be too specific and hard to apply when trying to define general support for context-awareness: e.g., how is it decided whether a type of information can be regarded as context if it is not listed in the definition, and how does a system handle new types of contexts if its design is based on a fixed set of types? Later, definitions of context became more general. Schmidt et al. [39] use the following definition: "[Context is defined as] knowledge about the user's and IT device's state, including surroundings, situation, and to a less extent location." Chen and Kotz [9] define context as "the set of environmental states and settings that either determines an application's behavior or in which an application event occurs and is interesting to the user." Similarly, Dey [15] provides the following definition: "Context is any information that can be used to characterize the situation of an entity. An entity is a person, place, or object that is considered relevant to the interaction between a user and an application, including the user and application themselves." These latter definitions are easier to apply than the former since information can be categorized as context on a per application basis, and they can be used as guidelines for designing general support for context-awareness.

C. Previous Models of Context

The different notions of context have led to a number of varied approaches to modeling and handling context in computer systems. Context defined as a fixed set of attributes can be represented by a hard-coded, optimized context model, whereas a more general definition will lead to more generic and flexible models. Typically, data structures include key/value pairs, tagged encodings, object-oriented models, and logic-based models [9]. In hypermedia, however, the prevalent way to model context has typically been as either composites serving as partitioning mechanisms on the global network or as key/value pairs—associated with links, nodes and other hypermedia objects—that describe parameters of the context the hypermedia objects belong to. Context modeled as composites is often a purely structural partitioning concept constraining browsing and linking to some kind of context given by the user explicitly or implicitly. Key/value pairs associated with objects are typically used to describe in which context objects are visible. Neptune's Contexts [13], Intermedia's Webs [46], and the Webvise Context composites [18] belong to the former category while FOHM's context objects [29] and Storyspace's guard fields [3] belong to the latter. Key/value pairs have also been used to model context in other context-aware systems e.g., Schilit's and Theimer's located objects [37] and the context-aware Web browser Mobisaic [40].

D. Conceptual Context Model

We view the definitions of context discussed above as a good starting point for a common context definition for hypermedia. When Chen and Kotz [9] describe context as

"[a] set of environmental states and settings" and Dey [15] uses the term "any relevant information" to define context, it sets some requirements for the design of a hypermedia framework. With context defined in general terms rather than specific entities, our framework must be able to handle context data in a very general way. Context must necessarily be modeled and defined as specific entities at some point, but this decision should generally be deferred to the design of concrete applications or perhaps even to run-time, where users should be free to specify the nature and format of context objects. In order to ease the categorization of context information and to facilitate designing for context, we adopt the terminology developed by Brynskov et al. [8] to help distinguish the use of "context" as a concept at different levels of system development. Context is classified into three domains: *physical*, *digital*, and *conceptual*. These domains ease the transition from traditional digital-only hypermedia models to models that also encompass a notion of physical objects and contexts:

Physical context includes the physical surroundings of an entity. This includes physical location, physical objects, physical interaction, absolute time and space, and other physical measurements. Computer systems may be aware of the physical context by using sensors.

Digital context includes computer models, infrastructure, protocols, devices, resources and services, logs, and relative time and space. Specifically, traditional hypermedia structures such as links, collections, guided tours, etc. correspond to objects that can be used to represent relationships in the digital context.

Conceptual context describes user activity, intention, focus, and understanding of surroundings.

Most entities may have both a physical and a digital representation, i.e., the physical phenomenon is modeled in the computer system. Ideally, the relationship between the physical entity and the digital model should correspond to the user's understanding of his or her current context (represented as the conceptual model or user model). The representation of physical entities in the digital model should thus reflect the conceptual context. As an example, consider a user using a Smartphone with a context-aware hypermedia system. Picture the user located in front of a building, and the system being aware of the user's position through the Smartphone's integrated GPS receiver. Based on this sensed data from the physical context, the hypermedia system can search its database (i.e., the digital context) for information pertinent to this particular physical context. In our example, the system finds a resource describing the building and presents it to the user. Furthermore, since additional resources annotate the displayed document, links to these resources are presented in the interface allowing the user to further explore the document's digital context. In this example, the user will have no problem coupling the physical building with the document describing it in the hypermedia system, and the conceptual context therefore corresponds to the relationship between the physical and digital contexts.

The main motive for adopting these three domains is not to enumerate a definitive list of parameters essential to the understanding of context, but rather to partition the design space. This allows developers to focus on the aspects they wish to support in their context-driven applications. The digital context corresponds to the classical problem domain of computer science and can be handled by well-known techniques from this field. Specifically, we can apply hypermedia models and structuring mechanisms to represent the relationships between digital entities. The physical context has traditionally not been a concern in computer science, but handled by other fields such as architecture, engineering, and (industrial) design. However, much work in the ubiquitous computing, pervasive computing, and context-aware computing fields has focused on capturing information about the physical world through sensory systems and making this information available to computer applications. Thus, we can leverage these techniques and apply them to hypermedia in order to handle the physical context. The conceptual context is the human factors—the internal context, which cannot easily be captured by sensors. It includes the user's task, goals, intent, and the user's internal model of the context. Again, this is something outside the computer, but research in educational systems and adaptive hypermedia systems [7] have developed techniques that try to build profiles of the user (user models), which can be used to adapt information and application behavior. These three domains and their supporting techniques are depicted in Figure 2.

III. DATA MODEL AND SOFTWARE ARCHITECTURE

Based on our conceptual framework, we now present our data model and our software architecture. Our notion of context is extensible by definition, and our data model must follow suit. We also discuss in this section how sensors for physical data are integrated in our architecture and how the data models handle the heterogeneous set of sensor data.

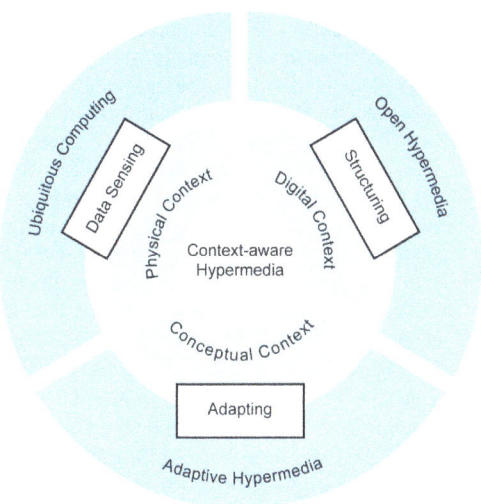

Figure 2. Three context domains: physical-, digital-, and conceptual context, and techniques that support them

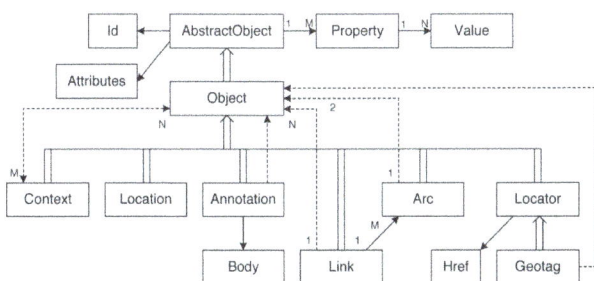

Figure 3. Overview of the Object oriented HyCon data model.

A. Context Model Implementations

In our work with the HyCon platform [5][24], we experimented with several ways to represent context. HyCon provides an object-oriented data model (shown in Figure 3) with context data that can be described by sub-classing an abstract data object.

These specialized objects can be associated with other data objects via a dedicated context composite. In addition to this approach, we also allowed key/value pairs to be associated with data objects to support the simple context tagging mechanism described above. During our evaluation of the HyCon framework [21] we found that while the composite mechanism is useful for modeling certain kinds of physical contexts, especially containment—such as locations where one location (a room) may be part of another location (a building)—this mechanism was seldom used in practice. And in addition, the object-oriented approach often resulted in minor changes to the existing framework services, so programmers preferred the simpler key/value pair model when new applications were created or old classes changed.

This observation led us to generalize the data model and develop the data model depicted in Figure 4. In this data model we use tagging as the fundamental paradigm for modeling data objects *and* their context. All resources are modeled as tag-able `BaseObjects`. If a resource has a certain property, we simply "tag" the resource with this property (as a typed property object). This simple mechanism allows us to model normal data objects, e.g, objects which have properties such as "`name`" or "`id`", but also their contexts e.g., adding GPS coordinates or a 2D barcode URL as properties of the object. The advantages of this structuring model is that we get a completely uniform data model, where it is just as easy to find an object with a given GPS coordinate as it is to find objects with a certain name or type, as all this information is represented by tagged properties.

This model also makes it very easy to handle heterogeneous sensor data. As depicted in Figure 4, a number of sensor inputs can be associated with a given resource and afterwards data from any of these sensors may be used to create a link to the resource, e.g., if a resource is tagged with both a 2D-barcode, an RFID tag, and a GPS position, the resource may be retrieved with data from any of these three sensor types. This mechanism is quite flexible, as it does not require special purpose structures for representing different kinds of physical contexts, but simply relies on object tagging and queries over the data to represent the digital-physical relationships.

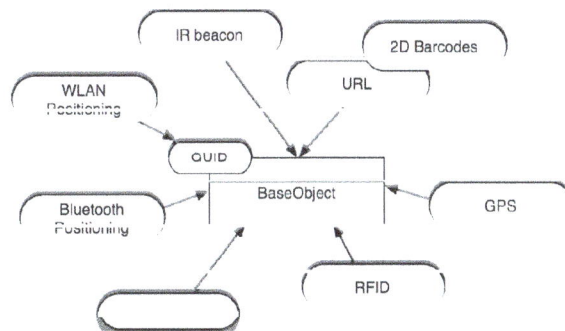

Figure 4. Context-aware hypermedia model as an extension of the notion of Web 2.0 tagging.

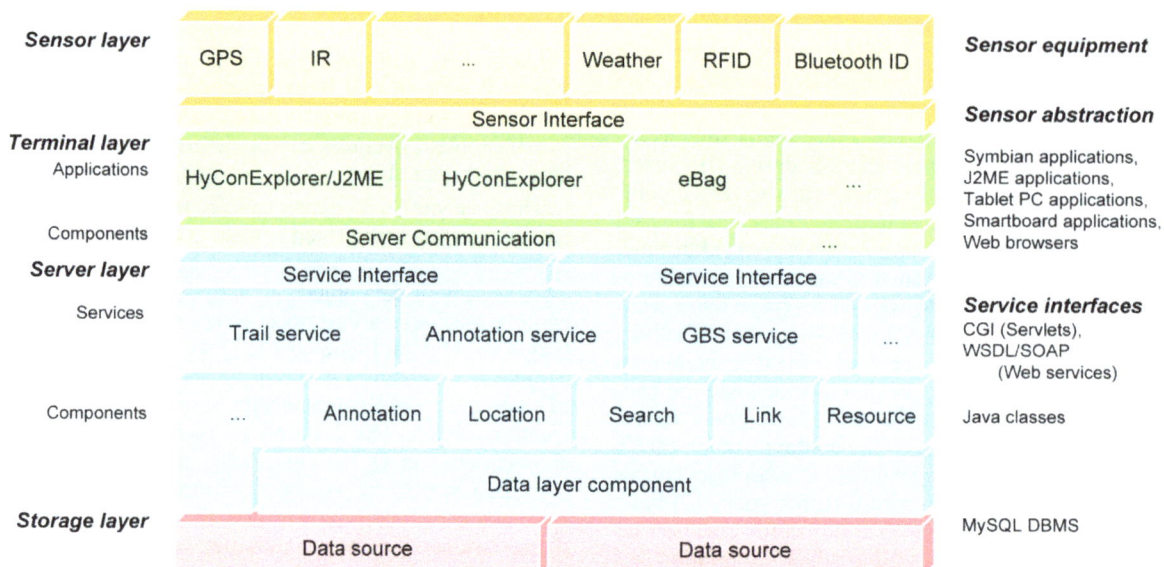

Figure 5.   The HyCon service framework architecture.

A.   HyCon Framework Architecture

The HyCon framework was developed to provide a general platform suitable for experiments with hypermedia mechanisms in a context-aware and mobile environment. HyCon encompasses an infrastructure for implementing context-aware services and applications and a framework, which can be used by applications programmers to build such services and applications. The HyCon framework was designed from a number of requirements in mind: The framework should support a data model combining both hypermedia structures and context objects and at the same time be extensible enough to support different types of context specified by application programmers. This requirement follows directly from our conceptual model and is realized by both data models described above. Having determined what constitutes context for a specific application, the framework should also support developers in implementing behaviors and features to take advantage of the context information in their applications. Furthermore, in order to support different aspects of project-based education as discussed later, a basic requirement for the design was to allow a large number of heterogeneous clients ranging from simple mobile phones to desktop based application, Smartboards, interactive floors and Web browsers to be used with the framework.

The HyCon service architecture is divided into four logical layers: Storage, Server, Terminal, and Sensor layer as illustrated in Figure 5. This architectural approach is quite similar to earlier hypermedia architectures, such as the Dexter architecture [20], the Open Hypermedia System Working Group's Open Hypermedia Model [34] and Construct [45]. However, key to the HyCon service architecture is the extra layer, the Sensor layer, dedicated to handle sensors and sensor information for terminal layer applications. The bottom layer, the Storage layer, handles persistent storage and retrieval of data, hypermedia structures and context structures produced in the system.

The Server layer is divided into core components implementing functionality for server layer applications i.e., services. The framework includes a number of pre-made components implementing functionality for the location, linking and annotation services. Applications in the Server layer can be developed by using one or several existing components from the framework and implementing specialized functionality, which is not available through the mixture of components e.g., computation on the output from different components. This design provides a mechanism for decoupling the responsibilities of the building blocks and not creating mutual dependencies between individual components in the framework.

Interfaces to the services in the Server layer are provided for communication with applications in the Terminal layer. The Terminal layer is not limited to applications running on a specific platform, but includes applications running on a variety of hardware platforms and software environments (mobile phones, Tablet PCs, laptops, Web browsers etc.). The component based design from the Server layer is also used in the Terminal layer: reusable components are e.g., implementing the communication to the server layer. The applications are implementing functionality and user interfaces appropriate for the terminal device in use. However, Terminal components are not reusable across platforms, but are tied to specific platforms (e.g., Java JME, Windows CE, or Symbian).

An innovation in the HyCon framework is the last layer, the Sensor layer. The purpose of the Sensor layer is to create a logical separation between application specific code and code concerned with acquiring context information for the application. The Sensor layer is implemented as an abstraction that specifies how sensor components integrate with and notify application of changes in the context. While this means that application programmers do have to write some code to handle sensors, it has the advantage that no complex sensor infrastructure need to be deployed to utilize sensors in the Sensor layer. Especially for mobile systems and field trip applications used for mobile learning, this can be a major benefit, as sensor infrastructures are not deployed in all cities or in nature areas. In such cases the sensor equipment can simply be brought along together with the mobile devices and accessed through the Sensor layer abstraction. Furthermore, the decoupling between application code and sensor code in the Sensor layer makes it easy to reuse sensor components across applications running on the same platform.

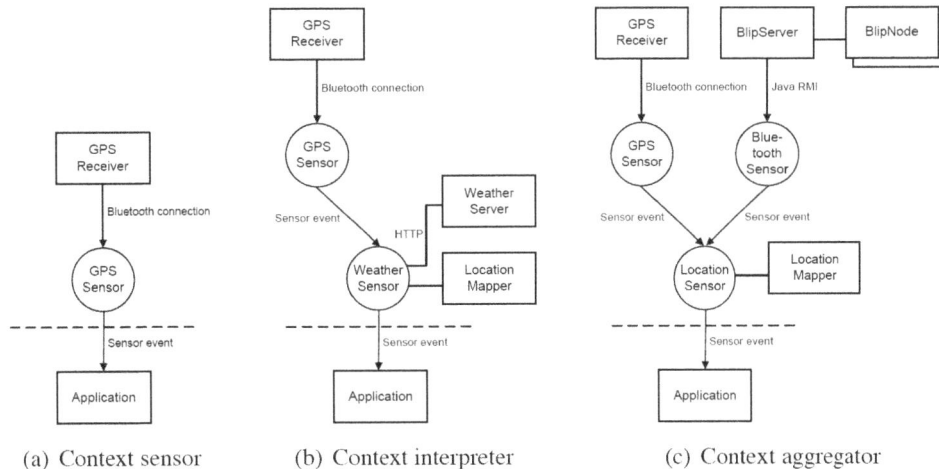

(a) Context sensor   (b) Context interpreter   (c) Context aggregator

Figure 6.   The HyCon sensor abstraction.

B. HyCon Sensor Abstraction

In HyCon, sensor information is accessed through tailored interfaces to components handling the parsing and computation of the raw sensor data. A HyCon sensor is a component that implements the sensor interface. A sensor component must allow applications to access context information and sensor history through both publish-subscribe and traditional poll mechanisms. If an application uses the publish-subscribe interface, it will receive a sensor event each time new context information is available. In this way, context information is made available in a similar fashion as explicit input from keyboards and pointing devices. Sensors can, through the sensor interface, make context information available to applications independently of how this information is sensed. There is no difference between accessing information from a sensor-implementation providing information directly from a physical sensor or from a purely virtual sensor, which aggregates information from a number of other sensor components.

Figure 6 illustrates three different HyCon sensor configurations. Figure 6(a) is a simple sensor that acts as an interface to a Bluetooth GPS receiver. The sensor implementation handles connections to the Bluetooth module and computation on the raw NMEA[1] data to provide context information as location objects delivered through sensor events to the application. Figure 6(b) is an interpreter used in the HyConExplorer to provide weather information and forecasts to the application. The weather sensor aggregates a GPS sensor. The location data from the GPS sensor is mapped to specific areas in the location mapper module, and this information can in turn be used to fetch weather data from a service on the Internet. In this configuration the data from the GPS sensor is interpreted to provide other interesting context information to the user. Figure 6(c) is an aggregated location sensor. The sensor aggregates two other sensors, the GPS sensor and a Bluetooth location system. The Bluetooth location system used here is based on an infrastructure of Bluetooth base stations (BlipNodes[2]) which make information available

about nearby Bluetooth devices through a dedicated server (BlipServer). The aggregating sensor uses a mapping between BlipNodes and their (x, y, z)-coordinates in 3D space to provide the application with uniform context information, independently of the sensor system used to acquire the data and thereby supports positioning both outdoors and indoors. As illustrated by the three configurations, the HyCon Sensor layer abstraction is quite flexible. In application scenarios that depend on a fully deployed context infrastructure, the HyCon sensor abstraction can still be used as a facade to this implementation in a similar way as the Blip infrastructure has been integrated in the third example above.

IV. APPLICATIONS FOR MOBILE LEARNING

We present in this section our systems as exemplars of our framework and as illustrations of context-aware computing support for mobile learning. Our systems have ranged in platform from Smartphones and tablet PCs, to stationary Smartboards in the school classrooms, and this has of course had implications for design and usability.

We discuss how browsing, searching, linking, annotation, and tagging can be augmented with context-aware techniques in order to support the three aspects of project-based education we outlined in the introduction. The applications that implement these techniques are not "stand-alone" applications, but rather an integrated set of applications built on top of our framework to support the different aspects of learning in and out of the classroom and collaboration in these different environments.

Project-based education has become increasingly important in Danish schools. By exploring the multi-faceted and multi-disciplined aspects of a theme chosen by the teacher, the students can apply many aspects of learning to synthesize an understanding.

Projects offer the possibility of combining multi-disciplinary in-class education with real-world experiences. By moving some of the education outside the classroom, lessons can be better learned, as they are experienced rather than just read about, e.g., the noise level of construction plant must be experienced in person in order to be appreciated. When learning moves outside, it becomes essential to bring the experience back into the classroom for further work and discussion. Students customarily bring notebooks and perhaps a camera with them

---

[1] Link: http://www.kh-gps.de/nmea-faq.htm

[2] The BlipNodes and BlipServer are part of a Bluetooth infrastructure developed by BlipSystems (http://www.blipsystems.com).

to document and record their experiences. Later, these experiences can be compiled, processed, and ultimately presented to the teacher and the rest of the class. Thus, project based learning can be described as a series of steps, involving planning, experience, recording, recall, reflection, and presentation. We have sought to support three main activities tied to project based learning with our system, as outlined in the introduction. These activities are:

Browsing with your feet: Much information is situated, i.e., it can be tied to specific locations and circumstances. By moving about in a real landscape, the student can simultaneously be browsing a virtual information landscape in which data pertinent to e.g. the current location automatically presents itself. This enables the presentation of information about the current location, but also opens the possibilities for on-site tests ("how old is the church that you are standing next to?).

Annotating the world: The information encountered above must be added to the system in some way. We have in our work focused mainly on human-entered data rather than automatically generated content (as seen in e.g., GIS). As the students or users of our system move in the world, they are free to inscribe their surroundings with multimedia annotations. This can be used for both factual information ("this church was built in the twelfth century") or digital graffiti ("Anne & Camilla say hi!").

Overview at a glance: While the two former activities support discovery and recording of information, recall is a crucial step in the learning process. Recall is improved by providing a context for the collected artifacts (e.g., notes, images, or other recordings), and this context is in our system provided by associating the knowledge artifacts with the place of creation, typically in the form of an annotated map.

In the following sections we discuss a number of Hy-Con prototypes that support these learning activities when used in combination during a project.

A. The HyConExplorer Prototypes

The mobile browsing and annotation system *HyConExplorer* was the first prototype built with the HyCon framework. Two versions of HyConExplorer have been developed as HyCon Terminal layer applications: one is designed to run on commercially available tablet PCs and the second is designed for Java enabled Smartphones. The goal of the prototypes were to support project based education in elementary schools and in particular, the process of collecting, producing and presenting information from heterogeneous. In order to support these activities, four different context-aware techniques were developed: *Context-aware browsing*, *Context-aware search*, *Context-aware annotation*, and *Context-aware linking*.

All four functions are derived from classical hypermedia techniques, but combined with context-awareness for use in context-aware hypermedia applications for mobile learning.

Context-aware browsing and search support the first aspect of project-based learning (browsing with your feet), by letting the user explore the digital context by changing parameters in the physical context, such as moving from one location to another. Context-aware annotation and linking support the second aspect (annotating the world), by letting users produce information and structures and

automatically tag these resources with available context information (i.e., contextual augmentation). The third aspect (overview at a glance) is supported through the HyCon presentation tools and is discussed in Section IV. F.

B. Context-aware Browsing

Navigating information resources by browsing is a core feature of most hypermedia systems and a typical way to quickly gain overview of the information space. By using hypermedia links, authors can contextualize information by linking to related material or other material that supports or contradicts the given information.

HyConExplorer applies context-awareness to browsing to let users navigate information stored in the digital context simply by changing parameters in the physical context (e.g., changing location). The prototypes support both implicit browsing of the digital context, where the user initiates browsing by moving around, as well as explicit user controlled browsing. These two ways of browsing can be described by adopting the terms direct physical navigation and indirect representational navigation from geo-spatial hypermedia [19]. Indirect representational navigation is browsing information related to an area without necessarily being present, e.g., by clicking on a map over a city or studying a list of museums. Direct physical navigation on the other hand is linked to time and space, as the information presented to the user is determined by the user's whereabouts. This approach has found use in a number of tourist guide systems, such as GUIDE [12], where information presented to the user depends on location (which attractions are nearby) and time (which of these are open). The HyConExplorer browser is shown in Figure 7.

Direct physical navigation lets the user browse information by changing parameters in the physical context. Physically walking or driving from one location to another will affect parameters such as time and location in the physical context, allowing the system to find associated information in the digital context. When engaged in this type of browsing, no direct user intervention is required. The users are "browsing information with their feet", simply moving about in the world. Behind the scenes, the system gathers contextual cues from the physical environment and maps it to tagged information (annotations, linked documents, and guided tours). This kind of browsing is useful especially while the user is in the field using small mobile devices with limited display capabilities and poor support for browsing through large collections (perhaps through multiple layers of menus or lists) of documents.

On the other hand, it is not desirable to require the user to physically move to a given location in order to access information associated with it. Indirect representational navigation allows the user to navigate information associated with remote locations by specifying a virtual location in the system. This creates a "what if I was there?"-scenario, where the system presents the information associated with the virtual context. This corresponds to investigating what Pascoe calls an imaginary world or a pretend context [30]. Systems such as GeoTags[3] and GeoURL[4] also fall into this category by letting users browse pages by their proximity to a given location (without actually being present at the location). In the HyConExplorer pro-

---

[3] Link: http://www.geotags.com/
[4] Link: http://www.geourl.org/

totype, the user can override the GPS sensor and simply click on the map to specify a virtual location for investigation e.g., when encountering a guided tour in the field, the user can browse through the tour stops without having to physically walk to the individual locations. Thus, the user can investigate the information before deciding whether following the tour is actually worthwhile.

Not all context-aware browsing systems allow both navigation modes e.g., early versions of GUIDE [10] only made information about nearby physical objects available and therefore constrained the users' ability to plan ahead or just browse. Similarly, following links in the Auld Linky server used in the Mack Room experiment [28] required users to actually move from the link source to its destination to follow a link. The HyConExplorer addresses such problems by focusing on (and distinguishing between) the digital and physical context and providing browsing modes for both.

C. Context-aware Search

Context-aware search is based on the general idea of using contextual information as part of the criteria to searches for information. This may help narrowing down the information, so the obtained results are more relevant for the user in the current context. The benefit of adding criteria based on context information is of course highly dependent on the specific search engine in use and how the context data can be formulated as an appropriate criteria.

To mobile users, an inherently interesting part of the context is the location. This single piece of contextual knowledge can be used to locate other resources associated with the location and thus bring location relevant information to the user. Ordinary Web searches based on keywords return results matching the topic of interest, but the pages will typically be completely unrelated to the geographic location of the user. However, many Web sites related to locations (such as landmarks and most businesses) include their postal addresses on their Web pages and this (textual) information is indexed by search engines. Thus, search criteria created as a combination of user supplied keywords and postal addresses can be used to find pages covering a specific topic and are related to a specific location.

The term geocoding refers in GIS systems to the process of assigning a latitude-longitude or (x, y, z)-coordinates to a piece of information, typically an address. Once a coordinate has been assigned, the address can be displayed on a map or otherwise used by the system. The term reverse geocoding refers to the reverse task of getting information about a location given a coordinate. The HyConExplorer prototype supports a novel reverse geocoding technique termed Geo Based Search (GBS). In essence, GBS implements a mechanism to augment search engine queries with information about the user's current location. The goal of GBS is to focus search results to pages covering both a topic of interest (specified by user supplied keywords) and the particular geographical area the user is located in. To our knowledge, GBS was quite novel when it was first implemented in 2002-2003. However, since then our concept has been validated through the success of services such as Google Local Search that have become commonplace. Nevertheless, we will describe the original concept below.

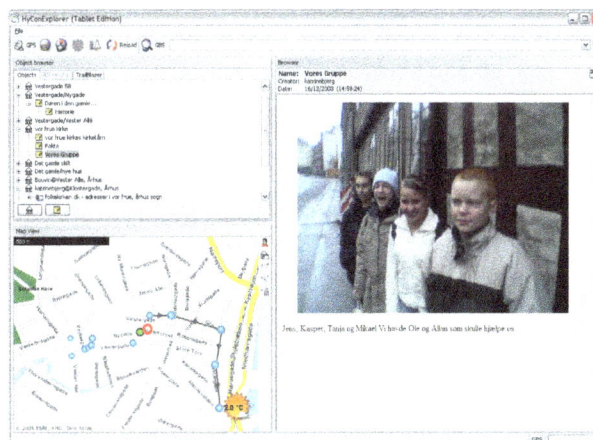

(a)

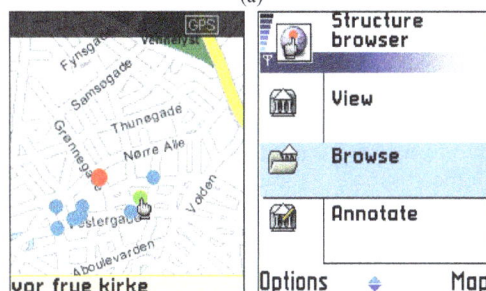

(b)

Figure 7. Browsing trails of annotations and linked Web pages with the HyConExplorer running on the Tablet PC (a) and on the mobile phone (b). Each dot on the map is a link marker to information associated with the physical location.

Raw (x, y, z)-coordinates from GPS sensors are not used directly in GBS, since very few search engines index Web pages by GPS coordinates. Instead, the GBS implementation transforms the GPS coordinates to postal addresses. A database of all public postal addresses in Denmark and their GPS coordinates is freely available from the Danish map provider KMS[5]. Using this database as the basis, mapping raw location information to textual postal addresses provides much more useful input to search engines. The current implementation of the GBS component uses Google as the back-end search engine and utilizes the Google Web APIs. When the user moves from location to location in the physical world the, GPS coordinates are continuously recorded from the GPS sensor. From the coordinates the names of every street within a fifty meters radius are determined and optionally combined with user supplied key words to formulate search strings. The matching Web pages contain the keywords and the postal address printed somewhere in the pages.

---

[5] KMS: Kort & Matrikelstyrelsen: http://www.kms.dk/

Currently, the first ten search results are returned for each street name with no further filtering. The search results are then plotted onto a map of the searched area to intuitively present the connection between geographical locations and Web pages.

Trying to determine the relationship between a Web page and its geographical location by analyzing the content of the page is what McCurley defines as content-based geo-parsing [27]. This may include determining the language of the page, finding names of geographical sites, telephone numbers, names of events that only occur certain places, names of people and locating them. Geo-parsing the content of Web pages will only be successful if the pages contain location information and can be parsed correctly. As GBS searches for postal addresses printed in the Web pages it naturally falls in the class of such content-based methods. In general, geo-parsing Web pages to determine a geographic association does not provide as much reliability and precision as the approach used by Geotags and GeoURL (mentioned in the previous section), where meta tags defining the precise geographic location are inserted into the Web pages. Including meta data in the resource ensures a very precise and reliable mapping, but relies on the authors of Web pages to embed the meta tags in their documents to enable indexing by the search engine. However, as these standards are currently poorly supported by Web page creators, geo-parsing is a viable alternative and functions especially well in urban areas, where businesses and societies supply the needed information on their Web pages. GBS demonstrated a novel technique of integrating existing unstructured information, which had not in advance been prepared for context-aware applications, into mobile, context-aware system such as HyConExplorer.

D. Context-aware Annotations

The browsing and searching techniques discussed in the previous sections present novel approaches to navigation of the information space based on context filtering. This kind of filtering proves especially useful when accessing information on small handheld devices with limited display and input capabilities such as Smartphones and PDAs. The annotation facilities in HyConExplorer are targeted toward user-created information and positions the prototypes as more than just context-aware browsers or navigation systems, but full context-aware hypermedia systems that support users as active producers of information and structures. Annotations created with the HyConExplorer are associated with context information captured by the client which makes it possible to locate the annotations later by browsing.

Combining mobile devices with built-in cameras and microphones (e.g., mobile phones or our prototype tablet PC) and cheap sensor equipment such as GPS receivers, RFID readers, or 2D barcode scanners (i.e., just the built-in camera in mobile phones) makes it possible to create photos, video, and audio documents and automatically tag them with context information captured by sensors. The annotation facilities in HyConExplorer utilize this combination to support linking of digital annotations to physical locations by linking the annotation documents to location objects.

Annotations can take several forms. A simple annotation consisting of a photo and a link to a given location can be used as a form of "digital graffiti" claiming "I was here" to other visitors later visiting the same location. More elaborate annotations featuring photos, videos, and textual descriptions can be used to document or comment on the location or situation in which they were created. Furthermore, annotations can also be linked to other existing annotations and thus support the creation of entire discussion threads which are anchored in the physical context they are concerned with.

E. Context-aware Linking

The World Wide Web presents a massive corpus of information and much of the information available through the Web may be relevant in a physical context. Thus, supporting links to existing Web pages is an effective means of integrating this information into the system. HyCon implements a link model based on XLink. The model supports simple links (2-ary links) linking external documents to objects in the linkbase and extended links (n-ary links) linking a collection of objects (typically locations) into a single trail through the objects.

Extended n-ary links are used to express trails or paths through a collection of objects. In mobile hypermedia, this typically means trails through a collection of locations, where the link represents a guided tour and the locations represent stops on the tour. So far we have only experimented with static link trails. However, adaptive trails could have some interesting applications as well. Registering the user's movement along the trail in a user model, could be used to gradually present and unfold the trail for the user or present different sub-trails depending on the user's preferences. In a school environment, this could be used when teachers prepare field trips for the children, by creating guided tours through an area the children are to investigate. Each stop on a tour may include a description of the site and additional linked material. This material can be divided into several different layers of information: history, archaeology, politics, environmental info, nature guidance, etc. Upon arriving at a stop the students are presented with one or more of the layers depending on their task and the state of their user model. Furthermore, the students can collect further material at the site and add annotations commenting on the material and on the teacher's predefined material through the linking and annotation mechanisms. At home in the classroom, the trail of information may serve as a way to present the results of the field trip for their classmates. In another scenario, the students may diverge from the predefined trail and create their own trail as they walk along. Other groups of students working simultaneously in the same area may bump into the newly created trail and instantly observe the results of diverging from the teacher's guided tour.

F. HyCon Presentation Tools

The third aspect of project work, *overview at a glance*, is supported by the HyConExplorer presentation module (presented in [4]) and the HyConBoard, HyConFloor, and HyConEditor prototypes illustrated in Figure 8. The HyConBoard and HyConFloor are awareness components designed for the classroom. The prototypes provide real-time overview of online HyCon users and the material they produce. This information is presented on interactive maps on Smartboards (HyConBoard) or projected onto floors (HyConFloor). Users can pan and zoom the map and select material from the map for further investigation. Both prototypes share the same interface (depicted in

Figure 8. (a)), but while the HyConBoard is controlled through a large, touch sensitive screen, the HyConFloor is controlled by tracking users position on a projected floor with a camera mounted over the floor. The users' positions are indicated by a round, semi-transparent cursor on the map and objects are selected by keeping the same position for a short while. The cursors will then change color to indicate that the user is about to select the object and if she does not move, a mouse click is sent to the object under the cursor. The HyConFloor is developed as part of other "floor interaction" applications in the Center for Interactive Spaces [33].

The HyConBoard and HyConFloor do not rely on sensors but poll the HyCon server for information about online users and their locations. This functionality is provided by the User Tracking Service, implemented atop the location component. The service relies on a number of attributes in the users' profiles: the user's online status, location, and a picture to show on the map. The users' status are assigned a time-to-live (TTL) so users will not appear as online if their location information is not up to date. When users are online, the HyConBoard and HyConFloor queries the annotation service for annotated location and plots available material and the users' position on the map. The HyConExplorer/J2ME provides this information and export the GPS sensor data to the HyCon User Tracking Service.

The awareness components were designed to provide teachers and students in the classroom with an instant, real-time overview of the whereabouts and activities of "field groups". This was used in several evaluations where the students produced online newspapers as part of the projects. Part of the class was appointed to be reporters, and sent out in the field to gather information, while the rest of the class remained in the classroom as part of the editorial team. The editors could follow the activities of field reporters and if dissatisfied with the gathered material they could immediately contact the reporter group and ask them to produce the needed information.

The HyConEditor prototype was also used during these evaluations to manipulate the produced material. HyConEditor is a Web based application supporting the creation of online journals, reports, and newspapers. Journals and journal entries are implemented on top of the annotation service. The HyConEditor implements a simple query interface for annotations and a WYSIWYG HTML editor that supports creation and editing of journal entries (see Figure 8. (b)). Each deployed instance of the editor is associated with a journal and through the interface, students are able to find annotations created by the field groups and write new journal entries based on this material.

Both the awareness components and the editor and journals support overview at a glance, but they differ with respect to temporal presentation: because the HyConBoard and HyConFloor applications are designed as awareness components, other users must be using one of the mobile applications and actively be producing annotations for the board and floor applications to display the users' positions and present newly created annotations. The HyConEditor, on the other hand, functions as a purely virtual browser and editor, that neither requires users nor the author to be co-located with the annotated resources or be working at the same time.

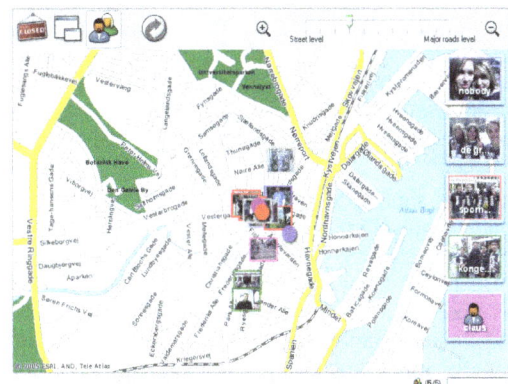

(a) Smart board and floor interface

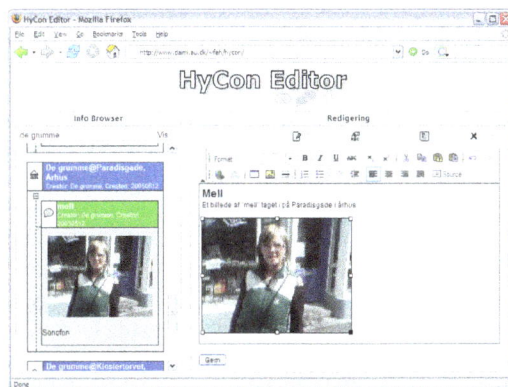

(b) Web authoring interface

Figure 8. Different interfaces for presentation of annotations. All applications are integrated through common services (built on the HyCon framework, in this case) and shared structures.

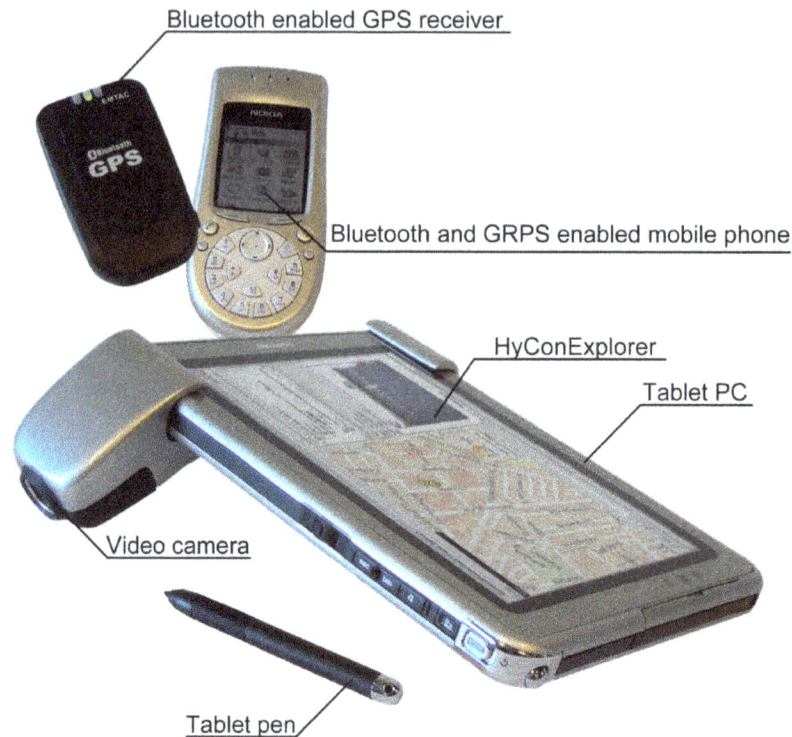

Figure 9.   The HyConExplorer hardware setup. All devices are needed for running the Tablet PC version of the system. The J2ME version only requires the Smartphone and GPS receiver.

V.   PROTOTYPE IMPLEMENTATION

A.   Requirements for a mobile learning tool

While supporting the activities described above were the basic requirements for our system, there are also a number of physical requirements and constraints to be considered.

The tools should be portable (i.e., lightweight, sturdy, and battery-powered), usable while standing or walking (eliminating ordinary laptop computers as potential platforms), support browsing (so a display and pointing device is necessary), support text input as well as possible (as text input is cumbersome without a keyboard), handle multimedia input (capturing a photograph or recording sound/video), preferably be connected to the Internet, and able to associate location and data in some fashion.

It became clear to us as we explored these constraints, that it would be difficult to find a single solution to accommodate all needs, as some of these needs are contradicting. High portability implies small size, which restricts screen size and hampers usability. Network connectivity and portability are at odds, as bandwidth will be limited in the field (as well as expensive and battery draining). Cost was also a consideration—ideally, schools should be able equip a number of groups or individual students with the necessary tools.

Since the initial prototypes of 2003, mobile computing has seen a tremendous rate of development, so it is possible today to have small devices that combine reasonable screen size, GPS, 3G connectivity, and full multimedia capabilities. The rate of development makes it likely that future ubiquitous learning applications need not rely on schools handing out specialized expensive mobile hand-

sets to the students, as the students' personal phones will be more than sufficient to handle the requirements.

B.   HyConExplorer Implementation

The first prototype of HyConExplorer was developed on the Java J2SE platform and was designed to run on commercially available tablet PCs. The second prototype was implemented on the Java micro edition (JME) platform and was designed to run directly on Java enabled Smartphones. The required hardware for both prototypes is illustrated in Figure 9.

The HyConExplorer tablet edition was designed to run on tablet PCs equipped with a mounted camera for capturing low-resolution images, video, and audio, and a Bluetooth enabled GPS unit for recording the user's physical location. The tablet PC prototype implements the full object-oriented HyCon data model. Communication to Server layer services is primarily through the Web Service SOAP interface, with data being sent as XML encoded Java objects. However, upload and download of shared multimedia files are done through the CGI interface. The prototype implements components handling the communication with the server applications, camera data manipulation, and sensor components acquiring location data from Bluetooth GPS receivers. The prototype also implements a map component handling retrieval of bitmap maps from various map providers (e.g., KMS and Esri[6]). Based on the user's position, the chosen map scale, and the geometry of the map view in the application, the map component retrieves maps and displays them in the interface. On top of the map, several layers of hypermedia structures and context information are displayed as link markers and other graphical representations.

---

[6] Link: http://www.esri.com/software/arcwebservices/

The second generation of HyConExplorer (HyConExplorer/J2ME) was designed to run on a much simpler hardware setup than the tablet PC version, namely directly on Java enabled Smartphones, which communicate with sensor equipment over Bluetooth and features built-in cameras, microphones, etc. The platform chosen for the implementation of HyConExplorer/J2ME was the Java micro edition (JME) with the MIDP 2.0 profile[7]. This platform is well suited for mobile, connected devices and supports network over HTTP, HTTPS, TCP sockets, and serial connections. To take advantage of the Smartphones cameras and Bluetooth implementations, we also required that the actual devices supported J2ME optional packages for Mobile Media[8] (JSR-135), and Bluetooth[9] (JSR-82). A number of devices met these requirements and HyConExplorer/J2ME has been developed and tested on Nokia Series 60 Symbian phones and SonyEricsson K750i phones. At the time of implementation, none of these devices supported Web services (through the J2ME Web services package[10] (JSR-172)) and it was therefore decided simply to communicate with the HyCon Server layer through the servlet CGI interface using a raw XML format instead of the Web service interface used by the tablet PC version. Like the tablet PC version, the HyConExplorer/J2ME also implements the full object-oriented data model, HyCon service components, a map component, and hardware integration components providing access to the Smartphones' camera and Bluetooth communication. As SVG support was quite poor on the J2ME platform, the J2ME implementation of HyConExplorer relies on the native double buffered GameCanvas class included in the MIDP-2.0 Game API.

VI. EVALUATION

We have evaluated our system in an educational setting and we report on our experiences in this section. The evaluation of the educational application HyConExplorer involved a project-based setting as outlined in the introduction.

A. Field testing

The HyConExplorer was first evaluated in the context of the NetWorking.Kids research project, headed by Christina Brodersen and Ole Iversen, which ran 2002-2003 [4]. The aim of the project was to enable students at fifth to seventh grade level (11-14 years old) to use technology proactively in the pursuit of project-based learning.

The majority of our tests were performed with a seventh grade class of 20 students from the Katrinebjerg School in central Århus; other schools in Århus were involved to a lesser extend.

We used a number of methods to learn about how teaching was accomplished, including field studies and design workshops with the students. See Iversen's PhD thesis [25] for a thorough discussion of all the aspects of the work with the students.

In order to evaluate how the HyConExplorer matched these requirements, we tested the HyConExplorer with twelve students and three teachers from three school in Århus over a single day. We had prior to this day together

with the teachers created a project based on the theme of "Our City", wherein students were to explore parts of Århus, document their findings, and later present them to their class. The students were divided into three groups of four persons. The teachers had, using the tablet based prototype, authored content, questions, and trails for their students to follow. This part of the evaluation went well: the teachers found the supplied tools easy to use and well suited for the task. They used the integrated web browser to locate additional information and further annotate the map and their trails.

With populated trails, we could then turn to the students. All groups had briefly been introduced to the prototype and its use. The groups were each given an area to cover in the city, and were accompanied by one researcher to recording the event and provide tech support, if needed. In order to explore the design space, we supplied one group with the Smartphone HyConExplorer/J2ME, and two groups with the tablet based version.

Using their allotted equipment, the groups proceeded to explore their part of the city. They made annotations, recorded interviews, and took pictures. On average, each group created five annotations. When they returned to the classroom, they used their gathered material to create more comprehensive presentations. In these, they would often expand on brief notes taken in the field that evidently served more as placeholders or reminders. The two groups with tablets continued to use their tablets, while the group with a Smartphone exported their collected material to a PC and used Powerpoint for their presentation. As all gathered information in HyConExplorer is geo-tagged, it was easy for the tablet users to recall the context of each collected data point. While the students found the software part of the HyConExplorer easy to use, there were some concerns over the physical aspects of the prototype. The tablet version of HyConExplorer consists of several separate devices, namely a tablet PC, a mobile phone for internet connectivity, and a GPS receiver for position. In contrast, the Smartphone version has only the mobile phone and the GPS receiver. As the GPS receiver is quite small, it can easily be carried together with e.g. the mobile phone. The students often divided the carrying among themselves, so that one would carry the tablet PC and another would carry the rest. If they moved too far apart, it was no longer possible to maintain a Bluetooth connection, and the Internet connectivity would falter. Additionally, the children found the tablet PC to be heavy and cumbersome, susceptible to rain, awkward to write or type on, and the Internet connection too intermittent. The Smartphone version received far fewer criticisms, as it was light, provided its own Internet connection, and was (to our surprise) not a text entry challenge to the texting savvy youths.

Based on the experiences we gained from the tests we found that the tablet PC software works well as a tool for the teachers in preparing and organizing information for the field trip and as a context-aware presentation tool that can be used by the students when they return back to the classroom. In the classroom, the overview and structuring mechanisms in the HyConExplorer are more important than the size and weight of the hardware; it is vital for the students' ability to assess their collection of project material. There is no doubt that the tablet provided a superior overview compared to the Smartphone's relatively small screen, which was also reflected in the teacher's clear

---

[7] JME: http://java.sun.com/javame
[8] JSR-135: http://www.jcp.org/en/jsr/detail?id=135
[9] JSR-82: http://www.jcp.org/en/jsr/detail?id=82
[10] JSR-135: http://www.jcp.org/en/jsr/detail?id=172

choice of the tablet over the Smartphone interface and the students' rejection of the Smartphone prototype when preparing their presentation. In the field, however, the simple phone setup turned out to be a much more elegant and usable solution than the tablet PCs. The strength of using a GPS-device is that the prototype automatically supplies the user with information relative to where they are; when the students moved through town and found an interesting spot, the HyConExplorer Smartphone prototype showed them the map and annotations centered around their current location. In the field, location takes priority over overview and the evaluation clearly supports this finding.

B. Implications for learning

From a learning perspective, the HyConExplorer prototypes provide both teachers and students with powerful tools for teaching and learning in the real world. The HyConExplorer prototypes provide teachers and students with means for extending the classroom into the real world as well as bringing the real world back into the classroom. Students should be taught in advance of their expedition, so that they can use this knowledge to inform them as they collect evidence and other data in the field, and finally, in the classroom, use their gained knowledge and data to form further hypotheses and present their findings.

Project work is supported on several different levels: taking advantage of the rich contextual information that is the real world; providing easy access to information related to the location you are currently in; easy capture of rich media (pictures, video, audio, text) annotations that are tied to your current location; providing an overview of the path you have traveled and the annotations created in the process, and finally the means to reflect on the experience and tie it all together in a presentation for the class. The HyConExplorer prototypes elegantly save collections of material that must ordinarily be retrieved from disparate sources and help the students in tying them together through the context in which they were experienced.

An interesting area, which we have not had the opportunity to explore is the interweaving of generated data over time – i.e., how would the accumulated data from previous years inform the students as they roamed the field? The ease with which digital artifacts can inscribed would lend itself to a continually expanding set of knowledge and experiences – something that would be more difficult where the knowledge artifacts in question physical rather than digital.

VII. RELATED WORK

Many systems have been developed to explore the possibilities of associating location (and more generally, context) to data, user generated or otherwise. Most of these systems have not been explicitly developed for learning purposes, yet they all have characteristics that are relevant and can inform future design of context or location aware systems.

Cyberguide [2][26] was one of the first projects to support both indoor and outdoor information browsing with a mobile, context-aware system. The aim of the Cyberguide was to replicate a human tour guide service through use of mobile and handheld technology and ubiquitous positioning and communication services. Several tour guide prototypes were built for environments within and around the

Georgia Tech campus, e.g., guided tours of the laboratories in the Georgia Tech GVU center using an infrared positioning system. Another prototype, called CyBARguide, was running on a Newton MessagePad equipped with a GPS receiver. With the CyBARguide users could locate points of interest that satisfied certain requirements (special offers, free parking, good ambiance, etc.). Cyberguide and CyBARguide were context-aware browsers with no support for user created information. However, the authors acknowledge the need for this kind of functionality and propose future support for a user modifiable information base. Abowd et al. [2] suggest that users should be able to leave comments and impressions for later visitors as "virtual graffiti" when visiting a location. This is similar to creating an annotation on a visited location in the HyConExplorer.

The more recent GUIDE project [12] was similarly developed as a context-aware tour guide and recommender system. Running on a handheld PC, GUIDE provided tourists with context-aware hypermedia information while they were visiting the city of Lancaster in England. Information could be related to the visitors current location or be links to nearby sites that might be of interest. The GUIDE system calculates its position relative to WiFi 802.11 base stations placed at Lancaster attractions. This makes for relatively coarse resolution, but provides high-speed data connection at the hot spots. The GUIDE system has been favorably tested in real use [11], supporting the concept of electronic tourist guides. However, like the Cyberguide, GUIDE only provides browsing capabilities and does not support users to actively create or structure tours and annotations while in the field.

Another project to employ WiFi positioning is Geonotes [17][32]. Geonotes is not designed as a guide system, but as a mass-scale location based annotation system that allows anybody to produce geo-spatial anchored notes. A main concern in the Geonotes project was aspects of social filtering. Realizing that a public annotation space would quickly become crowded and overloaded with information (e.g., spam or commercial advertisement) the designers introduced several collaborative filtering techniques in the system. Each note in the system is assigned metadata describing the popularity of the node, e.g., the number of comments on a note, the number of readings, and the age of the node. This data can be used to provide overview at a glance by only presenting the most popular notes at a given location. In addition, users can invite other users to be "Geonotes friends" to easily share notes among each other. The system supports both pull and push mechanisms for retrieving notes. Text search can be used to retrieve (or pull) notes from the Geonotes server which include specific words or phrases. Searches can also be saved as persistent queries that are continuously performed as the user moves about. When notes match a query, the user is notified with an audible alert. In addition to the WiFi positioning mechanism, Geonotes also allows users to manually specify an extra place label—a user defined location specification that describes the anchor point of the annotation. Even though the place label has to be supplied and interpreted by the user, it gives a much better description of the anchor point than the coarse grained measurement from the WiFi positioning system alone. Geonotes is a fine example of a geo-spatial annotation system with advanced filtering mechanisms. However, the systems services are built around a geographical data model, which may make it difficult to use the system for other context-aware applications that do not solely depend on location.

The Stick-e note architecture [6] is a framework developed at the University of Kent. It is not based on a distributed infrastructure like the previous systems, but aims at providing mechanisms for specifying relationships between application data and context information. The Stick-e note metaphor has been used intensively in mobile annotation systems designed for fieldwork [31]. Two field work applications were built on the Stick-e Notes platform: an application for giraffe observation [31] and a context-aware archaeological assistant [36]. The primary concerns for these applications were robustness of both devices and software while in the field and support for contextual augmentation i.e., automatic gathering and tagging of context information to collected material. Pascoe et al. [31] report that contextual augmentation proved to be almost indispensable during a giraffe observation in Kenya where a researcher recorded data (vegetation surveys and giraffe behavior observations) at a much faster rate than would have been possible using a traditional manual recording medium. Connectivity is also listed as an important aspect of the fieldwork applications. However, since no network infrastructure was readily available on the African savanna and the applications were developed on 1998 PalmPilots with no high-speed network support, data was simply collected on the device and later transferred to a PC back in the research laboratory. Thus, connectivity did not provide real-time collaboration support or sharing of material with co-workers while in the field, as supported by the HyCon framework and applications. But the connection to the PC allowed data to be transferred to a home-base repository that contained all collected material. When the researchers went back into the field, a subset of the home-base repository could be transferred back onto the PalmPilots to a project-dependent repository small enough to fit in the limited memory of the devices. In the field, notes from the project-dependent repository could be triggered and presented when the user's context matched the trigger conditions in the notes.

The Ambient Wood [35][42][43] also focuses on fieldwork, and in particular on supporting school children's education out of the classroom. The project aims at augmenting school children's learning experience in a specific woodland setting, by augmenting the wood with various sensors (RFID readers and location systems based on hidden radio transmitters) and actuators (speakers hidden in the trees). The system is based on a MUD (Multi User Dungeon) with "rooms" corresponding to places in the wood where radio transmitters (pingers) are placed and "players" corresponding to the groups of children in the wood. When the children enter a region in the wood, and are tracked by the pingers, the players position in the MUD is updated to the corresponding room, and the MUD issues an event that sets off audio playback on speakers close to the children. The children also carry moisture and light probes which can be used for taking readings from the environment. Such readings result in presentation of tailored information on the students' handheld computers. All events (the location of the children, probe readings, audio playback, etc.) are recorded in a system log for later reflection. The system also supports the creation of scenarios, which can be played out in the woodland. The scenarios contain pre-authored events that are triggered as the children move about. For example, to provoke further exploration of a site, the children could hear sounds from the hidden speakers and at the same time be presented with information on the handheld computer about specific kinds of animals that live in the grass nearby. After finishing a field trip, the students' exploration can be compiled into field trip journals based on the recorded data. However, unlike the journals created with the HyCon tools, the Ambient Wood journals are created from pre-authored snippets of text that are combined to reflect the specific path the children took through the wood. Thus, rather than trying to support the students as *producers* of material in the field, the Ambient Wood aims at orchestrating an ambient experience environment that provoke discussion and exploration of the surroundings. In evaluation, the authors found that the Ambient Wood system enabled children to connect class room theory with the field experiences. They could form and test hypotheses in the field based on their prior knowledge as well as knowledge imparted to them by the system in the field. Later, in the class room, their journals formed a solid basis for recollection and further work. Thus, the Ambient Wood system demonstrates the possibilities of ubiquitous learning to enhance the students' experience and learning process.

VIII. CONCLUSION

We have herein described the evolution of the HyCon platform, a framework designed to enable the seamless authoring and browsing of contextual information. The framework has explored the meaning of context and provides today a highly flexible and extensible context model as well as a multi-platform set of applications, ranging the very small (mobile phones) to the very large (large floor and wall displays).

The HyCon platform provides teachers and students with an extensible tool set that can support them in the planning, exploration, capture, recall, and presentation of project based education. It has been used in teaching settings on several occasions, and has been shown to work well.

A lesson of our evaluations has been that of choosing the correct tool for the job: when in a mobile setting, portability is key, whereas other settings (such as recall or presentation) favor larger devices. Still, across the various platforms supported by HyCon, the validity of relying on context as an anchoring point for browsing and authoring information has been shown.

The concept of using context as a mechanism for filtering through large data sets has since the conception of the HyCon platform in 2003 become widespread, and Smartphones now commonly feature mechanisms to provide information about the user's immediate vicinity. However, just as the Web moved from being largely a read-only media to today's user contributed "Web 2.0", so geo-based systems should allow its users to add content as they move through the world.

Learning does not occur in a void: it is by its very nature contextualized, and it is by linking and applying new knowledge and experience with the context of what is already known that learning is achieved. The strength of project-based learning is that by broadening the context of learning from one to several disciplines, the lessons learned can illuminate more.

ACKNOWLEDGEMENTS

This work was supported by ISIS Katrinebjerg, Center for Interactive Spaces. The authors wish to thank all our center colleagues. We also wish to thank all our HyCon collaborators over the years, in no alphabetical order: Christina Brodersen, Bent Christensen, Kaj Grønbæk, Ole Iversen, and Torben Pedersen.

REFERENCES

[1]   Abowd, G. D. 1999. Software engineering issues for ubiquitous computing. In *Proceedings of the 21ˢᵗ international conference on software engineering* (pp. 75–84). IEEE Computer Society Press.

[2]   Abowd, G. D., Atkeson, C. G., Hong, J., Long, S., Kooper, R., and Pinkerton, M. 1997. Cyberguide: a mobile context-aware tour guide. Wireless Networks, 3(5):421–433. (doi:10.1023/A:101919 4325861)

[3]   Bernstein, M. 2002. Storyspace 1. In Proc. of the 13th ACM Hypertext Conference, pages 172–181, College Park, Maryland, USA, 2002.

[4]   Bouvin, N. O, Brodersen, C., Hansen, F. A., Sejer, O. I., and Nørregaard, P. 2005: Tools of Contextualization: Extending the Classroom to the Field. In *Proc. of the 4ᵗʰ Int. Conf. for Interaction Design and Children (IDC 2005)*, June 2005, Boulder, Colorado, USA.

[5]   Bouvin, N. O., Christensen, B. G., Grønbæk, K., & Hansen, F. A. 2003. HyCon: A framework for context-aware mobile hypermedia. *The New Review of Hypermedia and Multimedia*, 9, 59–88. (doi:10.1080/13614560410001725310)

[6]   Brown, P., Bovey, J., and Chen, X. 1997. Context-aware applications: from the laboratory to the marketplace. IEEE Personal Communications, 4(5):58–64, Oct. 1997. (doi:10.1109/ 98.626984)

[7]   Brusilovsky, P. 2001. Adaptive hypermedia. User Modeling and User-Adapted Interaction, Ten Year Anniversary Issue 11p:87–110.

[8]   Brynskov, M., Kristensen, J. F., Thomsen, B., and Thomsen, L. L. 2003. What is context? Technical report, Department of Computer Science, University of Aarhus. Available from: http://www.daimi. au.dk/brynskov/publications/what-is-context-brynskov-et-al-2003.pdf.

[9]   Chen, G. and Kotz, D. 2000. A survey of context-aware mobile computing research. Technical Report TR2000-381, Dept. of Comp. Sci., Dartmouth College, Nov. 2000.

[10]   Cheverst, K., Davies, N., Mitchell, K., and Efstratiou, C. (2001). Using context as a crystal ball: Rewards and pitfalls. Personal and Ubiquitous Computing, 5(1):8–11. (doi:10.1007/s007790170020)

[11]   Cheverst, K., Davies, N., Mitchell, K., Friday, A., and Efstratiou, C. 2000. Developing a context-aware electronic tourist GUIDE: some issues and experiences. In Proceedings of the SIGCHI conference on Human factors in computing systems, pages 17–24. ACM Press.

[12]   Cheverst, K., Mitchell, K., and Davies, N. 2002. The role of adaptive hypermedia in a context-aware tourist GUIDE. Communications of the ACM, 45(5):47–51. (doi:10.1145/506218. 506244)

[13]   Delisle, N. M. and Schwartz, M. D. 1987. Contexts-a partitioning concept for hypertext. ACM Transactions on Information Systems, 5(2):168–186, 1987. (doi:10.1145/27636.27639)

[14]   Dey, A. K. 1998. Context-aware computing: The CyberDesk Project. In Proc. of the AAAI 1998 Spring Symp. on Intelligent Environments (AAAI Tech. Report SS-98-02), pages 51–54, Stanford, CA, 1998.

[15]   Dey, A. K. 2001. Understanding and using context. Personal and Ubiquitous Computing, 5(1):4–7, 2001. (doi:10.1007/s00779017 0019)

[16]   Dey, A. K., *et. al* 2001. A conceptual framework and a toolkit for supporting the rapid prototyping of context-aware applications. *Human-Computer Interaction (HCI) Journal*, 16(2-4):97–166., 2001.

[17]   *Espinoza, F.,* et al. 2001. Geonotes: Social and navigational aspects of location-based information systems. In *Proc. of the 3ʳᵈ in-ternational conference on Ubiquitous Computing*, pages 2–17, London, UK. Springer-Verlag.

[18]   Grønbæk, K., Sloth, L., and Bouvin, N. O. 2000. Open hypermedia as user controlled meta data for the Web. In Proc. of the 9th Intl. World Wide Web Conference, pages 553–566, Amsterdam, Holland, May 2000.

[19]   Grønbæk, K., Vestergaard, P. P., and Ørbæk, P. 2002. Towards geo-spatial hypermedia: Concepts and prototype implementation. In Anderson et al. (2002), pages 117–126.

[20]   Halasz, F. G. and Schwartz, M. 1994. The Dexter hypertext reference model. Communications of the ACM, 37(2):30–39. (doi:10.1145/175235.175237)

[21]   Hansen, F. A. 2005. Representing Context in Hypermedia Data Models. In Proc. of the Intl. Workshop on Context in Mobile HCI, Salzburg, Austria, Sep. 2005. Available from: http://mobilehci. icts.sbg.ac.at/context/papers.htm.

[22]   Hansen, F. A. 2006. Ubiquitous annotation systems: technologies and challenges. In *Proceedings of the 17ᵗʰ Hypertext Conference* (Odense, Denmark, August 22 - 25, 2006). ACM Press.

[23]   Hansen, F. A. and Grønbæk, K. 2008. Social Web Applications in the City: A Lightweight Infrastructure for Urban Computing. In *Proceedings The Nineteenth ACM Conference on Hypertext and Hypermedia*, Pittsburgh, PA, USA 19 - 21 June, 2008.

[24]   Hansen, F. A., et al. 2004. Integrating the Web and the World: Contextual trails on the move. In *Proc. of the 15ᵗʰ ACM Hypertext Conf.*, pages 98–107, Santa Cruz, CA, USA.

[25]   Iversen, O.S. 2006: Participatory Design beyond Work Practices - Designing with Children, Ph.D. dissertation, Dept. of Computer Science, University of Aarhus. Available from: http://www.daimi. au.dk/~sejer/Publications%20_files/dissertation.pdf

[26]   Long, S., Kooper, R., Abowd, G. D., and Atkeson, C. G. 1996. Rapid prototyping of mobile context-aware applications: the Cyberguide case study. In Proceedings of the 2nd annual international conference on Mobile computing and networking, pages 97–107. ACM Press.

[27]   McCurley, K. S. 2001. Geospatial Mapping and Navigation of the Web. In Proceedings of the 10th International World Wide Web Conference, pages 221–229, Hong Kong. W3C, ACM Press.

[28]   Millard, D. E., Roure, D. C. D., Michaelides, D. T., Thompson, M. K., and Weal, M. J. 2004. Navigational hypertext models for physical hypermedia environments. In Proceedings of the 15th ACM conference on Hypertext & hypermedia, pages 110– 111. ACM Press.

[29]   Millard, D. et. al. 2002. Beyond the traditional domains of hypermedia. In the Workshop on Open Hypermedia Systems, pages 26–32, College Park, MD, 2002.

[30]   Pascoe, J. 1997. The stick-e note architecture: extending the interface beyond the user. In Proceedings of the 2nd international conference on Intelligent user interfaces, pages 261–264. ACM Press.

[31]   Pascoe, J., Ryan, N. S., and Morse, D. R. 1998. Human Computer Giraffe Interaction: HCI in the Field. In Johnson, C., editor, Workshop on Human Computer Interaction with Mobile Devices, GIST Technical Report G98-1. University of Glasgow. Available from: http://www.cs.kent.ac.uk/pubs/1998/617.

[32]   Persson, P., Espinoza, F., and Cacciatore, E. 2001. GeoNotes: social enhancement of physical space. In CHI '01 Extended Abstracts on Human Factors in Computing Systems, pages 43–44. ACM Press.

[33]   Petersen, M. G., Krogh, P. G., Ludvigsen, M., and Lykke-Olesen, A. 2005. Floor interaction hci reaching new ground. In CHI '05: CHI '05 extended abstracts on Human factors in computing systems, pages 1717–1720, New York, NY, USA. ACM Press. (doi:10.1145/1056808.1057005)

[34]   Reich, S., Wiil, U. K., Nürnberg, P. J., Davis, H. C., Grønbæk, K., Anderson, K. M., Millard, D. E., and Haake, J. 1999. Addressing interoperability in open hypermedia: the design of the open hypermedia protocol. The New Review of Hypermedia and Multimedia, 5:207–248. (doi:10.1080/13614569908914714)

[35]   Rogers, Y., Price, S., Randell, C., Fraser, D.S., Weal, M., and Fitzpatrick, G. 2005. Ubi-learning Integrates Indoor and Outdoor Experiences. Communication of the ACM, 48:1:2005, pp. 55-59.

[36] Ryan, N. S., Pascoe, J., and Morse, D. R. 1998. Enhanced reality fieldwork: the context-aware archaeological assistant. In Computer Applications in Archaeology, Oxford, Oct. 1998.

[37] Schilit, B. N. and Theimer, M. 1994. Disseminating active map information to mobile hosts. IEEE Network, 8:22–32, 1994. (doi:10.1109/65.313011)

[38] Schilit, B., Adams, N., and Want, R. 1994. Context-aware computing applications. In Proc. of the Workshop on Mobile Computing Systems and Applications, pages 85–90, Santa Cruz, CA, USA, Dec. 1994.

[39] Schmidt, A., et. al. 1999. Advanced interaction in context. LNCS, 1707, 1999.

[40] Voelker, G. M. and Bershad, B. N. 1994. Mobisaic, an information system for a mobile wireless computing environment & engineering. In Proc. of the Workshop on Mobile Computing Systems and Applications, pages 85–90, Santa Cruz, CA, USA, Dec. 1994.

[41] Want, R. et. al. The active badge location system. Technical Report 92.1, Olivetti Research Ltd, Cambridge, UK, 1992.

[42] Weal, M. J., Michaelides, D. T., Thompson, M. K., and DeRoure, D. C. 2003a. The AmbientWood Journals: Replaying the Experience. In Proceedings of the 14th ACM conference on Hypertext and hypermedia, pages 20–27, New York, NY, USA. ACM Press.

[43] Weal, M. J., Michaelides, D. T., Thompson, M. K., and Roure, D. C. D. 2003. Hypermedia in the Ambient wood. Hypermedia, 9(1):137–156. (doi:10.1080/13614560410001725347)

[44] Weiser, M. (1991, February). The computer for the 21st century. Scientific American, 265(3), 66–75.

[45] Wiil, U. K., Hicks, D. L., and Nürnberg, P. J. 2001. Multiple open services: A new approach to service provision in open hypermedia systems. In Davis et al. (2001), pages 83–92.

[46] Yankelovich, N., Haan, B. J., Meyrowitz, N. K., and Drucker, S. M. 1988. Intermedia: the concept and the construction of a seamless information environment. Computer, pages 81–96, January 1988. (doi:10.1109/2.222120)

AUTHORS

Frank Allan Hansen is Post. Doc. at Aarhus University (Center for InteractiveSpaces, Department of Computer Science, email: fah@cs.au.dk). He received his PhD and MS degrees in computer science from Aarhus University in 2006 and 2003, respectively.

Niels Olof Bouvin is Associate Professor at Aarhus University (Center for InteractiveSpaces, Department of Computer Science, bouvin@cs.au.dk). He received his PhD and MS degrees in computer science from Aarhus University in 2001 and 1996, respectively.

20

A Collaborative Webbased Framework with Optimized Mobile Synchronisation: Upgrading to Medicine 2.0

P.L. Kubben

Maastricht University, The Netherlands

Abstract—**Magazines like BusinessWeek and Time have confirmed "The Power of Us" as we collaborate in the Web 2.0. However, current medical practice still relies on Medicine 1.0. The use of webbased collaborative frameworks ("wiki's") combined with optimized output and synchronisation for mobile devices can help us upgrading to Medicine 2.0. This serves as an aid for community-driven clinical decision support systems increasing medical safety. NeuroWiki.com will be demonstrated as a useful concept in this philosophy, and its technical background will be described.**

Index Terms—**mobile, pda, handheld, wiki, Medicine 2.0**

Figure 1.   BusinessWeek cover of June 20, 2005

I. INTRODUCTION

The number of peer reviewed journals almost doubled over forty years, and the number of citations yearly added to MEDLINE increased four fold (table 1) [1]. Although reviews and meta-analysis resources may serve as floating aids in a sea of knowledge, it remains doubtful whether they let us reach the shore. Accurate numbers on the amount of articles having clinical impact, are lacking. Citation indices are neither capable of measuring the quality of publications, nor the clinical impact [2]. For all kinds of meta-analysis resources and databases little evidence exists supporting its effectiveness in enhancing their uptake or changing clinician behaviour [3]. As evidence does not speak for itself, it seems worthwile to expand the horizon.

TABLE I.
MEDLINE / PUBMED SUMMARY INDEXING STATISTICS

	1965	2005
Number of journals indexed in Index Medicus	2,241	4,279
Number of citations added to MEDLINE	151,635	606,000

II. MEDICINE 2.0

A. Lessons learned from the web

In 2004 the term Web 2.0 was defined [4], thereby promoting the Internet as a platform of services in which the user controls and possibly shares his own data to harness collective intelligence. This philosophy has lead to cover articles in magazines like BusinessWeek (figure 1)

[5] and Time [6] emphasizing the importance of collaborative online communities. The same principles can be applied by clinicians to install "Medicine 2.0" [7]. A collaborative webbased framework ("wiki") has proven to be accurate [8] and may overcome the problems mentioned in the first paragraph by stimulating active participation of the user group. Content experts can share and maintain knowledge, add new articles to the database, and discuss different opinions in an open, accessible form.

B. Considering success predictors

To be successful this system should meet three criteria. First, it should be comprehensible to use, not only for technologists, but mainly for content experts in all fields. Second, especially in a field where professionals do not have a fixed workplace, such a framework should allow access to this knowledge without a desktop computer available. As safety regulations in many hospitals prohibit the use of cell phones or wireless networks, the framework should also allow access to the data without continuous Internet access. Third, content quality control is mandatory. In a collaborative system, peer-review is a continuous and active process. Information that is generally accepted as valid could be shown as the "stable version" of the article, with a separate editable addition of new conclusions that are still under debate [8].

III. MOBILE COMPUTING

A. Current situation

Mobile computing has an important place in modern clinical practice where it supports medical students, physicians and patients [9-12]. Physicians frequently use decision support software (drug references, calculators, textbooks) that may change management in up to 30% of cases [13]. Most patients realize that there is too much medical information to be remembered by the physician, and 50% reported an increased confidence in a physician using a personal digital assistant (PDA), whereas only 5% reported decreased confidence [13]. Besides commercially available software, physicians add their own medical notes to their PDA. However, not only is data sharing difficult this way, data loss due to hardware or software failures and slow manual data entry are main disadvantages of this technique [14].

B. Improving existing technology

The author developed a simplified webbased collaborative framework (NeuroWiki.com, figure 2) that allows content experts to create short notes, which can be shared and exported to mobile devices [15]. Information is automatically converted for optimal display on the small PDA screen (figure 3), and hyperlinks between articles are reconstructed to work even without continuous web access. The next section will provide some technical background information.

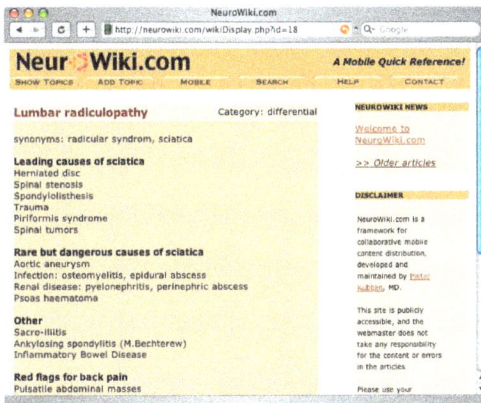

Figure 2: NeuroWiki.com on desktop computer running Mac OS X

IV. TECHNICAL ASPECTS OF NEUROWIKI.COM

The website is running on a Linux Apache server, and has been programmed using open source software (PHP scripting with MySQL database). Cascading Style Sheets (CSS) provide a universal interface throughout the website. Additional markup has been applied with XHTML and HTML. All techniques are approved by the World Wide Web Consortium (W3C) [16], thereby offering maximal cross-platform compatibility on all major desktop operating systems: Windows, Linux and Mac OS X. A separate script prepares all pages for optimal display on handheld computers. It does not only change the text markup, but also recreates hyperlinks between wiki articles: this provides access to other topics in the database without the necessity for an active Internet connection. Cross-platform synchronisation for handheld computers is performed using AvantGo [17], currently targeting the major mobile operating systems like Windows Mobile, Palm OS, Symbian Series 60 / UIQ, and Blackberry. This synchronisation can be performed manually, but also completely automatic while backing up the handheld's data to a desktop computer at any time. As the current small screens of PDAs makes data input a cumbersome task, the developed framework requests database maintenance currently on desktop operating systems only.

V. FUTURE DIRECTIONS

Besides improvement of the usability and technical options of the current system, future developments need to focus at interactivity. Mobile computers should not be degraded to electronic books. However, as the system is as weak as the weakest link, providing accurate and updated content is the primary concern. Wiki-based techniques have proven to be a useful approach to achieve this goal, and should also be considered in future developments.

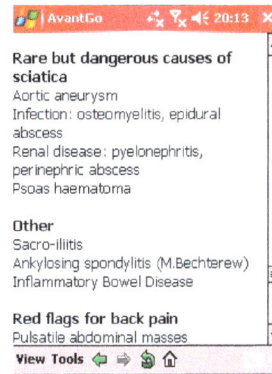

Figure 3:
NeuroWiki.com on handheld
computer running Windows Mobile

VI. CONCLUSION

Evidence does not speak for itself, emphasizing the need for clinical decision support systems. Webbased collaborative frameworks offer a method for active participation of content experts, and can contribute to a relevant and frequently updated source of information. This article demonstrates such a framework with an extension for optimal synchronisation on mobile devices, allowing access even without wireless Internet connection. The concept provides a basis for future directions where Medicine 2.0 is applied using interactive applications built on community-driven information resources.

ACKNOWLEDGMENT

My sincere thanks go to dr. Henk Hoogland, prof. Albert Scherpbier, prof. Cees van der Vleuten and prof. E. Beuls for their support of mobile computing.

REFERENCES

[1] www.nlm.nih.gov. "Summary Indexing Statistics: 1965-2005", 2007.

[2] Doring TF. "Quality evaluation needs some better quality tools." *Nature* 2007;**445**(7129):709.

[3] Wyer PC, Rowe BH. "Evidence-based Reviews and Databases: Are They Worth the Effort? Developing Evidence Summaries for Emergency Medicine." *Acad Emerg Med* 2007, in press.

[4] www.oreilly.com. "O'Reilly -- What Is Web 2.0", 2007.

[5] www.businessweek.com. "The Power Of Us", 2007.

[6] www.time.com. "Time's Person of the Year: You – TIME", 2007.

[7] blogs.nature.com. "Spoonful of Medicine: Nature Medicine 2.0", 2007.

[8] Giles J. "Internet encyclopaedias go head to head." *Nature* 2005;**438**(7070):900-1.

[9] Baumgart DC. "Personal digital assistants in health care: experienced clinicians in the palm of your hand?" *Lancet* 2005;**366**(9492):1210-22.

[10] Honeybourne C, Sutton S, Ward L. "Knowledge in the Palm of your hands: PDAs in the clinical setting." *Health Info Libr J* 2006;**23**(1)**:**51-9.

[11] Garrett BM, Jackson C. "A mobile clinical e-portfolio for nursing and medical students, using wireless personal digital assistants (PDAs)." *Nurse Educ Today* 2006;**26**(8)**:**647-54.

[12] Greenberg R. "Use of the personal digital assistant (PDA) in medical education." *Med Educ* 2004;**38**(5)**:**570-1.

[13] Rudkin SE, Langdorf MI, Macias D, Oman JA, Kazzi AA. "Personal digital assistants change management more often than paper texts and foster patient confidence." *Eur J Emerg Med* 2006;**13**(2)**:**92-6.

[14] Barrett JR, Strayer SM, Schubart JR. "Assessing medical residents' usage and perceived needs for personal digital assistants." *Int J Med Inform* 2004;**73**(1)**:**25-34.

[15] www.neurowiki.com. "NeuroWiki.com", 2007.

[16] www.w3.org. "World Wide Web Consortium", 2007.

[17] my.avantgo.com. "AvantGo: Home", 2007.

AUTHOR

P.L. Kubben, MD, is a resident at the Department of Neurosurgery, Maastricht University Hospital, The Netherlands. Besides he is with the Faculty of Health, Medicine and Life Sciences, Maastricht University, The Netherlands, for development of mobile computing and clinical decision support systems. (e-mail: pieter@kubben.nl). For more information, visit http://www.kubben.nl .

21

Developing a Mobile Application via Bluetooth Wireless Technology for Enhancing Student-Instructor Communication

Dr. Sahar Idwan

The Hashemite University, Zarqa, Jordan

Abstract—A Bluetooth mobile application is developed to enhance student-instructor interaction during delivery of university courses. The paper presents Mobile application via Bluetooth wireless technology (MAvBT) consists of a server installed on a Bluetooth enabled computer, mobile application installed on student's mobile phones, and a website that enables the instructors to edit materials and the demonstrator to obtain appropriate feedback and relevant reports. Students' mobile phones are connected to the server which utilize the MAvBT application implemented to facilitate communication with instructors outside office hours as well as information retrieval with minimum time and from anywhere ranging from 100m up to 1 km. Experimental results show that the proposed system offers fast and low-cost interaction with minimum administrative offers.

Index Terms—Bluetooth, MAvBT, mobile application, wireless technology

I. INTRODUCTION

Bluetooth is a wireless communication protocol used to communicate two or more other Bluetooth-capable devices. It is similar to any other communication protocol such as HTTP, FTP, SMTP, or IMAP in a way that it has a client-server architecture where the one who initiates the connection is the master and the other who receives the connection is the slave. There are two communication protocols available in the mobile device or PDA depending on the cost of establishing the connection. The first one is a non free connection protocol that needs to pay for every loaded and uploaded byte between the client and the server. It is similar to the HTTP protocol that uses GPRS profile. The other one is a free connection protocol that establishes the connection between server and client without any need to be connected to a large provider like the mobile communication companies. The second protocol is used with our system. MAvBT allows the students and the instructors to communicate with each other in different ways without limitation such as capacity of labs, costs, type or numbers of recipients within the domain. The instructor is able to publish notes, exam, and homework's. The students establish the connection by using the Bluetooth and receiving the required information based on his/her request. The student is able to retrieve the evaluation form of the instructor fill it and send it back by using the MAvBT.

This paper is organized as follows. In Section II, we briefly review the related work via Bluetooth connectivity. The developed system is introduced in Section III. The implementation of the system by using the Java standard edition (J2SE) for the server application and Java micro (J2ME) edition for the mobile application is presented in Section IV. Section V provides the conclusion of the paper.

II. RELATED WORK

Reference [1] presented Bluetooth as an industrial specification for wireless Personal Area Networks that provides a way to connect and exchange information between devices, such as mobile phones, laptops, PCs, printers and digital cameras via a secure, low-cost and globally unlicensed short-range radio frequency

Building a wireless information system by using the Bluetooth wireless technology described in [2]. A Bluetooth-enabled client–sever system is under development to conduct assignment during the lectures.

In [3], they introduced a multimedia message transmitter tool that provides businesses with the opportunity to use Bluetooth wireless connectivity delivering the relevant information to relevant people in the right places and time. It empowers any business to market their services or products for unlimited time with a one-time fee.

The opportunity of using cellular phone for educational purposes specifically for teaching and learning mathematics examined in [4]. The students can update or share their learning process without any needs to be in the same place by using emails, logging, and MMS. This approach is expensive since we need to pay the cost of the sending and receiving messages.

A mobile web services solution to support education activities is proposed in [5]. Their work utilizes the mobile web technologies and SMS.

The use of classroom response to improve the classroom interactivity examined in [6].

In this paper, we introduced a communication tools for sharing data through the mobile phone by using the Bluetooth ID. Our system provides benefits to both students and professors. It reduces the pressure on the labs by allowing the students to access their information at any time, anywhere (in the range of his/her department), and from his own tool (mobile phone device).

III. THE PROPOSED MAvBT SYSTEM

In the revolution of mobile and wireless technology and using it in different environment such as universities, schools and marketing, we developed MAvBT. Fig. 1 presents the architecture of the system and the sequence of the operation. Our system consists of one application server, one client application and a website that enables the instructors to edit materials and share their own course documents and information at any time and the demonstrator to obtain appropriate feedback and relevant reports. The application server that utilizes the MAvBT is dedicated for handling client requests and responding to them. The client application is designed to be stored on the student's mobile phones, which enable them to access the required information via MAvBT. Bluetooth Radio Technology represents the transmission media between server and client.

The student uses the mobile application to send a message via the Bluetooth to the server that utilizes the MAvBT application to trigger the appropriate function. The MAvBT in the server side receives the message analyzes it, and returns the required information to the mobile application.

MAvBT is considered as an enhancement for the student portal system and solves the following problems:

1. When the labs are busy, it will be easy for the students to reach their information such as exam table, or classroom number through his/her phone device and within a range of the faculty.

2. Receiving notes with or without the instructor's absences.

3. Receiving shared files that the instructor need to distribute through their phone devices.

4. Ability to retrieve their general information such as grades system or current courses, and the general announcements.

5. The ability to make the instructor assessment through the Bluetooth by using a stand alone mobile application.

A. Security

In order to provide information confidentially, the Bluetooth system provides security measures at both the application layer and the link layer that presented in [7] and [8]. These measures are appropriate for a peer environment such that in each Bluetooth unit, the authentication and encryption routines are implemented in the same way. Four different entities are used for maintaining security at the link layer: a public address that is unique for each user, two secret keys, and a random number that is different for each new transaction. Ref. [9] discussed that the Bluetooth is a highly secure system for the following reasons:

1. Every Bluetooth device is given a 48-bit address that uniquely identifies the device. Every Bluetooth device on earth will have a unique address.

2. When one device wants to communicate with another device, the second device is authenticated.

3. Data on the channel is encrypted so that only the intended recipients can receive it.

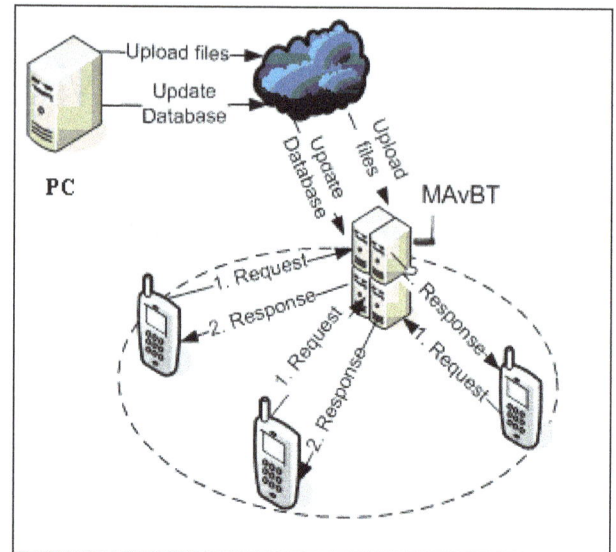

Figure 1. Architecture of the MAvBT

4. Every Bluetooth device has a random number generator; these numbers are used for authentication.

5. A frequency-hopping scheme provides built in security. Only those devices that know the hopping sequence can decode the data sent by the master.

The MAvBT used these technologies to send secure message that reduce the possibility of intruding the mobile with viruses.

IV. IMPLEMENTATION

Our system was implemented by using the Active Server Pages (ASP.NET) in [10] with C-Sharp (C#.NET) Technologies in [11] and [12] to implement the application servers, and Java 2 Micro Edition (J2ME) Technology in [13], [14], and [15] to implement the client application. The web administration page which is designed for the instructor's use to publish any information for his/her students' runs in the command prompt program (Win-32 application) or DOS-based application. Fig. 2 presents sample from the admin page, this page is used to publish an assignment to a specific course or course section.

We download the client application in the phone that allows the student to access the required information after inserting the user name and the password. Then the main list in Fig. 3 contains several options such as:

- Instructors: Where the student can get any related information about the instructor.

- Announcements: Where he/she can find any announcement dedicated to him/her, or to a specific course section.

- Search Mode: If he/she could not find which instructor instructs a specific course.

- Student: he/she can access the information related to him/her.

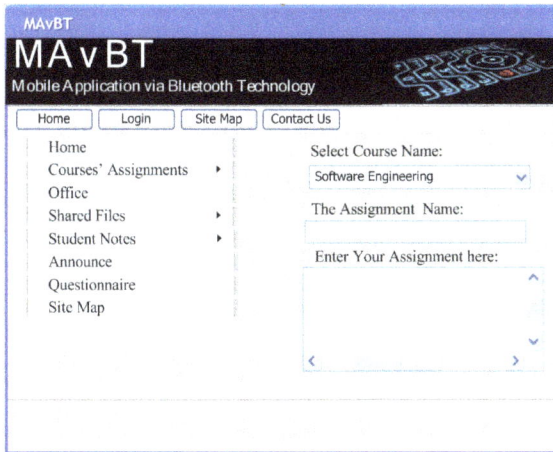

Figure 2. Admin Page

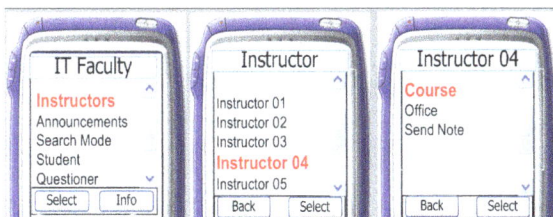

Figure 3. Client Application

As appears in Fig. 3, after selecting the instructors, a list of the instructor InstructorList appears so you can choose any instructor. After you select the instructor a new window appears contains the following selections:

- Course: To find out all courses for the selected instructor.
- Office: To find out the information related to the instructor's office (Office Hours, Office Room).
- Send Note: To send a note or complain for this instructor.

MAvBT helps to evaluate the instructor. The evaluation form can be downloaded into the client application. After answering all the questions MAvBT asks the student to press the submit button, then the application will establish a connection to the server to do the following:

- Submitting the answers to the Server's Database.
- Pointing the selected instructor to be evaluated once by the student, avoiding the reassessment of the instructor by the same student.
- Retrieve the rest of the instructor's assessment forms, in case the student wants to continue evaluating other instructors.

A. Testing

Table 1 shows the Bluetooth dongles information that is used to test the MAvBT. A "dongle" is a small hardware device that helps the computer to authenticate a piece of software or the identity of a user [16]. It connects with a computer's USB port and enables it to communicate with the other Bluetooth-enabled devices. It is available in one of three classes: Class1 covers up to 100 meter, Class2 covers up to 10 meter, and Class3 covers up to 1 meter. In our experiments, we used Class1 and Class2.

TABLE I.
BLUETOOTH DONGLE INFORMATION

Bluetooth Dongle name	Instant Connections	Class	Range	No. of sending Files
SAM sync	Up to 7	Class 2	10 meter	1
MSI	Up to 7	Class 1	100 meter	5

Two kinds of questionnaire are distributed to the students and instructors to evaluate the MAvBT based on enhancing the education and improving the learning outcomes. The first one distributed among two hundred different students contains a variety of questions such as: age, gender, owning mobile with Bluetooth technology, how frequently using the Bluetooth technology within the university, can you use the Bluetooth technology, did the mobile support the java applications technology, the availability of the labs, and are you able to contact the instructors in their office hours. Fig. 4 shows the results to selected questions of the first questionnaire. We found that the percentage of student's phone that supports Bluetooth technology is 89%, while the rest does not. When we searched about the percentage of the students who use Bluetooth technology frequently we found, 71% do, while 29% don't. The mobile phones that support Java applications technology which is the core of the client side is 67%. When we asked about the availability of the labs we found that only 7% said they are always available, and 93% said they are rarely or sometimes available.

64% of the students that contact the instructors (A lot, Sometimes, And Rarely) find it hard to contact their instructors in their office hours.

The second questionnaire distributed among thirty different instructors containing several of questions such as: did you use the e-mail or the paper to send the information to the students, did the students get the most recent information, and did you notice an increase in the number of students that do the evaluation for the instructors by using MAvBT.

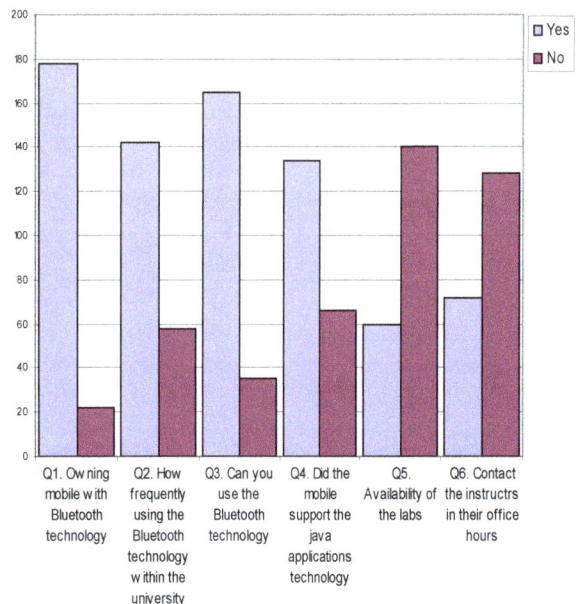

Figure 4. Evaluation of MAvBT based on Student Responses

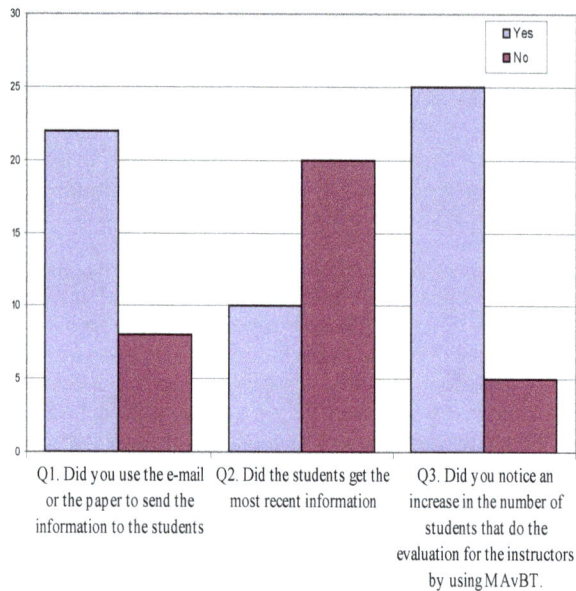

Figure 5.   Evaluation of MAvBT based on Instructors Responses

Fig. 5 shows the results of the second questionnaire. Most of instructors use the e-mails and the paper to communicate with their students. They were excited about the MAvBT, because it was efficient, effortless, and cheap for the student and it increased the number of students that can be reached by the instructors.

V. CONCLUSION

A wireless system which may enhance student-instructor interaction through utilization of Bluetooth technology is presented. The proposed system was based on Java standard edition (J2SE) for the server application and Java Micro (J2ME) edition for the mobile application. The MAvBT was implemented as a tool for interaction between students and educators as it offers a low-cost solution since it avoids use of complex network infrastructure. The proposed system enabled students to retrieve information at any time from anywhere. A simple user-friendly administrator web page, designed for instructor's use only, was used to assist the instructor to publish information to his/her students. Test results showed that the system provides flexible, easy and a viable solution for using mobile technology as it fulfill needs of students as well as the educators.

REFERENCES

[1]   Hopkins, B. and Antony, R. (2003) *Bluetooth for Java*. Berkeley, CA: Apress

[2]   Davidrajuh, R, "Exploring the Use of Bluetooth in Building Wireless Information Systems, " International Journal of Mobile Communications, Vol. 5, No. 1, pp. 1-10, 2007 (doi:10.1504/IJMC.2007.011486)

[3]   Idwan, S., Alramouni, S., Al-Adhaileh, M. and Al-Khasawneh, A., "Enhancing Mobile Advertising via Bluetooth Technology," International Journal of Mobile Communications, v. 6, n. 5, p. 587-597, July 2008 (doi:10.1504/IJMC.2008.019323)

[4]   Yerulshalmy, M., Ben-Zaken, O. (2004), "Mobile Phones in education: the case of mathematics," The Institute for Alternatives in Education, University of Haifa, available at: http://construct.haifa.ac.il/michalyr/cellular%20report.pdf , 2004.

[5]   M. Lytras, E. Sakkopoulos, and A. Tsakalidis, "Adaptive Mobile Web Services Facilitate Communication and Learning Internet Technologies," in IEEE Transactions on Education, special issue on "Mobile Technology for Education" edited by Keng Siau, Vol 42, No 2, pp. 208-215, 2006

[6]   Siau, K., Sheng, H. and Nah, F., "Use of Classroom Response System To Enhance Classroom Interactivity," *IEEE Transactions on Education*, Vol. 49, No. 3, pp. 398-403.,2006 (doi:10.1109/TE.2006.879802)

[7]   Direct Marketing Association (UK) Ltd , "Mobile Marketing Best Practice Guidelines," http://www.dma.org.uk/content/Pro-BestPractice.asp, 2005

[8]   Miller, M. (2001), *Discovering Bluetooth*, SYBEX Inc.

[9]   Dreamtech Software Team, *WAP, Bluetooth, and 3G Programming: Cracking the Code*, Hungry Minds, Inc., 2002

[10] Francis, Rob, and Jamsa, Kris Rob Francis, *Rescued by Active Server Pages and Asp.Net*, Onword Pr, 2002

[11] Jason Price ,and Mike Gunderloy, *Mastering Visual C# .NET*, Sybex; 1st edition (August 20, 2002)

[12] Mickey Williams , *Microsoft Visual C# .NET (Core Reference)*, Microsoft Press; (March 27, 2002)

[13] Eric Giguere, *Java 2 Micro Edition: Professional Developer's Guide*, John Wiley & Sons

[14] O'Connor , *Java 2 Micro Edition Programming*, Wiley Hungry Minds, 2002

[15] Michael Kroll, and Stefan Haustein, *Java 2 Micro Edition (J2ME) Application Development*, Pearson Education; 1st edition (June 25, 2002)

[16] SafeCom Technologies Limited, Bluetooth, User's Manual Class 1 Bluetooth USB Dongle, 2005

AUTHORS

Sahar Idwan is with the Computer Science Department in Prince Al-Hussein Bin Abdullah II of Information Technology, Hashemite University, P.O. Box 150459, Zarqa 13115, Jordan (E-mail: sahar@hu.edu.jo; sahar700@hotmail.com)

22

Computer-Based Wireless Advertising Communication System (CBWACS)

Yahya S. H. Khraisat, Anwar Al-Mofleh

Al-Balqa Applied University/ Al-HusonUniversity College, Amman, Jordan

Abstract—In this paper we developed a computer based wireless advertising communication system (CBWACS) that enables the users to advertise the information they want to display from their own offices to the screen in front of the customer via wireless communication system. This system consists of two PIC microcontrollers, transmitter, receiver, LCD, serial cable and antenna. The main advantages of the system are the wireless structure and the digital communication techniques. So this system is less susceptible to noise and other interferences.

Index Terms—Computer; Microcontroller; Amplitude Shift Keying; Transmitter; Receiver; antenna.

I. INTRODUCTION

The idea of CBWACS came from the need to visualize some certain information in public by typing it in the user's display and sending it via wireless system to main display. The system provides a kind of dialog and interaction between the company and its customers.

Banks for example, may want to improve their customer services by putting a large, common and well seen monitor to display and inform their clients of when the bank closes, opens, what services are available and the newest released offers.......etc.

From what had been described we can see that CBWACS shows us a clear view of what the future would look like; by its capability of establishing an interaction between companies and their clients, that save time and money.

Systems based on analogue communication techniques are more sensitive to the interferences. So we designed our system to have more handling capability to noise by modulating the base-band signals to band-pass signals and transmit the data using digital techniques. The advantages of digital communications are:

1. Digital signals are easily represented (ones or zeros) compared with analogue ones.
2. Digital circuits are less subject to distortion and interference unlike analogue circuits, because binary digits operate in one of two cases i.e. fully on or fully off.
3. More reliable and can be produced at lower cost than analog circuits.

A. Structure of CBWACS

The structure of a CBWACS system consists of the following essential components: two PIC microcontrollers, transmitter, receiver, LCD, serial cable, and antenna.

We will discuss each component in the next sessions.

CBWACS structure consists of two parts:

1. transmitting part (PC, MAX 232C, PIC, transmitter, antenna)
2. receiving part (antenna, receiver, PIC, LCD)

Figure 1 represents the block diagram of our wireless advertising system

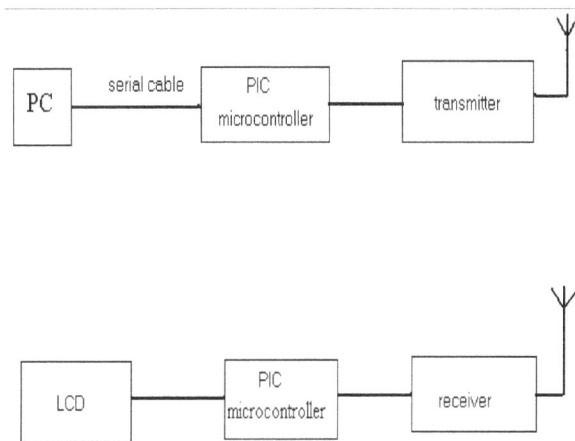

Figure 1. Block diagram of Wireless Advertising System

B. Data Processing Overview

The system has two parts transmitting and receiving. We shall view the data processing for each element in the system [1]:

The transmitting part of CBWACS does the following:

1. Receive the textual message from the serial cable in ASCII code (0-12 volt).
2. MAX232C interfaces the incoming signals to PIC (0-5 volt).
3. The PIC microcontroller converts the ASCII bits into pulse width modulating (PWM).
4. The output of the PIC is fed to the transmitter.
5. The transmitter sends the modulated data to the receiving part of the system through the antenna.

The receiving part of CBWACS does the following:

1. When the RF signal is picked up from space by a receiving antenna, a voltage is induced into the antenna (a conductor). The RF voltages induced into the receiving antenna are then passed into the receiver and converted back into the transmitted RF information.
2. The receiver received band-pass signals that were transmitted and recovered the base-band signal. This signal is then fed to the PIC microcontroller.

3. The PIC microcontroller converts the incoming PWM back to their original forms.

4. The PIC microcontroller then sends the message signal in ASCII code to the LCD.

5. The LCD displays the received data.

In other words the first PIC microcontroller programmed to convert its input serial data from ASCII into PWM, and send it to the transmitter. The second PIC microcontroller is programmed to convert the received waveform back into its basic pulse representation (PWM) then its code contains some instructions to control the LCD.

The PIC in the transmitting part converts the incoming serial data in ASCII to a PWM. Every letter in the textual message has its own separate identifying width so as to minimize the bit error probability in the detection process at the receiving part of the system. Then those PWM sequences are sent to the receiving part of the system through the transmitter after modulating it to a band-pass signal in the form of ASK –amplitude shift keying-.

The receiving part of our system which has other PIC microcontroller receives whatever the transmitter sends. The demodulation/detection process takes place to reconstruct the original message signal. Then the PIC converts the incoming serial bits into its original form (extracting the letters). The letters representing the user textual message are fed to the LCD screen.

II. TRANSMITTING PART DATA PROCESSING

The transmitting part of the system consists of the computer, MAX232C, PIC, and the transmitter. The transmitter processes the data through its terminals from input to output in four separated stages: computer-MAX232C, MAX232C-PIC, PIC-transmitter and transmitter-antenna).

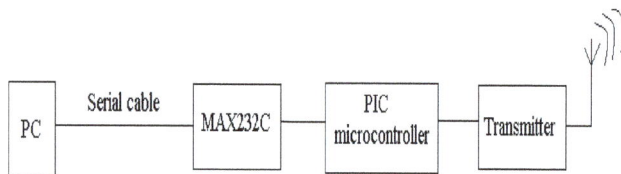

Figure 2.   Transmitting part components

The primary subject in this session is the processing of the message signal. The block diagram which represents the stages at which the data would be flown within the transmission part of the CBWACS system is shown in figure 3.

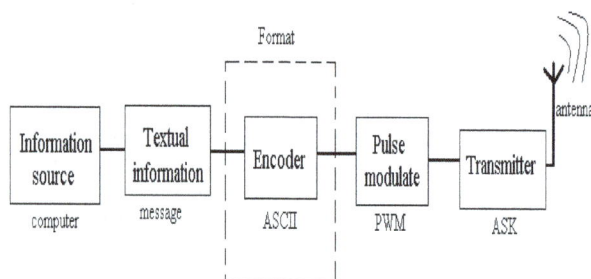

Figure 3.   Overall data processing in the transmitting part

A.   Formatting Textual Data (Character Coding)

The goal of the first essential signal processing step is to insure that the message (or source signal) is compatible with digital processing. Textual information is transformed into binary digits by the computers' CPU.

The computers' keyboard is the information source of the system that provides a textual data entrance to the computer system. It could be encoded by the ASCII code into binary sequence [2].

The computer sees the user's message as a sequence of binary ones and zeros formed of every letter ASCII code representation. The USART terminal of the MIKROBASIC attaches that sequence to the computer transmitting unit that is the computers serial port. This port sends the message binary sequence through the serial cable with amplitudes of 0-12 volts.

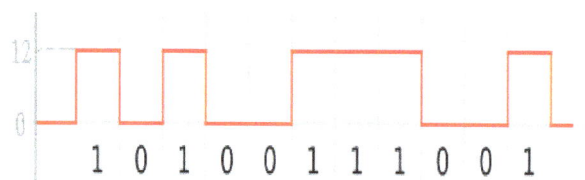

Figure 4.   USART terminal

Then the sequence goes to the MAX232C, which is the interface of the PIC microcontroller. MAX232C converts the incoming sequence amplitudes from (0-12 volts) to (0-5 volts) so as PIC could handle and process it.

Figure 5.   Pulse modulation

Conversion from a bit stream to a sequence of pulse waveforms takes place in the block labelled as pulse modulation.   As the message flows in the system, arriving to the first PIC; the sequence is read in the following manner: the PIC microcontroller reads the sequence letters letter by letter. This message stored in the PIC microcontroller, RAM in parallel. Then the PIC microcontroller reads and processes the sequence by its algorithm. The algorithm was designed to give the letters that represent the message a unique width. In other words the PIC microcontroller's algorithm is designed to do a kind of modulation known as PWM, involving varying each letter duty cycle to discriminate it from other letters in the message.

Figure 6.   Pulse-Width Modulation [3]

The width that every letter occupies is affected by the baud rate that the transmitter used. After the PIC microcontroller had modulated the message pulses, it would be sent to the transmitter serially. The PIC microcontroller final duty is to perform a parallel to series conversion.

B. Data Transmission

The transmitter is an electronic device which with the aid of an antenna propagates an electromagnetic signal. The transmitter converts incoming base-band signal to a band-pass signal which will be transmitted to the receiver by a kind of band-pass modulation technique known as ASK –amplitude shift keying. ASK is a form of band-pass modulation which represents digital data as variations in the amplitude of a carrier wave [3].

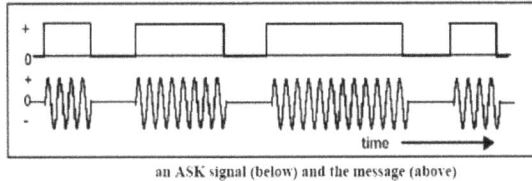

an ASK signal (below) and the message (above)

Figure 7. Amplitude Shift Keying

The amplitude of an analog carrier signal varies in accordance with the bit stream (modulating signal), keeping frequency and phase constant. The level of amplitude can be used to represent binary logic 0s and 1s. We can think of a carrier signal as an ON or OFF switch. In the modulated signal, logic 0 is represented by the absence of a carrier, thus giving OFF/ON keying operation and hence the name (OOK) was given. The transmitter gives a maximum baud rate of 40 kbps, but as we sent a textual data we satisfied with a less baud rate than what the transmitter gives. If we use transmitter at 1 kbps, the letter "a" will need (0-1) ms to be transmitted. Adding some more width to the next letter say 0.5 ms, the letter "z" have the longest time that is approximately around 0.03 second. Assuming that the message has a hundred letters all of them are "z", then the message needs three seconds to appear on the receiving part display unit. This is a very long time that the users are unhappy to see. So we used the transmitter baud rate at a higher range. If we make the transmitter baud rate 20 kbps, the letter "a" needs a 0,5 microseconds, and thus the letter "z" needs 0.13 microseconds plus some interval gab to reduce the errors, say 0.14 microseconds. So if the message contains a hundred letters all of are "z", it needs 14 microsecond to be transmitted, which is better from the fist one.

Finally, the transmitter sends its band-pass modulating signal to the antenna. The antenna is used either for radiating electromagnetic energy into space or for collecting it from space. The use of an antenna improves the range of the RF link and reduces the susceptibility to noise. The choice of antenna and its positioning can affect the performance of the RF link. The best place to mount the antenna is outside the box away from all sources of interference, metal items and ground planes.

III. Receiving Part Data Processing

The receiving part of the system has the following parts:

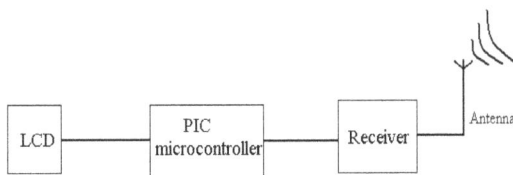

Figure 8. Receiving part components

We are interested in recovering the original message signal. Since band-pass model of the detection process is virtually identical to the base-band model, the final detection step takes place, according to the equivalence theorem which states:

Performing band-pass linear signal processing followed by heterodyning the signal to base-band, yield the same result as heterodyning the band-pass signal to base-band, followed by base-band linear signal processing [4].

The term "heterodyning" refers to a frequency conversion or mixing process that yields a spectral shift in the signal.

The received signal will be processed in the receiving part of the system as shown in the following block diagram.

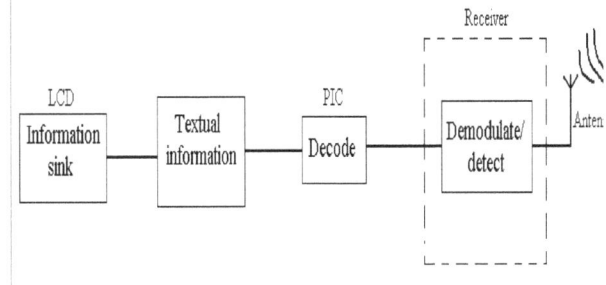

Figure 9. The over all data processing in the receiving part

When the RF signal is picked up from space by a receiving antenna, a voltage is induced into the antenna. The RF voltages induced into the receiving antenna are then passed into the receiver and converted back into the transmitted RF information. In other words on the receiving part, electromagnetic energy is converted into electrical energy by the antenna and fed into the receiver.

A. Demodulation and Detection

The channel typically causes the received pulse sequence to suffer from inter symbol interference (ISI) and thus appears as smeared signal, not quite ready for sampling and detection. That is the receiver signal is distorted, and the receiver will deal with that smeared signal. The receiver has to fix and sample it in a right way so as to recover the original message signal.

The goal of the demodulator (receiving filter) is to recover a base-band pulse with the best possible of signal-to-noise ratio (SNR), free of any ISI. Equalization is a technique used to help accomplishing this goal. The equalization process is not required for every type of communication channel.

However, since equalization embodies a sophisticated set of signal-processing techniques, making it possible to compensate for channel induced interface, it is an important area for many systems.

During a given signalling interval T, a binary base-band system will transmit one of two waveforms, denoted g1 (t) and g2 (t). Similarly, a binary band-pass system will transmit one of two waveforms, denoted s1 (t) and s2 (t).

Since the general treatment of demodulating and detection are essentially the same for base-band and band-pass systems, we used si (t) here as a generic designation for a transmitted wave form.

Then any binary transmitted signal over a symbol interval (0, T) is represented by

$$Si(t) =$$
$$s1(t) \quad , 0 < t < T \quad \text{for binary 1}$$
$$s2(t) \quad , 0 < t < T \quad \text{for binary 0}$$
$$(1)$$

The received signal r (t) is degraded by the noise n (t) and by the impulse response of the channel hc (t). r (t) is written as

$$r(t) = si(t) * hc(t) + n(t) \qquad (2)$$

Where n (t) is here assumed to be a zero mean additive white Gaussian noise AWGN process, and * represents the convolution operation.

Figure 10 shows the typical demodulation/detection function of a digital receiver.

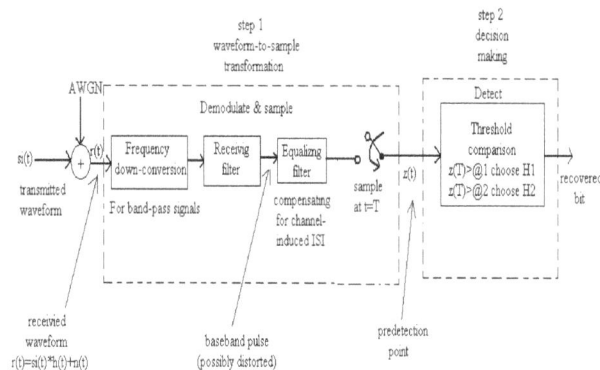

Figure 10. The two basic steps in the demodulation/detection of digital signals

We define demodulation as recovery of a waveform (to an undistorted base-band pulse). We designated detection as the mean of decision-making process of selecting the digital meaning of the waveform.

The frequency down-conversion block, shown in the demodulator portion of the figure 10, performs frequency translation for band-pass signals operating at some radio frequency (RF).

The receiving filter (essentially the demodulator) performs waveform recovery in preparation for the next important step-detection. The filtering and the channel at the transmitter typically cause the received pulse sequence to suffer from ISI, and thus it is not quite ready for sampling and detection.

The goal of the demodulator (receiving filter) is to recover a base-band pulse with the best possible signal-to-noise ratio (SNR), free of any ISI. The optimum receiving filter for accomplishing this is called a matched filter or correlator [5]. An optional equalizing filter follows the receiving filter. This filter is only needed for those systems where channel induced ISI can distort the signals.

The received filter and the equalizer filter are shown separately to indicate the importance of their functions in demodulating the signals. In most cases, however, when an equalizer is used, a single filter would be designed to incorporate both functions and thereby compensate for the distortion caused by both the transmitter and the channel. Such composite filter is sometimes referred to simply as the equalizing filter or the receiving and equalizing filter.

In the demodulation/detection process there are two steps namely step1 and step 2. In step 1, the waveform-to-

sample transformation, is made up of the demodulator followed by a sampler. At the end of each symbol duration T, the output of the sampler yield a sample Z(T) called the test statistics. Z (T) has a voltage value proportional to the energy of the received symbol and that of the noise [4].

In step 2 a decision (detection) is made regarding the digital meaning of that sample. We assume that the input noise is a Gaussian random process and that the receiving filter in the demodulator is linear. A linear operation performed on a Gaussian random process produces a second Gaussian random process thus; the filter output noise is Gaussian. The output of step 1 yields the test statistics

$$Z (T) = ai (T) + n0 (T) \qquad i = 1, 2 \qquad (3)$$

Where ai (T) is the desired signal component, and n0 (T) is the noise component. The noise component is a zero mean Gaussian random variable, and thus z (T) is a Gaussian random variable with a mean of either a1 or a2 depending on whether a binary one or zero was sent. The probability density function (PDF) of the Gaussian random noise n0 is p(z/n0), like the PDFs of s1 and s2.

The last two conditional pdfs are illustrated in the next figure. The right most conditional PDF, p (z/s1) called the likelihood of s1 which illustrates the probability density function of the random variable z (T), given that symbol s1 was transmitted. Similarly, the left most conditional PDF, p(z/s2) called the likelihood of s2, illustrate the probability density function of the random variable z (T), given that symbol s2 was transmitted.

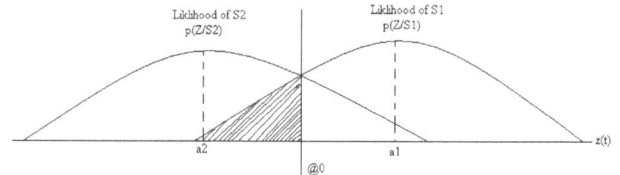

Figure 11. Conditional probability density functions: p (Z/S1) and p (Z/S2)

After the received wave form has been transformed to a sample, the actual shape of the wave form is no longer important. All waveform types that are transformed to the same value of z (T) are identical for detection purpose. Later it will be shown that the matched filter (receiving filter) maps all signals of equal energy into the same point z(T), therefore the receiving signal energy not its shape is the important parameter in the detection process. This is why the detection analysis for base-band signals is the same as that for band-pass signals. Since z (T) a voltage signal that is proportional to the energy of the received symbols, the larger the magnitude of z (T), the more error frees the decision-making process.

In step 2, detection is performed by choosing the hypothesis that result from the threshold measurement

$$Z (T) > @ \qquad \text{choose H1} \qquad (4)$$

$$Z (T) < @ \qquad \text{choose H2}$$

When H1 and H2 are the two possible (binary) hypotheses, the inequality relationship indicates that hypothesis H1 is chosen if z (T)>@, and hypothesis H2 is chosen if z (T) <@, if z (T) =@, the decision can be an arbitrary one. Choosing H1 is equivalent to deciding that s1 (t) was sent and hence a binary one is detected. Similarly, Choosing H2 is equivalent to deciding that s2 (t) was sent and hence a binary zero is detected.

1) The Matched Filter

A matched filter [5] is a linear filter designed to provide the maximum signal-to-noise power ratio at its output for a given transmitted symbol waveform. Consider that a known signal s(t) plus AWGN n(t) are the inputs to a linear, time invariant (receiver) filter followed by a sampler, as shown in figure 10. At time t=T, the sampler output z (T) consists of a signal component ai and a noise component n0. Manipulation for the impulse response of the matched filter yields the following

$$H(t) = \begin{cases} Ks(T-t) & 0 \leq t \leq T \\ 0 & elsewhere \end{cases} \qquad (5)$$

Thus, the impulse response of a filter that produces the maximum output signal-to-noise ratio is the mirror image of the message signal s (t), delayed by the symbol time duration T. Without the delay the response s (-t) is unrealizable because it describes a response as a function of a negative time. Figure 12 illustrates the matched filter's basic property: the impulse response of the filter is a delayed version of the mirror image (rotated on the t = 0 axis) of the signal waveform.

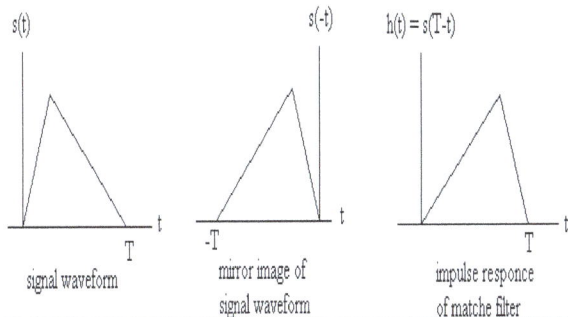

Figure 12. Matched filter characteristics

B. Decoding the Recovered (Base-Band) Signal

The original base-band signal (Textual information) is sent to the PIC microcontroller bin serially and stored in the PIC microcontroller RAM. Then PIC microcontroller (No.2) deals with the arriving signal with its algorithm in the following manner:

The algorithm was written in order to decode the signal from its binary form to its basic meaning. The sequence that is letters, and sends them to the display unit of our system, that is the LCD. The PIC microcontroller converts the serial message signal into parallel so as to deal with each letter separately.

As the ASCII code gives a unique pulse representation for each letter, we had modulated those pulses in the PIC microcontroller with the PWM. ASCII code gives each letter a unique width to discriminate it from the other letters in the message in order to reduce the probability of errors and ISI. Now the PIC microcontroller algorithm recovers each letter with its determined width in the transmitting part plus extra small width. To insure that we recovered the exact transmitted letter without any errors at all. We could consider that process as a form of error immunity for our system.

C. Passing Textual Information-to-Information Sink

After the PIC microcontroller's algorithm had finished its duty, the original message combining all the letters would be recovered. The algorithm sends that message to its output bin serially to the LCD. The mikrobasic is software that we used to program the PICs.

IV. CBWACS COMPONENTS CONNECTION

In this paragraph we represent the necessary hardware which we used. We will talk about the necessary connection to support the system software. In part one, we showed that the system contains two parts, transmission part and receiving part. Now we will demonstrate the components of each part.

A. Transmission Part

The transmission part has the following components

1. Power supply
2. Oscillator
3. Rest
4. RS 232 interface
5. PIC 16F877A
6. Transmitter

1) Power supply

The power supply is needed in the two parts (transmission and receiving). The input power supply is applied to systems that have voltage range of V equal 15Vor less than 13V and its maximum current is 20mA. We can choose any value in this range to get 5V from the regulator, as shown in figure 12 [6].

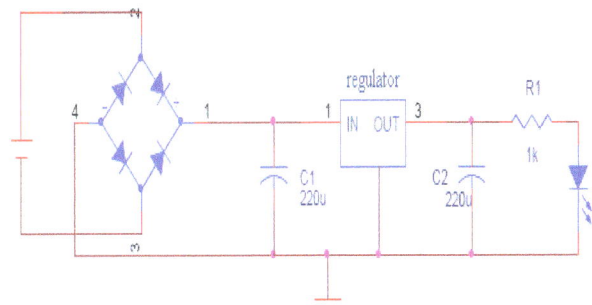

Figure 13. Power Supply Circuit

Generally microcontrollers are based on CMOS technology.

The operating voltage is connected to Vdd (pin #11).

In the transmitter the operating voltage 5V is connected to pin #4, in max232 the 5V is connected to pin #2, and pin #16. Pin #2 is connected to a capacitor to keep the voltage stable.

2) Oscillator

We used crystal oscillator for the PIC16F877A at 4 MHZ. The oscillator used as an external clock. Increasing the clock speed is proportional to increasing the size of the incoming data. The steps needed to complete an instruction are 4 clocks. The 4 clocks are equal to one cycle. So to get the instruction speed, the four clocks are divided by 4MHZ equals 1us (1us is the duration to complete the instruction) [7].

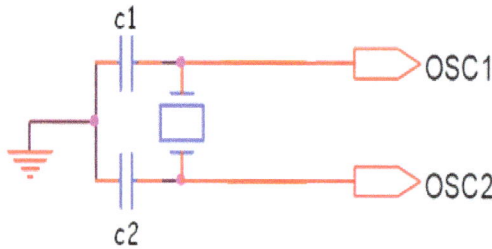

Figure 14. Oscillator circuit

The smoothing capacitors C1 and C2 are used to reject any attenuation.

3) Reset

Reset is used to restart the downloaded program in PIC, in other words it is used to initiate the external clock. A click switch is connected to the _MCLR pin (1st pin in the PIC). After polling it up with a 10KΩ resistance and smoothed by 0.1uF capacitor. R2 is used to reduce the back current form capacitor. It is required that you consider the Reset button in our design.

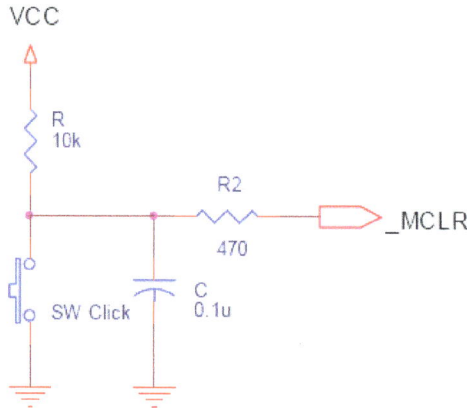

Figure 15. Reset circuit

4) RS 232 Interface

The binary pulses which came from the serial port had the amplitudes 0 and +12 volts. The PIC microcontrollers don't deal with 0 and +12 volts, they are dealing with 0 and +5 volt. So we used max232c to convert the incoming pluses to 0 and +5 volts. So RS 232 interface, fig15, supports the USART module.

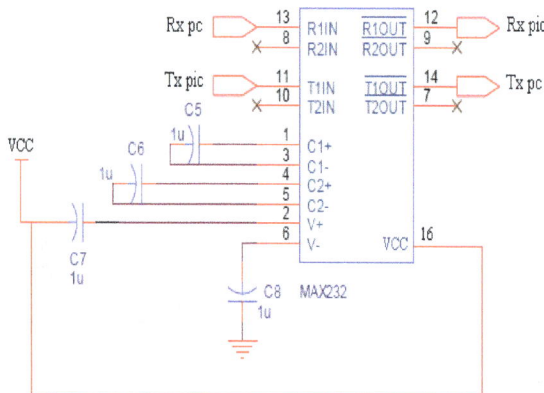

Figure 16. The (RS232 interface) circuit

5) PIC 16F877A

PIC is a family of RISC microcontrollers made by Microchip Technology. PIC stands for "Peripheral Interface Controller" Or "Programmable Intelligent Computer". We are interesting in the 16F877A type of the wide range of the PIC microcontrollers, which is amid range family of a various families of PIC microcontroller that consist of 40 pins. The F in the name 16F877A generally indicates that the PIC microcontroller uses flash memory and can be erased electronically. The number 16 in front of a name shows the family range that this PIC belongs to (low, medium, high). The last three numbers refer to a certain features the PIC have. The last letter A shows the category of improvement that was done to the PIC.

6) Transmitter

In this work we used a 434 MHZ OOK (on-off keying) AM transmitter with external antenna. Fig 16 illustrates the connection between transmitter and PIC.

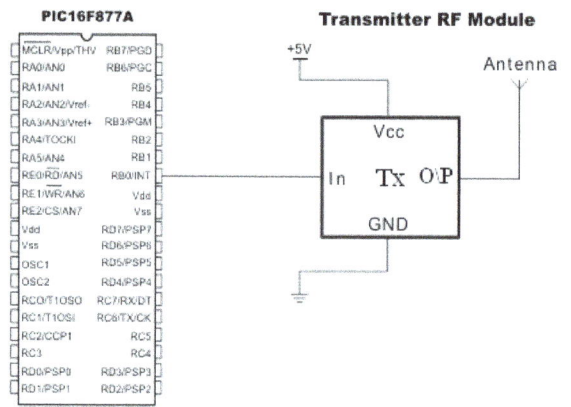

Figure 17. Transmitter connection

7) Transmitting Part Overall Connection

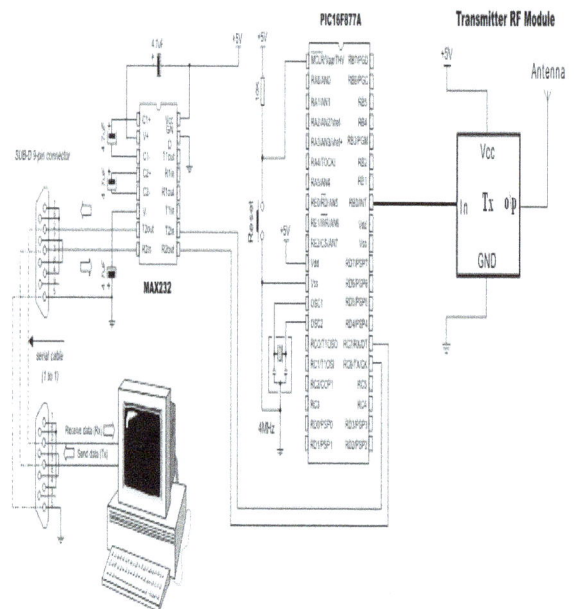

Figure 18. Transmitting part

B. Receiving Part

This part has the following components:

1. Power supply
2. Oscillator
3. Reset
4. PIC 16F877A
5. Receiver
6. LCD (16x2)

The power supply circuit, oscillator circuit, and Reset circuit are the same as what we have discussed in the transmission part.

1) Receiver

Fig 19 illustrates the hardware connection of this receiver.

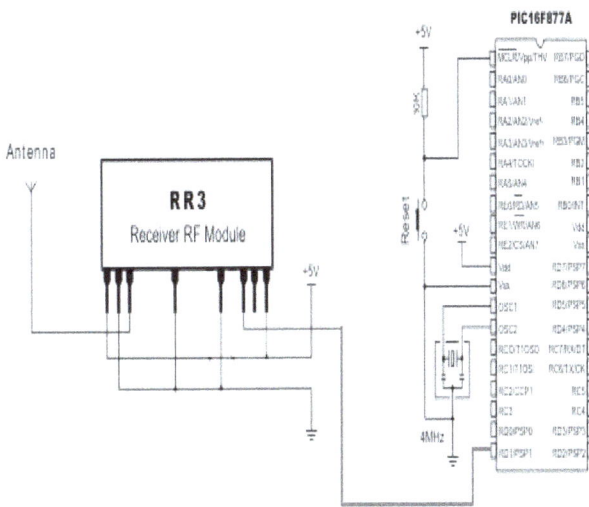

Figure 19. Receiver connection

2) LCD (Liquid Crystal Display)

A liquid crystal display (LCD) is a thin, flat display device made up of any number of arrayed pixels. It is prized by engineers because it uses very small amounts of electric power. LCD contains a small processor and small memory that is controlled by the LCD. There are two kinds of LCD. The first uses a serial data bus and the other uses a parallel data bus. In our work we used parallel data bus because it's simple and fast read\write. We will use the LCD to view the output. The LCD that we used is similar to the one in the previous picture. It has two rows and 16 characters, as shown below.

Figure 20. A general purpose alphanumeric LCD, with two lines of 16 characters

The hardware connection of LCD with PIC is shown in figure 21.

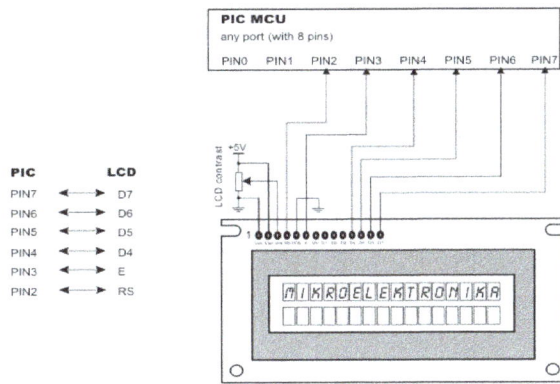

Figure 21. LCD connection

We can write and read from LCD by using two modes; eight bit mode and four bit mode. The general mode to be displayed is eight bits long and is sent to the LCD either four or eight bits at a time. If four bits mode is used, two nibbles of data are sent to make up a full eight bits transfer (sent high four bits and then low four bits with an E clock pulse with each nibbles). This mode has the disadvantage of making errors. So we used the four bit mode that is a reliable method to accomplish our goal. For this work there is no reason to read from LCD, therefore the R/W usually tied to ground. We used the potentiometer wired as a voltage divider to specify the contrast of the character on the LCD screen.

3) Receiving Part Overall Connection

The general circuit for the receiving part is shown in figure 23.

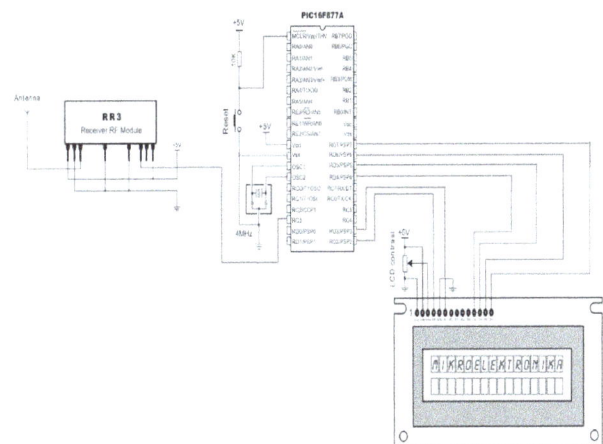

Figure 22. Receiving Part

REFERENCES

[1] Yahya S. H. Khraisat, Anas Al-Kurdi, Nimer Al-Shalabi and Salah-Eddin Mahasneh, " Local Area Mobile - System Communication (LAMS)", 3rd International conference on Interactive Mobile and Computer Aided Learning IMCL 2008, Princess Sumaya University for Technology, 16-18 April 2008, Amman, Jordan.

[2] Thomas L. Floyd, Digital Fundamentals, Prentice Hall, Eight edition, 2003.

[3] Haykin, S., Communication Systems, John Wiley & sons, Inc., New York, 2000.

[4] Sklar, B., Digital Communications Fundamentals and Applications, Prentice Hall P T R, New   Jersey, 2001.

[5] Merrill I. Skolnik, Introduction to Radar System, 3rd edition. New York: McGraw-Hill, 2001

[6]   Thomas L. Floyd, Electronic Devices, Prentice Hall, Six edition, 2003.

[7]   Neaman, Microelectronics Circuit Analysis and Design, Mc Graw Hill Inc, 2007.

[8]   Millman, J., Microelectronics, Mc Graw Hill Inc, 1987.

[9]   AUR.EL, Wireless System, Italy, available: www.aurel.it.

AUTHORS

Yahya S. H. Khraisat is an associated professor of the Electrical and Electronics Department, Al-Balqa Applied University/ Al-Huson University College, Jordan, P.O. Box 1375, Irbed 21110. (E-mail: yahya@huson.edu.jo).

Anwar Al-Mofleh is an engineer of the Electrical and Electronics Department, Al-Balqa Applied University/ Al-Huson University College, Jordan, P.O. Box 1375, Irbed 21110. (E-mail: Anwar1971@yahoo.com).

23

Mobile Education: Towards Affective Bi-modal Interaction for Adaptivity

E. Alepis[1], M. Virvou[1] and K. Kabassi[2]

[1] University of Piraeus, Piraeus, Greece

[2] TEI of the Ionian Islands, Zakynthos, Greece

Abstract—One important field where mobile technology can make significant contributions is education. However one criticism in mobile education is that students receive impersonal teaching. Affective computing may give a solution to this problem. In this paper we describe an affective bi-modal educational system for mobile devices. In our research we describe a novel approach of combining information from two modalities namely the keyboard and the microphone through a multi-criteria decision making theory.

Index Terms—Educational systems, affective computing, mobile systems, adaptive systems .

I. INTRODUCTION

In the last decade, there is a growing interest in mobile technology and mobile networks. As a result, a great number of services are offered to the users of mobile phones including education. In the fast pace of modern life, students and instructors would appreciate using constructively some spare time that would otherwise be wasted [12]. For example, when they are travelling or even when they are waiting in queue. Students and instructors may have to work on lessons at any place, even when away from offices, classrooms and labs where computers are usually located. Assets of mobile interaction include device independence as well as more independence with respect to time and place in comparison with web-based education using standard PCs.

Mobile education may be quite impersonal since the presence of a human instructor and human co-students may not be available. A remedy to this kind of problem may be given by providing affective interaction based on the user's emotional state. The recognition of emotions can lead to affective user interfaces that take into account the users' feelings and can adapt their behavior according to these feelings. Regardless of the various emotional paradigms, neurologists/psychologists have made progress in demonstrating that emotions play an important role in the process of decision making and action deciding [1]. Moreover, the way people feel may play an important role in their cognitive processes [2]. Recently, significant research effort has been put in the recognition of emotions of users while they interact with software applications. Picard points out that one of the major challenges in affective computing is to try to improve the accuracy of recognizing people's emotions [3].

Improving the accuracy of emotion recognition may imply the combination of many modalities in user interfaces. Indeed, human emotions are usually expressed in several ways. Human faces, people's voices, or people's actions may all show emotions.

Ideally, evidence from many modes of interaction should be combined by a computer system so that it can generate as valid hypotheses as possible about users' emotions [4]. It is hoped that the multimodal approach may provide not only better performance, but also more robustness [5]. As it is stated in [6], although the benefit of fusion (i.e., audio-visual fusion, linguistic and paralinguistic fusion, multi-visual-cue fusion from face, head and body gestures) for affect recognition is expected from engineering and psychological perspectives, our knowledge of how humans achieve this fusion is extremely limited.

In previous work, the authors of this paper have implemented and evaluated with quite satisfactory results emotion recognition systems, incorporated in educational applications ([7], [8]). As a next step we have extended our affective educational system by providing mobile interaction between the users and the system. In many situations this means that learning may take place at home or some other site, supervised remotely and asynchronously by a human instructor but away from the settings of a real class.

The main characteristic of the proposed mobile system is that it combines evidence from two modes, namely the mobile device's microphone and the keyboard, in order to identify the users' emotions. The results of the two modes are combined through a multi-criteria decision making method. More specifically, the system uses Simple Additive Weighting (SAW) [9] for evaluating different emotions, taking into account the input of the two different modes and selects the one that seems more likely to have been felt by the user. In this respect, emotion recognition is based on several criteria that a human tutor would have used in order to perform emotion recognition of his/her students during the teaching course.

In view of the above, in this paper we describe a novel mobile educational system that incorporates bi-modal emotion recognition through a multi-criteria theory. The two modes of interaction are the interaction through the mobile device's keyboard and the interaction through the users' voice. Users may use their mobile device in order to read parts of the theory about a particular lesson, as well as to take tests. In the case of the sort examinations about particular lessons, users may write their answers directly to their mobile device through the keyboard, or in other cases they may use the mobile device's microphone as a mode of interaction. The proposed system collects evidence from the two modes of interaction and analyses it in terms of criteria for emotion recognition. The main focus of the paper is on presenting the empirical studies that are essential for the application of a multi-criteria model, which is used for making final assumptions about the recognition of one or more than one emotional states.

II. AFFECTIVE INTERACTION IN MOBILE DEVICES

After a thorough investigation in the related scientific literature we found that there is a shortage of educational systems that incorporate multi-modal emotion recognition. Even less are the existing affective educational systems with mobile facilities. In [10] a mobile context-aware intelligent affective guide is described, that guides visitors touring an outdoor attraction. The authors of this system aim mainly at constructing a mobile guide that generates emotions. On the contrary our proposed system aims at recognizing the users' emotions through their interaction with a mobile device rather than generating emotions. As a second related approach we found that Yoon et al. [11] propose a speech emotion recognition agent for mobile communication service. This system tries to recognize five emotional states, namely neutral emotional state, happiness, sadness, anger, and annoyance from the speech captured by a cellular phone in real time and then it calculates the degree of affection such as love, truthfulness, weariness, trick, and friendship. In their approach only data from the mobile device's microphone are taken into consideration, while in our research we investigate a mobile bi-modal emotion recognition approach. Moreover, our proposed system is incorporated in an educational application and data pass through a linguistic and also a paralinguistic stage of analysis. This derives from the fact that in an educational application we should take into consideration how users say or type something (such as low or high voice, slow or quick typing speed), as well as what users say or type (such as correct answers or mistakes).

III. OVERVIEW OF THE SYSTEM

The main architecture of the mobile bi-modal emotion recognition system is illustrated in figure 1. Participants were asked to use their mobile device and interact with a pre-installed educational application. Their interaction could be accomplished either orally (through the mobile device's microphone) or by using the mobile devices keyboard and of course by combining these two modes of interaction. All data are captured during the interaction of the users with the mobile device through the two modes of interaction and then transmitted wirelessly to the main server. All the input actions are used as trigger conditions for emotion recognition by the emotion detection server. Finally all input actions as well as the possible recognized emotional states are stored in the system's database.

The discrimination between the participants is done by the application that uses the main server's data base and for each different user a personal profile is created and stored in the data base. In order to accomplish that, user name and password is always required to gain access to the mobile educational application.

A snapshot of a mobile emulator, operated by a participant is illustrated in figure 2. Users may answer questions and take tests using the mobile system. They may write their answers through the mobile device's keyboard, or alternatively give their answers orally, using

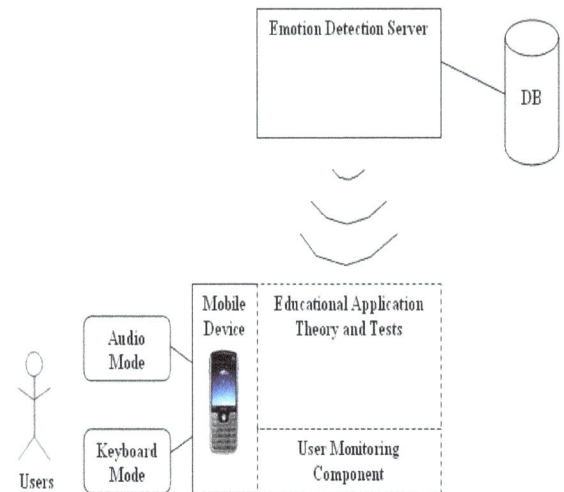

Figure 1.   Architecture of the mobile emotion detection system.

Figure 2.   A user is answering a question of a test either using the keyboard or orally through the mobile device's microphone.

their mobile device's microphone. In both cases, the data from the two possible modes of interaction are stored in the main system's database (emotion detection server), in order to be processed for emotion recognition purposes. When participants answer questions, the system tries to perform error diagnosis in cases where the participants' answers have been incorrect. Error diagnosis aims at giving an explanation about a participant's mistake taking into account the history record of the participant and the particular circumstances where the error has occurred.

I. EMPIRICAL STUDY

The empirical study that we have conducted concerns the audio-lingual emotion recognition, as well as the recognition of emotions through keyboard evidence. The audio-lingual mode of interaction is based on using a mobile device's microphone as input device. The empirical study aimed at identifying common user reactions that express user feelings while they interact with mobile devices. As a next step, we associated these reactions with particular feelings.

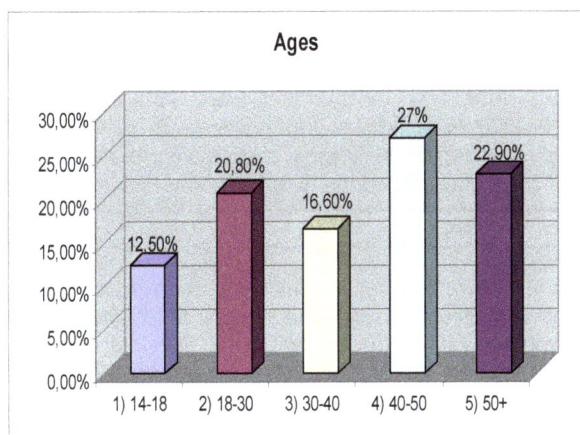

Figure 3. Distribution of the ages of the participants

The empirical study involved a total number of 200 male and female users of various educational backgrounds, ages and levels of familiarity with computers. Individuals' behavior while doing something may be affected by several factors related to their personality, age, experience, etc. Figure 3 illustrates the distribution of participants in the empirical study in terms of their age. In particular there were 12,5 % of participants under the age of 18, approximately 20% of participants between the ages of 18 and 30. A considerable percentage of our participants was over the age of 40.

The participants were asked to use a mobile educational application which incorporated a user monitoring component. The user monitoring component that we have used can be incorporated in any application, since it works in the background, recording silently each user's input actions. Our aim was not to test the participants' knowledge skills, but to record their oral and keyboard behavior. Thus, the educational application incorporated the monitoring module that was running unnoticeably in the background. Moreover, users were also videotaped while they interacted with the mobile application.

After completing using the educational application, participants were asked to watch the video clips concerning exclusively their personal interaction and to determine in which situations they were experiencing changes in their emotional state. Then, they associated each change in their emotional state with one of the six basic emotion states in our study and the data was recorded and time-stamped.

As the next step, the collected transcripts were given to 20 human expert-observers who were asked to perform audio and keyboard emotion recognition with regard to the six emotional states, namely happiness, sadness, surprise, anger, disgust and neutral. All human expert-observers possessed a first and/or higher degree in Psychology. These experts where first asked to analyze the data corresponding to the audio-lingual input only. To this end, they were asked to listen to the video tapes without seeing them. They were also given what the user had said in printed form from the computer audio recorder. The human expert-observers were asked to justify the recognition of an emotion by indicating the weights of the criteria that they had used in terms of specific words and exclamations, pitch of voice and changes in the volume of speech.

In the case of the keyboard mode, human expert-observers analyzed the data corresponding to the keyboard input separately. Thus, they were asked to watch the video tape and were also given a print out of what the user had written as well as the exact time of each event as it was captured by the user monitoring component. Human observers were asked to justify the recognition of an emotion by indicating the weights of the criteria that they had used in terms of specific changes in the speed of typing, in the absence of typing as well as to the written context. Correct answers, wrong answers as well as the consequence of these events were analyzed as long as these events involved only the keyboard mode of interaction. Frequent use of backspace and other basic keyboard buttons was also recorded and associated with specific emotional states.

Finally after processing the data from both the human experts and the monitoring component we came up with statistical results that associated user input action through the mobile keyboard and microphone with emotional states. More specifically, considering the keyboard we have the following categories of user actions: a) user types normally b) user types quickly (speed higher than the usual speed of the particular user) c) user types slowly (speed lower than the usual speed of the particular user) d) user uses the "delete" key of his/her mobile device often e) user presses unrelated keys on the keyboard f) user does not use the keyboard. These actions are considered as criteria for the evaluation of emotion with respect to the user's action in the keyboard.

Considering the users' basic input actions through the mobile device's microphone we have 7 cases: a) user speaks using strong language b) users uses exclamations c) user speaks with a high voice volume (higher than the average recorded level) d) user speaks with a low voice volume (low than the average recorded level) e) user speaks in a normal voice volume f) user speaks words from a specific list of words showing an emotion g) user does not say anything. These seven actions are considered as criteria for the evaluation of emotion with respect to what the user says.

Concerning the combination of the two modes in terms of emotion recognition we came to the conclusion that the two modes are complementary to each other to a high extent. In many cases the human experts stated that they could generate a hypothesis about the emotional state of the user with a higher degree of certainty if they had taken into account evidence from the combination of the two modes rather than one mode. For example, when the rate of pressing the mobile device's "deletion key" of a user increases, this may mean that the user makes more mistakes due to a negative feeling. However this hypothesis can be reinforced by evidence from speech if the user says something bad that expresses negative feelings.

II.   THE MULTI-CRITERIA MODEL

The input actions that were identified by the human experts during the empirical study provided information for the emotional states that may occur while a user interacts with an educational system. These input actions are considered as criteria for evaluating all different emotions and selecting the one that seems more prevailing. More specifically, each emotion is evaluated first using only the criteria (input actions) from the keyboard and then only the criteria (input actions) from the microphone. In cases where both modals (keyboard and microphone) indicate the same emotion then the probability that this emotion has occurred is increased significantly. Otherwise, the mean of the values that have occurred by the evaluation of each emotion is calculated and the one with the higher mean is selected.

For the evaluation of each alternative emotion the system uses SAW for a particular category of users. This particular category comprises of the young (under the age of 19) and novice users (in computer skills). The likelihood for a specific emotion (happiness, sadness, anger, surprise, neutral and disgust) to have occurred by a specific action is calculated using the formula below:

$$\frac{em_{1e_{1}1} + em_{1e_{1}2}}{2} \quad \text{where}$$

$$em_{1e_{1}1} = w_{1e_{1}k1}k_1 + w_{1e_{1}k2}k_2 + w_{1e_{1}k3}k_3 + w_{1e_{1}k4}k_4$$
$$+ w_{1e_{1}k5}k_5 + w_{1e_{1}k6}k_6 \quad (1)$$

$$em_{1e_{1}2} = w_{1e_{1}m1}m_1 + w_{1e_{1}m2}m_2 + w_{1e_{1}m3}m_3 + w_{1e_{1}m4}m_4$$
$$+ w_{1e_{1}m5}m_5 + w_{1e_{1}m6}m_6 + w_{1e_{1}m7}m_7 \quad (2)$$

$em_{1e_{1}1}$ is the probability that an emotion has occurred based on the mobile keyboard actions and $em_{1e_{1}2}$ is the probability that refers to an emotional state using the users' input from the mobile device's microphone. These probabilities result from the application of the decision making model of SAW and are presented in Eqs.(1) and (2), respectively. $em_{1e_{1}1}$ and $em_{1e_{1}2}$ take their values in [0,1].

In Eq.(1) the k's from k1 to k6 refer to the six basic input actions that correspond to the keyboard. In Eq.(2) the m's from m1 to m7 refer to the seven basic input actions that correspond to the microphone. These variables are Boolean. In each moment the system takes data from the bi-modal interface and translates them in terms of keyboard and microphone actions. If an action has occurred the corresponding criterion takes the value 1, otherwise its value is set to 0. The w's represent the weights. These weights correspond to a specific emotion and to a specific input action and are acquired by the constructed database.

In order to identify the emotion of the user interacting with the mobile system, the mean of the values that have occurred using Eqs.(1) and (2) for that emotion is estimated. The system compares the values from all the different emotions and determines whether an emotion is taking effect during the interaction. As an example we give the two formulae with their weights for the two modes of interaction that correspond to the emotion of happiness when a user (under the age of 19) gives the correct answer in a test of our educational application. In case of $em_{1e_{1}1}$ considering the keyboard we have:

$$em_{1e_{1}1} = 0.4k_1 + 0.4k_2 + 0.1k_3 + 0.05k_4 + 0.05k_5 + 0k_6$$

In this formula, which corresponds to the emotion of happiness, we can observe that the higher weight values correspond to the normal and quickly way of typing. Slow typing, often use of the backspace key and use of unrelated keys are actions with lower values of stereotypic weights. Absence of typing is unlikely to take place. Concerning the second mode (microphone) we have:

$$em_{1e_{1}2} = 0.06m_1 + 0.18m_2 + 0.15m_3 + 0.02m_4 + 0.14m_5 + 0.3m_6 + 0.15m_7$$

In the second formula, which also corresponds to the emotion of happiness, we can see that the highest weight corresponds to m6 which refers to the 'speaking of a word from a specific list of words showing an emotion' action. The empirical study gave us strong evidence for a specific list of words. In the case of words that express happiness, these words are more likely to occur in a situation where a novice young user gives a correct answer to the system. Quite high are also the weights for variables m2 and m3 that correspond to the use of exclamations by the user and to the raising of the user's voice volume. In our example the user may do something orally or by using the keyboard or by a combination of the two modes. The absence or presence of an action in both modes will give the Boolean values to the variables k1…k6 and m1…m7.

A possible situation where a user would use both the keyboard and the microphone could be the following: The specific user knows the correct answer and types in a speed higher than the normal speed of writing. The system confirms that the answer is correct and the user says a word like 'bravo' that is included in the specific list of the system for the emotion of happiness. The user also speaks in a higher voice volume. In that case the variables k1, m3 and m6 take the value 1 and all the others are zeroed. The above formulae then give us $em_{1e_{1}1} = 0.4 * 1 = 0.4$ and

$$em_{1e_{1}2} = 0.15 * 1 + 0.3 * 1 = 0.45.$$

In the same way the system then calculates the corresponding values for all the other emotions using other formulae. For each basic action in the educational application and for each emotion the corresponding formula have different weights deriving from the stereotypical analysis of the empirical study. In our example in the final comparison of the values for the six basic emotions the system will accept the emotion of happiness as the most probable to occur.

III. EVALUATION

The 200 participants that were involved in the first phase of the empirical study in section 3 were also used in the second phase of the empirical study for the evaluation of the multi-criteria emotion recognition mobile system. In this section we present and compare results of successful emotion recognition in mobile audio mode, mobile keyboard mode and the two modes combined. For the purposes of our study the whole interaction of all users with the mobile educational application was video recorded. Then the videos collected were presented to the users that participated to the experiment in order to perform emotion recognition for themselves with regard to the six emotional states, namely happiness, sadness, surprise, anger, disgust and the neutral emotional state. The participants as observers were asked to justify the recognition of an emotion by indicating the criteria that s/he had used in terms of the audio mode and keyboard actions. Whenever a participant recognized an emotional state, the emotion was marked and stored as data in the system's database. Finally, after the completion of the empirical study, the data were compared with the systems' corresponding hypothesis in each case an emotion was detected.

TABLE I.
RECOGNITION OF EMOTIONS USING THE MULTI-CRITERIA THEORY.

Using Stereotypes and SAW			
Emotions	Audio mode recognition	Recognition through keyboard	Multi-criteria bi-modal recognition
Neutral	17%	32%	46%
Happiness	52%	39%	64%
Sadness	65%	34%	70%
Surprise	44%	8%	45%
Anger	68%	42%	70%
Disgust	61%	12%	58%

Table 1 illustrates the percentages of successful emotion recognition of each mode after the incorporation of modes' weights and the combination through the multi-criteria approach.

The results in table 1 indicate that the application of the multi-criteria model lead our system to noticeable improvements in its ability to recognize emotional states of users successfully.

IV. CONCLUSIONS

In this paper, we described a mobile affective educational application that recognizes users' emotions based on two modes of interaction, namely the mobile device's microphone and the keyboard. The proposed system uses an innovative approach that combines evidence from the two modes of interaction based on a multi-criteria decision making theory to improve the system's accuracy in recognizing emotions. Real users participated in the first experiment as well as human experts and aimed at revealing the criteria that are taken into account for emotion recognition in mobile devices.

For the evaluation of the affective mobile system a second experimental study with the participation of human experts was conducted and its results revealed noticeable improvements in the proposed system's ability to recognize emotional states of users.

The results of these experiments also showed that the criteria used for emotion recognition in mobile devices were similar to the criteria that were identified by other experimental studies [7] for emotion recognition in personal computers. As a result, these findings may also be used by other systems that perform emotion recognition in other domains. However, what may differ in the application of these criteria in other systems is their weight of importance in emotion recognition. In such a case, the setting of the second experiment should be repeated for the new domain.

REFERENCES

[1] E. Leon, G. Clarke, V. Gallaghan, F. Sepulveda, "A user-independent real-time emotion recognition system for software agents in domestic environments", Engineering applications of artificial intelligence, 20 (3): 337-345, 2007. (doi:10.1016/j.engap pai.2006.06.001)

[2] D. Goleman, "Emotional Intelligence", Bantam Books, New York 1995.

[3] Picard, R.W., "Affective Computing: Challenges", Int. Journal of Human-Computer Studies, Vol. 59, Issues 1-2, 55-64, 2003.

[4] C. Busso, Z. Deng, S. Yildirim, M. Bulut, C. M. Lee, A. Kazemzadeh, S. Lee, U. Neumann, and S. Narayanan, "Analysis of emotion recognition using facial expressions, speech and multimodal information", In Proceedings of the 6th international Conference on Multimodal interfaces (State College, PA, USA, October 13 - 15, 2004). ICMI '04. ACM, New York, NY, 205-211, 2004.

[5] M. Pantic, L.J.M. Rothkrantz, "Toward an affect-sensitive multimodal human-computer interaction", Vol. 91, Proceedings of the IEEE 1370-1390, 2003.

[6] Z. Zeng, M. Pantic, G. I. Roisman, and T. S. Huang, "A survey of affect recognition methods: audio, visual and spontaneous expressions", In Proceedings of the 9th international Conference on Multimodal interfaces (Nagoya, Aichi, Japan, November 12 - 15, 2007). ICMI '07. ACM, New York, NY, 126-133, 2007.

[7] E. Alepis, M. Virvou, and K. Kabassi, "Knowledge Engineering for Affective Bi-modal Human-Computer Interaction", Sigmap, 2007.

[8] E. Alepis, and M. Virvou, "Emotional Intelligence: Constructing user stereotypes for affective bi-modal interaction", Lecture Notes in Computer Science, Volume 4251 LNAI - I, Pages 435-442, 2006.

[9] P.C. Fishburn, "Additive Utilities with Incomplete Product Set: Applications to Priorities and Assignments", Operations Research, 1967.

[10] M.Y. Lim, and R. Aylett, "Feel the difference: A guide with attitude!", Lecture Notes in Computer Science, Volume 4722 LNCS, Pages 317-330, 2007.

[11] W.J. Yoon, Y.H. Cho, and K.S. Park, "A study of speech emotion recognition and its application to mobile services", Lecture Notes in Computer Science, Volume 4611 LNCS, Pages 758-766, 2007.

[12] M. Virvou, and E. Alepis, "Mobile educational features in authoring tools for personalised tutoring" Computers and Education, 44 (1), 2005, Pages 53-68, 2005.

AUTHORS

E. Alepis is with the University of Piraeus, Department of Informatics, Piraeus 18534, Greece. He is a Ph.D student in the Department of Informatics. He received a first degree in Informatics in 1998. He has authored over 25 articles, which have been published in international conferences, books and journals. He has served as reviewer in international conferences. His research interests are in the areas of Affective Computing, E-learning, M-learning, User Modeling and Human Computer Interaction (e-mail: talepis@ unipi.gr).

M. Virvou, is with the University of Piraeus, Department of Informatics, Piraeus 18534, Greece. She is an Associate Professor in the Department of Informatics, University of Piraeus, Greece. She received a degree in Mathematics from the University of Athens, Greece (1986), a M.Sc. (Master of Science) in Computer Science from the University of London (University College London), U.K.(1987) and a D.Phil. from the School of Cognitive and Computing Sciences of the University of Sussex, U.K.(1993). She is the sole author 3 computer science books ("Object Oriented Software Engineering", "Object Oriented Analysis and Design" and "Introduction to Compilers"). She has authored or co-authored over 150 articles, which have been published in international journals, books and conferences. The international journals and conferences where she has published articles are in the areas of Knowledge based Software

Engineering, Web-based information systems, Computers and Education, Personalization Systems, Human Computer Interaction, Student/User Modeling. She has served as a member of Program Committees and/or reviewer of International journals and conferences. She has supervised or currently supervising 12 Ph.D.s in the areas of web-based information systems, knowledge-based human computer interaction, personalization systems, software engineering, e-learning, e-services and m-services. She has served or is serving as the project leader and/or project member in 15 R&D projects in the areas of e-learning, computer science and information systems. In the year 2003-2004 she received the first research award from the Research Center of her University as the member of faculty (among 200 colleagues of hers) having the highest number of research publications in high quality research journals. Many articles of hers have been among the top 5 most downloaded articles of the respective journals where they were published. She is a member of the IEEE. (e-mail: mvirvou@unipi.gr).

K. Kabassi is with the Technological Educational Institute of the Ionian Islands, Zakynthos, Greece. She is a Lecturer in the Department of Ecology and the Environment, Technological Educational Institute of the Ionian Islands. She received a first degree in Informatics (1999) and a D.Phil. (2003) from the Department of Information, University of Piraeus (Greece). She has authored over 40 articles, which have been published in international journals, books and conferences. She has served as a member of Program Committees and/or reviewer of International conferences. Her current research interests are in the areas of Knowledge based Software Engineering, Human Computer Interaction, Personalization Systems, Multi-Criteria Decision Making, User Modeling, Web-based Information Systems and Educational Software. (e-mail: kkabassi@teiion.gr).

24

Let's Meet at the Mobile –
Learning Dialogs with a Video Conferencing
Software for Mobile Devices

Hans L. Cycon[1], Thomas C. Schmidt[2], Gabriel Hege[2], Matthias Wählisch[2, 3], and Mark Palkow[4]

[1] FHTW Berlin, Berlin, Germany, [2] HAW Hamburg, Hamburg, Germany,
[3] link-lab, Berlin, Germany, [4] daViKo GmbH, Berlin, Germany

Abstract—**Mobile phones and related gadgets in networks are omnipresent at our students, advertising itself as** *the* **platform for mobile, pervasive learning. Currently, these devices rapidly open and enhance, being soon able to serve as a major platform for rich, open multimedia applications and communication. In this report we introduce a video conferencing** *software*, **which seamlessly integrates mobile with stationary users into fully distributed multi-party conversations. Following the paradigm of flexible, user-initiated group communication, we present an integrated solution, which scales well for medium-size conferences and accounts for the heterogeneous nature of mobile and stationary participants. This approach allows for a spontaneous, location independent establishment of video dialogs, which is of particular importance in interactive learning scenarios. The work is based on a highly optimized realization of a H.264 codec.**

Index Terms—**Mobile video-based learning, H.264/MPEG-4 AVC software codec, mobile conferencing, peer-to-peer group communication, distributed SIP conference management.**

I. INTRODUCTION

Remote and distance learning aims at facilitating a process of immersive qualification and life-long personal development, and does not primarily focus on a supplementary aid for memorizing facts. Neither the simple, archaic look-up model of Wikipedia, nor the schoolmasterly presentation of instructional films does advance an active, involved process of learning, which is demanding and seldom arises without incentives. Nevertheless, the motivation of students may be stimulated by problem oriented tasks and imaginative environments.

The discursive acquisition of knowledge and understanding requires dialogs, which significantly gain intensity by the synchronous video-based presence among participants. Live Q&A sessions based on mobiles enable students in an inexpensive way to talk with experts as presented in [18]. Mobile phones and their wide usage allow to get in contact with almost all professions and thereby grant a "real-life" experience to active learners.

It is important to enrich discussion processes by group orientation, but unnatural to restrict interaction to two-party dialogs. The idea of video conferencing in groups is around for some time, now, but only the flexibility of the Internet generated a noticeable deployment. As compared to audio, video processing places significantly higher demands on end system and network transmission capabilities. The rapid evolution of networks and processors have paved the way for realistic group conferences conducted at standard personal computers, combining about a dozen visual streams of Half-QVGA (240 x 160 pixel @ 15-30 fps) resolution.

Mobile phones and networked consumer portables are now on the spot to deliver sufficient performance for rich multimedia applications and communication, as well. Videoconferencing though, which requires simultaneous decoding and encoding in real-time, poses still a grand challenge to the mobile world. Limited and expensive wireless channels on the one hand, high consumer demands on visual quality on the other, advise applications to take advantage of the latest standard for video coding H.264/AVC [1].

H.264/AVC provides gains in compression efficiency of up to 50 % over a wide range of bit rates and video resolutions compared to previous standards. While H.264/AVC decoding software has been successfully deployed on handhelds, high computational complexity still prevented pure software encoders in current mobile systems. There are, however, also fast hardware implementations available, which give rise to an increasing offer of device- and operator-bound video services.

In this work we first introduce a pure *software* solution for real-time video communication on standard smartphones in section 2. These mobile clients extend a lightweight, feature rich conferencing application developed for an infrastructure compliant use on standard PCs. This software includes ad hoc group building capabilities, as well as an application sharing facility. It is the vision that future learning environments will seamlessly integrate the conferencing software, so that students can collaborate on topics and tasks while at home or moving, or professional technicians, physicians, etc. can discuss problems 'in the field' and receive support from back-office experts at difficulties encountered. It is expected for many cases, that the ability to transmit visual streams from any location will greatly simplify distant collaboration and personal understanding. At the present of technical emergence, though, it is too early to report on practical real-life experiences, which is reserved for the future time of deployment. However, in the next section we will give some indications about future learning on mobiles.

In the second part of this paper we present the underlying peer-to-peer group communication scheme, which performs well for medium-size conferences and accounts for the heterogeneous nature of mobile and stationary participants, cf. section 2 and 3. This includes on the one hand SIP [2] standard compliant session signaling with respect to group communication, and on the other hand efficient, serverless media distribution, self-adjusting to the actual network infrastructure support. Conclusions and an outlook on future work follow in the final section.

II. FUTURE LEARNING WITH MOBILE PHONES

Mobile phones are widely spread. They are continuously carried by children, adults as well as ancients, who all are familiar with its usage. The well-known employment of mobiles as well as their tight integration makes these devices a promising tool for upcoming learning scenarios. In the following, we leave aside technical issues as low battery power or poor network connectivity, and sketch two straightforward learning scenarios for mobile phones equipped with a video transmission component.

In contrast to desktop PCs or notebooks mobile phones possesses the advantage of a very small chassis, which let them inconspicuous and thus applicable to experiments based on observations. For example, the biology subject in elementary schools can be enhanced, when pupils use their mobile phone to keep track of domestic animals. Mobiles will be distributed over the schoolyard or in specific vivariums. Based on the integrated cam, the video stream will be transmitted to a classroom without disturbing the natural behavior of the animal.

Mobile phones displace increasingly dedicated cameras in the context of tourism. Keeping in mind that mobiles are almost online, recorded data of sights can be sent immediately to online portals or web pages. Embedding such information in encyclopedia sites, may provide the learner with an up-to-date view of the location under consultation. Recalling the social change from privacy to publicity, e.g., while people upload holiday photos to portals like Flickr, it is not unlikely to assume that tourists are frank to share their current experiences. Video conferencing enforce the liveliness between the learner and the people on site.

III. THE DAVIKO VIDEOCONFERENCING SOFTWARE

In this section we give an overview of our reference implementation, a digital audio-visual conferencing system, realised as a serverless multipoint video conferencing software without MCU developed by the authors [3]. It has been designed in a peer-to-peer model as a lightweight Internet conferencing tool aimed at email-like friendliness of use.

The system is built upon a fast H.264/MPEG-4 AVC standard conformal video codec implementation [4] called DAVC. By controlling the coding parameters appropriately, the software permits scaling in bit rate from 48 to 1440 kbit/s on the fly.

Audio data is compressed using a 16 kHz speech-optimized variable bit rate codec [5] with extremely short latencies of 40 ms (plus network packet delay). All streams can be transmitted by unicast as well as by multicast protocols. Within the application, audio streams are prioritized over video since user experience is usually more sensitive to losses in audio packets than those of video packets, which both may result from transmission errors or network congestions.

An application-sharing facility is included for collaboration and teleteaching. It enables participants to share or broadcast not only static documents, but also any selected dynamic PC actions like animations. All audio/video-streams including dynamic application sharing actions can be recorded on any site. This system is equally well suited to intranet and wireless video conferencing on a best effort basis, since the audio/video quality can be controlled to adapt the data stream to the available bandwidth.

The daViKo conferencing system is available for desktop computers running MS-Windows or Linux and on handhelds with Windows Mobile operating system.

A. The DAVC Codec

DAVC codec, the core of the videoconferencing system, is a fast, highly optimized H.264/MPEG-4 AVC standard implementation. It realizes a Baseline Profile, optimized for real-time encoding and decoding by means of a fast motion estimation strategy including integer-pel diamond search as well as a fast sub pel refinement strategy up to ¼ pel motion accuracy. Motion estimation includes the choice of several different macroblock (MB) partitions and multiple reference frames, as permitted by the H.264/MPEG-4 AVC standard. For choosing between different MB partitions for motion-compensated (i.e. temporal) prediction and MB-based intra (i.e. spatial) prediction modes, a fast rate-distortion (RD) based mode decision algorithm with early termination conditions has been employed.

In comparison to the well-known open source H.264/MPEG-4 AVC encoder implementation of x264 [6], our DAVC encoder implementation achieves up to 0.5 dB PSNR better RD performance and a considerable increase in encoding speed when using comparable encoder settings. For selected RD points we measured 284 encoded frames per second (fps) as compared to 210 fps for x264 on an Akiyo CIF test sequence. Comparing with the H.264/MPEG-4 AVC Joint Model (JM) reference software implementation (Baseline profile settings, using a high-complexity RD-based mode decision strategy neglecting any real-time constraints) our DAVC achieves only about 0.5 dB PSNR less on a speed of 284 fps. Similar results were also achieved for other test sequences.

The DAVC codec along with the H.264/AVC design also includes some suitable mechanisms to quickly recover from video packet loss.

B. Mobile Video Codec

In ongoing work, the DAVC codec has been adapted to sustain real-time performance on mobile devices. The mobile codec version operates at reduced complexity for motion compensation with a highly optimized code base for the target platform. Motion compensation has been limited to work on 16 x 16 pixel blocks, only. The code tuning includes the efficient use of the wireless MMX instruction set available at the target system. Portability is sustained by an ANSI compliant C version, to be augmented incrementally by platform specific injections.

Figure 1. The Mobile Video Application

The application was tested on a 520 MHz Xscale processor built in an Asus P735 system. Thereon it can reliably encode and decode a QCIF video stream in parallel at 5/10 fps, without CPU exhaustion or frame dropping. Real-time encoding rate increases up to 10 fps for moderate video complexity. QCIF @15 fps is the maximal image feed that can be obtained from the front camera in our test equipment. Performance values for the mobile encoder are moderately lower as compared to the results for the full DAVC.

Reduced coding complexity results in an enhanced data rate send by the mobile, but the gross total rate for a bidirectional video exchange at 10 fps complies to 3GPP/UMTS bandwidths constraints. A sample picture of the mobile conferencing application is shown in figure 1. For further information we refer to [19].

IV. DISTRIBUTED POINT-TO-MULTIPOINT CONFERENCING

Our application aims for simple, flexible, and cost-efficient ad-hoc conferencing functions, which scale appropriately well, but avoid any infrastructure assistance. Such a solution requires group session management and media distribution at peers, which for the sake of standard compliance we realize with group conferencing functions in SIP, cf. [7-9]. Implemented as pure software on standard personal devices, user agent peers are exposed to severe restrictions in real-world deployments: Often they are located behind NATs and firewalls with network capacities confined to asymmetric DSL or wireless links. Capacity constraints and resilience to node failures require peer-managed ad-hoc conferences to organize in a distributed multi-party model. As a key component, the heterogeneity of clients must be accounted for, whereas the range of scalability is limited to about a dozen parties in videoconferences.

A. P2P Adaptive Architecture

A peer-to-peer conferencing system faces the grand challenge to be robust with respect to the infrastructure. The role a user agent is able to attain in a distributed scenario needs to be adaptively determined according to constraints of its device and current network attachment. In a simplified scenario, clients may be divided into two groups, distinguished by their ability to act as a SIP conference focus or not. A focus must be globally addressable and have access to necessary processing and network resources.

This elementary adaptation scheme can be based on individual decisions of user agents and gives rise to a hybrid architecture of super peers, chosen from potential focus nodes, and remaining leaf nodes. To decide on its potential role of building a focus, a client at first needs to determine NATs and firewalls. Aside from address evaluation, this is done by a simple probe packet exchange. As the implementation is CPU-type aware, processing restrictions are easily evaluated, as well. However, an á priori judgement on available network bandwidth is not easily obtained. An evaluation of the local link capacity is frequently misleading, as wireless devices may be located behind wired transmitters of lower, asymmetric capacity such as in ADSL. Current experiments to quickly retrieve reasonable estimates of up- and downstream capacity are ongoing on the basis of variable packet size, nonintrusive estimators, cf. [10]. Note that network capacity detection is of vital use for temporal adaptation of the video codecs, as well.

Leaf nodes attach to super peers in subordinate position, whereas potential focus nodes may be assigned to be super peers or leaves. Super peers provide global connectivity among each other and NAT traversal assistance to leaves (Fig. 3), while leaf nodes experience super peers in different roles: A leaf nodes sees its next hop super peer as the conference focus, while the remote super peers act as proxies on the path to the leaves behind.[1] This set-up corresponds to the well known architecture of Gnutella 0.6 and successive hybrid unstructured peer-to-peer systems, cf. [11]. Despite its architectural analogy, our routing layer for real-time group applications follows a different, next-hop design.

The daViKo video conferencing implementation is designed along the above hybrid network architecture with some additional "low number of peers features". It works as a pure peer-to-peer network as long as there are less than 4 peers in a conference. There is a super peer which works in this case only as address reference in a LDAP server scenario [15, 17]. If there are 4 or more peers the super peer will channel the audio/video streams in a multicast manner (see Fig. 2). If all peers however are located within an intranet and connected with sufficient bandwidth, the super peer will, connect all participants in a peer-to-peer topology. A SIP stack implemented in the super peer provides additional daViKo connectivity to customary SIP based video conferencing systems.

V. CONCLUSIONS & OUTLOOK ON FUTURE WORK

We have presented a hybrid peer-to-peer software for high-quality videoconferencing on mobiles, admitting utmost flexibility with respect to end systems, operators and network provisioning. To the best of our knowledge, this was the first software implementation of an H.264 video encoder that operates in real-time on mobile phones at the time it was presented on CeBIT (March 2008). An adaptive, fully distributed conference management scheme with SIP has been developed as part of the multi-party scenario. This hybrid peer-to-peer model accounts

for client capabilities as well as network attachment, and does scale well beyond standard use.

We are currently experimenting with an early deployment in teaching and learning scenarios. Thereby we target on augmenting the Hypermedia Learning Object System (hylOs) [12] by mobile and group communication tools. It is our vision that eLearning Objects accessed from this eLearning Content Management can be shared and discussed in video conferencing, while user presence can be equally supported on desktops and handhelds [13].

We are also about to extend our wired scenarios on distributed asynchronous and synchronous learning system on mobiles. An asynchronous mobile learning system has been deployed already using our so called Nibbler server [16] (see Fig. 4). This is a rapid deployment system for mobile learning content. Technically it is a multiformat transcoding system which semi automatically transcodes video and audio content into a large variety of different formats for web cast and streaming or download and display on almost any mobile device. Thus video coded learning content can be provided for pull on demand on mobiles. (a feature we may call lectures2go).

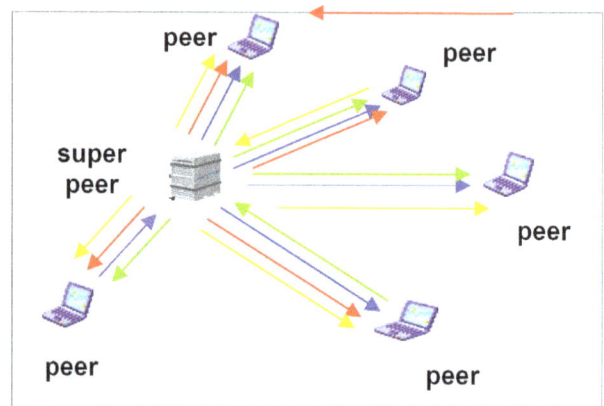

Figure 2. Super Peer-centric Multipoint Videoconferencing

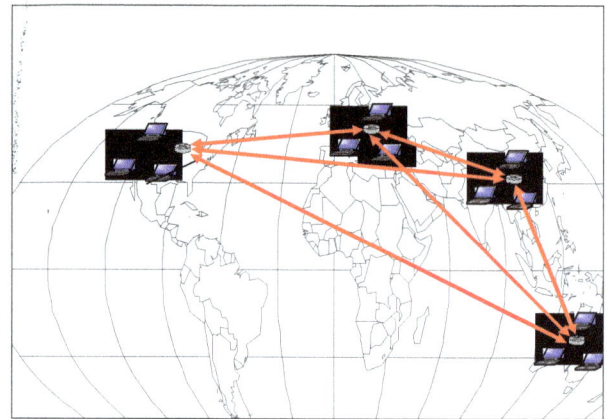

Figure 3. Global Hybrid P2P Network

REFERENCES

[1] ITU-T Recommendation H.264 & ISO/IEC 14496-10 AVC, Advanced Video Coding for Generic Audiovisual Services, ITU, Tech. Rep., 2005, draft Version 3.

[2] J. Rosenberg, H. Schulzrinne, G. Camarillo, A. Johnston, J. Peterson, R. Sparks, M. Handley, and E. Schooler, SIP: Session Initiation Protocol," IETF, RFC 3261, June 2002.

[3] M.   Palkow,   The   daViKo   homepage,"   2008, http://www.daviko.com.

[4] J. Ostermann, J. Bormans, P. List, D. Marpe, N. Narroschke, F. Pereira, T. Stockhammer, and T. Wedi, Video Coding with H.264/AVC: Tools, Performance and Complexity, IEEE Circuits and Systems Magazine, vol. 4, no. 1, pp. 7-28, April 2004. (doi:10.1109/MCAS.2004.1286980)

[5] The Speex projectpage," http://www.speex.org, 2008.

[6] VideoLan: x264   -   a   free   h264/avc   encoder, http://www.videolan.org/developers/x264.html, 2007.

[7] A. Johnston and O. Levin, Session Initiation Protocol (SIP) Call Control – Conferencing for User Agents, IETF, RFC 4579, August 2006.

[8] R. Mahy, R. Sparks, J. Rosenberg, D. Petrie, and A. Johnston, A Call Control and Multi-party usage framework for the Session Initiation Protocol (SIP)," IETF, Internet Draft - work in progress 9, November 2007.

[9] T. C. Schmidt and M. Wählisch, Group Conference Management with SIP, in SIP Handbook: Services, Technologies, and Security, S. Ahson and M. Ilyas, Eds. Boca Raton, FL, USA: CRC Press, pp. 123-158, November 2008, on invitation.

[10] R. Prasad, C. Dovrolis, M. Murray, and kc claffy, Bandwidth Estimation: Metrics, Measurement Techniques, and Tools, IEEE Network, vol. 17, no. 6, pp. 27-35, November-December 2003. (doi:10.1109/MNET.2003.1248658)

---

[1] This architecture relies on the presence of at least one globally addressable, sufficiently powerful peer. As there are many scenarios, where this is likely to fail, we advise for and offer a permanently deployed 'silent' relay-peer at some unrestricted place.

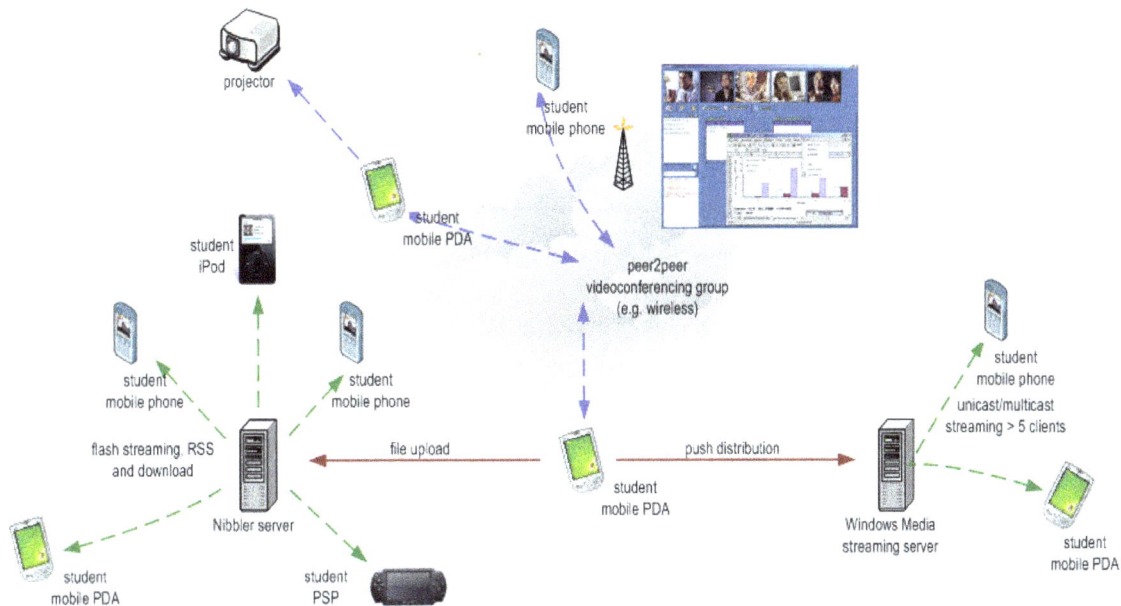

Figure 4. Joint Synchronous and Asynchronous Mobile Learning

[11] R. Steinmetz and K. Wehrle, Eds., Peer-to-Peer Systems and Applications, ser. LNCS. Berlin Heidelberg: Springer-Verlag, 2005, vol. 3485.

[12] The hylOs Homepage, http://www.hylOs.org, 2008.

[13] A. Hildebrand, T. C. Schmidt, and M. Engelhardt, Mobile eLearning Content on Demand, International Journal of Computing and Information Sciences- (IJCIS), vol. 5, no. 2, pp. 94-103, August 2007. [Online]. Available: http://www.ijcis.info/Vol5N2/Vol5N2pp94-103.pdf

[14] M. Wählisch and T. C. Schmidt, Between Underlay and Overlay: On Deployable, Efficient, Mobility-agnostic Group Communication Services, Internet Research, vol. 17, no. 5, pp. 519 - 534, November 2007. (doi:10.1108/10662240710830217)

[15] H.L.Cycon, M.Palkow, D. Marpe, T.C.Schmidt, M.Wählisch: A Fast Wavelet-Based Video Codec and its Application in an IPv6-ready serverless Video conferencing , , Proc. Third International Conference on Wavelet Analysis and its Applications (ICWAA'03), vol. 2, pp. 577-583,(2003).

[16] Hans L. Cycon, Anja C. Wagner, Fabian Topfstedt, Henrik Regensburg: Distribution & Communication Tools for Video based m-Learning, Proceedings WCI 2007, pp.139-149, October 2007.

[17] T. C. Schmidt, M. Wählisch, H. L. Cycon, and M. Palkow, "Global Serverless Video Conferencing over IP", Future Generation Computer Systems, vol. 19, no. 2, pp. 219–227, February 2003. (doi:10.1016/S0167-739X(02)00148-6)

[18] T. McNeal and M. van't Hooft, Anywhere, anytime: Using mobile phones for learning, Journal of the Research Center for Educational Technology, vol. 2, no. 2, November 2006. http://www.rcetj.org/?type=art&id=79575

[19] Hans L. Cycon, Thomas C. Schmidt, Gabriel Hege, Matthias Wählisch, Detlev Marpe, Mark Palkow, Peer-to-Peer Videoconferencing with H.264 Software Codec for Mobiles, In: WoWMoM08 - The 9th IEEE International Symposium on a World of Wireless, Mobile and Multimedia Networks - Workshop on Mobile Video Delivery (MoViD), (Ramesh Jain, Mohan Kumar Ed.), IEEE, pp. 1-6, Piscataway, NJ, USA:IEEE Press, June 2008.

AUTHORS

Hans L. Cycon is with the University of Applied Sciences (FHTW) Berlin, Treskowallee 8, 10318 Berlin, Germany (e-mail: hcycon@fhtw-berlin.de).

Thomas C. Schmidt is with the University of Applied Sciences (HAW) Hamburg, Department. Informatik, Berliner Tor 7, 20099 Hamburg, Germany. (e-mail: t.schmidt@ieee.org).

Gabriel Hege is with the University of Applied Sciences (HAW) Hamburg, Department Informatik, Berliner Tor 7, 20099 Hamburg, Germany. (e-mail: hege@fhtw-berlin.de).

Matthias Wählisch is with link-lab Hönower Str. 35, 10318 Berlin, Germany. He is also with the HAW Hamburg, Dept. Informatik and with the Freie Universität Berlin, Institut für Informatik. (e-mail: waehlisch@ieee.org).

Mark Palkow is with the daViKo GmbH, Am Borsigturm 40, 13507 Berlin, Germany (e-mail: palkow@daviko.com).

25

Simulation and Proposed Handover Alert Algorithm for Mobile Communication Networks

Muzhir Al-Ani[1] and Wael Al-Sawalmeh [2]

[1]Amman Arab University, Amman, Jordan
[2]Philadelphia University, Amman, Jordan

Abstract—This paper deals with a novel approach to realize the handover process of a mobile systems network. It concentrates on the existing challenges of Global System for Mobile communication (GSM) networks and how to overcome these challenges. The proposed algorithm extracts the received signal information features in order to track the significant coverage cell. The presented algorithm distinguishes between real problems and false alarms. One of the important contributions of the algorithm is how it predicts the adequate signal of the effective coverage cell.

Index Terms—Handover, Coverage Area Measurements, Mobile Positioning & Digital Cellular System

I. INTRODUCTION

During the early 1980s, analog cellular telephone systems were experiencing rapid growth in Europe, particularly in Scandinavia and the United Kingdom, but also in France and Germany. Each country developed its own system, which was incompatible with everyone else's in equipment and operation.

In 1989, GSM responsibility was transferred to the European Telecommunication Standards Institute (ETSI), and phase I of the GSM specifications were published in 1990. Commercial service was started in mid-1991, and by 1993 there were 36 GSM networks in 22 countries. Although standardized in Europe, GSM is not only a European standard. Over 200 GSM networks (including DCS1800 and PCS1900) are operational in 110 countries around the world. In the beginning of 1994, there were 1.3 million subscribers worldwide, which had grown to more than 55 million by October 1997. With North America making a delayed entry into the GSM field with a derivative of GSM called PCS1900, GSM systems exist on every continent, and the acronym GSM now aptly stands for Global System for Mobile communications.

The developers of GSM choose an unproven digital system, as opposed to the then-standard analog cellular systems like AMPS in the United States and TACS in the United Kingdom. They had faith that advancements in compression algorithms and digital signal processors would allow the fulfillment of the original criteria and the continual improvement of the system in terms of quality and cost.

The revolution of the telecommunication technology opens new doors in many countries all over the world to start development and investigation over mobile system fields starting from the analog systems and then to digital systems that allow the subscriber mobility and comfort.

The cellular system as implemented in most countries demonstrates that the world is a small village. On the other hand, a substantial growth in the number of subscribers makes it necessary to evaluate large and extended networks to overcome the growth of mobile networks [1], thus GSM becomes the most popular network. Nowadays there are many different and potential types of GSM with different features and applications. The popularity of mobile phones and the number of mobile devices users is continuously increasing, so manufacturers are continually trying to introduce new features and services to attract new customers [2].

GSM and 2G networks have been started in many Arab countries during 1990s, while the transition from 2G to 3G started at the beginning of 2000s. Due to some financial problems, some countries have not evolved to 2G or 3G technologies. Therefore, our work is concentrated on improving the GSM network performance, since it is still widely used.

II. DIGITAL COMMUNICATION SYSTEM

The usage of mobile communication systems keeps expanding, leading to large number of studies related to GSM services, techniques and algorithm improvement. A fundamental goal of mobile telephone network operators is to provide seamless service for their subscribers [3] & [4].

The report of the GSM Association in 2004 indicates that the number of GSM networks exceeds 500 in more than 180 countries worldwide; the Latina American and Asian Pacific regions have the largest share of the highest rates of growth in both number of subscribers and number of networks lunched [5].

As in [6], developing and proposing a new algorithm for receiving GSM broadcasted data is important because it allows subscribers to obtain the identification parameters and carry out an analysis of mutual interference effects.

Cellular Systems are moving from 2G to 3G, while 2G already brings high speed data transmission enabling wideband multimedia applications. The next generation of cellular systems will be increasingly similar to a data communication system, which not only transfers voice and multimedia, but will also be integrated with WLAN to access Internet [7].

Great effort has been devoted towards the design and development of mobile architectures and multiservice networks. High performance modeling and evaluation of General Packet Radio Service (GPRS) are adapted via

GSM cell involving both voice and multiple class data services under a complete partitioning scheme [8].

GPRS provides packet switched data transfer to efficiently utilize the radio resources. GPRS is considered the main development step of GSM networks toward the next generation of mobile communication system like UMTS. GPRS allows a single mobile station to transmit data using multiple time slots to increase the transmission rate [9].

Existing GSM/GPRS base station site utilization can be sped up 3G wide band code division multiple access (WCDMA) deployment. It is possible to provide full coverage even for bit rates higher than 144 or 384 kbps when 3G WCDMA use GSM 1800 sites [10].

As operators continue to roll out networks, it is obvious that providing the depth of coverage equal to the mass-market population coverage achieved with GSM 900/1800 will not be possible for many years. Early release of UMTS enables circuit-switched users to access similar services via GSM or UMTS [11].

Detecting and explaining fully states in complex telecommunications systems such as GSM networks are mentioned as challenging tasks. Mobile networks are hierarchical cell-based systems with complex dynamics influenced by the stochastic user demand. The network operating characteristic such as hand over algorithms, carrier frequency and the non-stationary influence of the environment [12] are considered as main areas of this field.

III. DIGITAL CELLULER SYSTEM

Digital cellular systems are implemented in numerous countries all over the world. This system consist of three main parts: mobile switching center (MSC), base station system (BSS) and mobile station (MS). Each division is divided into other entities. The overall system is connected to the normal public service telephone network (PSTN) as shown in Figure 1 [13]. In cellular systems, the coverage area of an operator is divided into cells. These cells are normally hexagonally shaped, but in practice, because of the influence of the terrain, the shape is irregular as shown in Figure 2. The size of cells varies from a few hundred meters to many tens of kilometers, depending on the coverage area and the number of subscribers and transmitting power [14].Each base station (BS) is allocated to a different carrier frequency and each cell has a usable bandwidth associated with this carrier. The cellular radio permits the use of a limited part of the radio spectrum, so the available number of carrier frequencies is limited [15].

Therefore, to provide sufficient channels for more subscribers, it is necessary to re-use the available frequencies many times, which creates possible interference between cells carrying the same frequencies. To avoid interference it is preferable to isolate the cells by employing different frequencies [1].

The cells with different carrier frequencies form a section which identifies the re-use distance, which can depend on many factors, including the number of co-channel cells in the vicinity of the center cell and the geography of the terrain and the transmitted power within each cell [16].

Mobile network providers installing thousands of base stations in each country implies using small cells, although it is more expensive than using big cells.

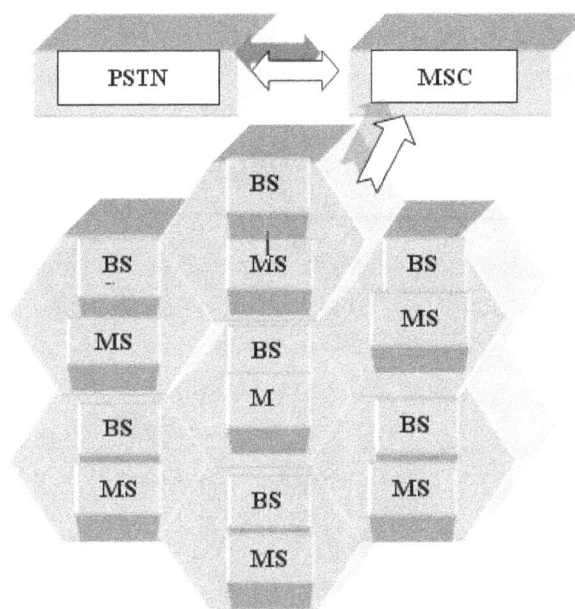

Figure 1. Main parts of Cellular System

The advantages of cellular systems with small cells are:

- Higher capacity, they are limited to fewer possible users per km^2.

- Less transmission power, a receiver near the base station requires less power.

- Only local interference occurs.

- Robustness, if one cell fails, the problem occurs in small area.

Small cells also have disadvantages:

- They need a complex infrastructure to connect all base stations.

- They require a fast handover because of mobility between cells.

- Frequencies have to be distributed carefully to avoid interference between transmitters.

IV. GSM SYSTEM ARCHITECTURE

GSM became popular rapidly because of the quality improvement and its use of international standards. The GSM communication network is divided into three main groups; the mobile station (MS), the base station subsystem (BSS) and the network subsystem (NS), which is illustrated in Figure 3.

The MS or mobile equipment (ME) includes the subscriber identity module (SIM). Each SIM card has an identification number called international mobile subscriber identity (IMSI). Each MS is assigned to a hardware identification called international mobile equipment identity (IMEI). Besides providing transmission and reception of voice and data, the mobile performs other tasks such as authentication, handover encoding and channel encoding [17] & [18].

The BSS consists of the base station controller (BSC) and the base transceiver station (BTS). The BTS is used to connect the mobiles to a cellular network.

Their tasks include channel coding/decoding and encryption/decryption. A BTS is comprised of radio trans-

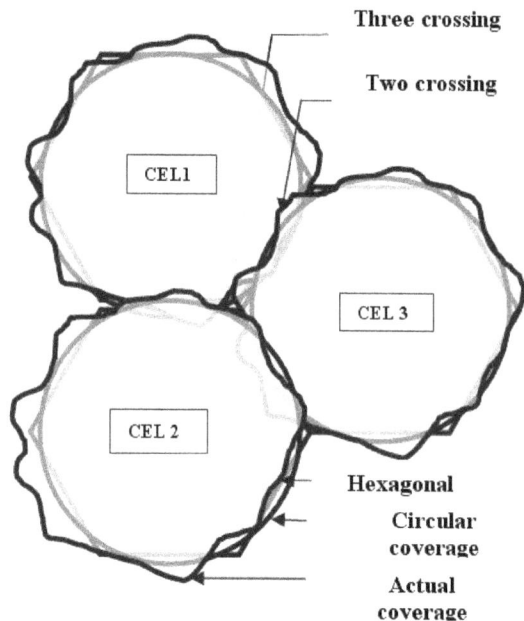

Figure 2.   Effects of crossing on the coverage

mitters and receivers, antennas and an interface to the PCM facility ...etc. A group of BTS are connected to a particular BSC. The main function of the BSC is call maintenance. The MS sends a report of their received signal strength to the BSC, so the BSC decides to initiate hand over to other cell [19].

The network subsystem includes the mobile switching center (MSC), the home location register (HLR), the visitor location register (VLR), the authentication center (AUC) and the equipment identity register (EIR). The main functions of MSC are registration, authentication, location updating, handovers and call routing to a roaming subscriber. MSC also has a gateway function for communication with other networks. HLR is a data base used for management of mobile subscribers; it stores the IMSI, mobile station ISDN number (MSISDN) and current visitor location register (VLR) address.

VLR contains the current location of the MS and selected administrative information from the HLR, necessary for call control and the provision of subscriber services, for each mobile currently located in the geographical area controlled by the VLR, which is connected to one MSC. AUC is a protected database that holds a copy of the secret key stored in each subscriber's SIM card which is used for authentication and encryption. The EIR is the database that contains a list of all valid mobile station equipment within the network, where each mobile station is identified by its IMEI. The operation and management center (OMC) is a management system which maintains the operation of the GSM network. The OMC is responsible for controlling and maintaining the MSC, BSC and BTS.

V.   GEOGRAPHICAL AREAS OF GSM NETWORK

In order to work properly, a cellular system must verify the following main conditions:

- The power level of a transmitter within a single cell must be limited in order to reduce the interference with the transmitters of neighboring cells. The interference will not produce any damage to the system if

a distance of about 2.5 to 3 times the diameter of a cell is reserved between transmitters.

- Neighboring cells can not share the same channels. In order to reduce the interference, the frequencies must be reused only within a certain pattern.

- In GSM, five main functions must be defined: Transmission, Radio Resources management (RR), Mobility Management (MM), Communication Management (CM), Operation, and Administration and Maintenance (OAM).

The geographical area of a GSM network is illustrated in Figure 4. In a GSM system, a cell is identified by its Cell Global Identity number (CGI), corresponding to the radio coverage of a base transceiver station. A Location Area (LA), identified by its Location Area Identity (LAI) number, is a group of cells served by a single MSC/VLR. A group of location areas under the control of the same MSC/VLR defines the MSC/VLR area. A Public Land Mobile Network (PLMN) is the area served by one network operator.

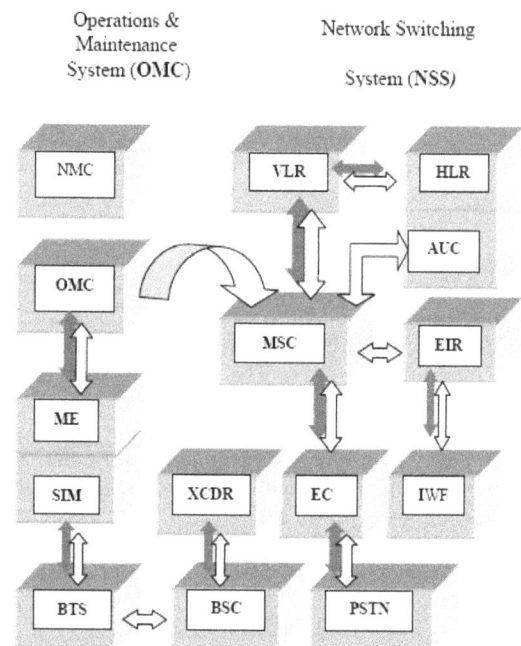

Figure 3.   GSM System Architecture

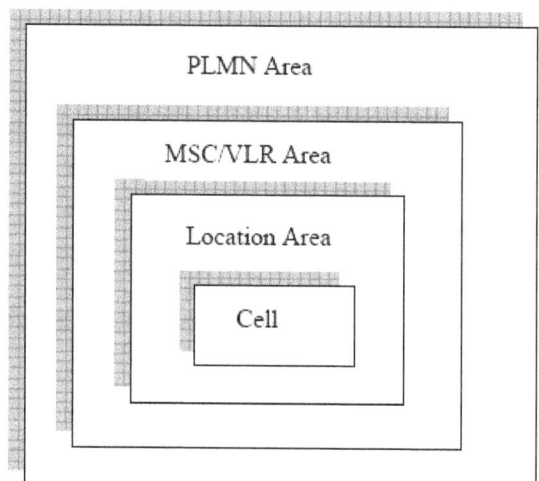

Figure 4.   GSM Network Areas

VI. Signal Power

The selected area/zone for measurements is a multilevel geographical area/zone that needs a special solution for each case. Because an isotropic antenna has unity gain for both transmitter and receiver, hence the power received by such antenna is given by:

$$Pr = Pt(\lambda / 4\pi d)^2 \qquad (1)$$

Where: Pr is the MS received power signal, Pt is the BS transmitted power signal, d is the distance between the BS and the MS and λ is the wavelength of the signal.

The above expression indicates that the attenuation is inversely proportional to the square of the distance between BS and MS [19] & [20]. In practice of real system as illustrated in Figure 5 exist direct, diffracted and reflected paths. In this system a small difference between these three signals is mentioned. The received power can be written as:

$$Pr = Pt(\lambda / 4\pi d)^2 * |1 + \rho e^{j\phi}|^2 \qquad (2)$$

where ρ is the reflection coefficient which is equal to -1 for an ideal reflector and ϕ is the phase difference between direct and reflected path.

Let the height of the BS and MS are HBS & HMS respectively. The phase difference can be calculated using the following equation,

$$\phi = 4\pi H_{BS} H_{MS} / \lambda d \qquad (3)$$

So, the received power can be written as,

$$Pr = Pt(\lambda / 4\pi d)^2 * |1 - e^{j\phi}|^2 \qquad (4)$$

Because the value of ϕ is very small, the second term is approximately equal to ϕ

Therefore $|1 - e^{j\phi}| = \phi \qquad (5)$

Substituting equations 3 & 5 in equation 4 so

$$Pr = Pt(H_{BS} H_{MS} / d^2)^2 \qquad (6)$$

This equation is used in the frequency re-used calculation [18]&[20].

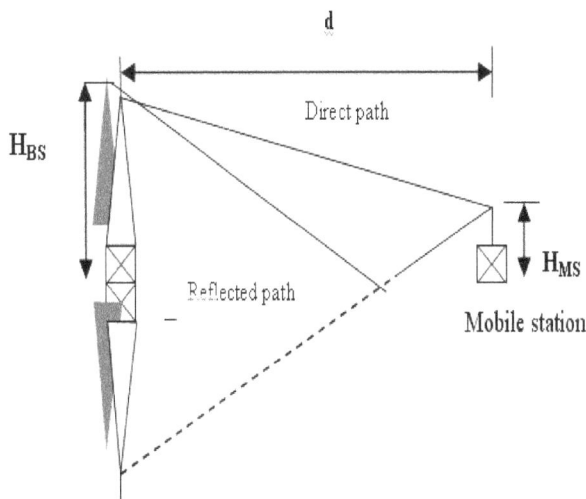

Figure 5. Multiple paths of GSM

VII. Proposed Handover Algorithm

The subscribers' movement across the cells results in the need to change the allocated channels, which also happens when interference occurs. These events affect the quality of the communication so it is necessary to change the resources; defined as handover (HO) that is mainly controlled by the MSC.

The MS continuously monitors the strength of received signal strength either from its BS or the surrounded BSs, which depends on the list of all cells that must be monitored by its BS. This will help in deciding the destination of the MS, based on the best transmission of the surrounding cells in order to maintain the quality of the communication network.

The unwanted case (the worst case) situation occurs when the mobile is placed on the boundary of its serving cell. In this case the interference is produced by all overlapping surrounding cells. This case causes the strength and the quality of the signal to fall under a certain threshold. Therefore, HO must happen to maintain the call continuity.

In the GSM architecture shown in Figure (6) there are four possible types of HO which involve transferring a connection between two BSs:

- HO1: Channels in the same cell named Intra-cell handover
- HO2: Cells in the same BSC named Inter-cell, Intra-BSC handover
- HO3: Cells in the different BSC named Inter BSC handover
- HO4: Intra MSC handover and cells in the different MSC named Inter MSC handover

Most cellular systems that employ GSM aim at maximizing HO for 60ms. The proposed algorithm depends on the continuous measurement at any point of the received power and the carrier to interference ratio. Depending of the refused values, a trade off process will be used to choose the best suitable value between these two measurements. The scanning process is necessary to check the mentioned values concerned with the received signal around all the nodes related to the MS. These processing, are used to verify various situations of handover problem, whenever it occurs during the crossing cells or MSC. When the problem is defined correctly, as indicated in the first phase, then the processing can be pass to the second phase that deals with the keeping continuity of the call during a correct decision.

The proposed algorithm is summarized by the following:

1. Management of received power measurements
2. Verification of the received power measurements
3. Management of carrier to interference ratio measurements
4. Verification of carrier to interference ratio measurements,
5. Comparison of measured and reference values
6. Selection of the optimal base station to connect the mobile
7. Handover decision depends on performance and the average value of the received signal

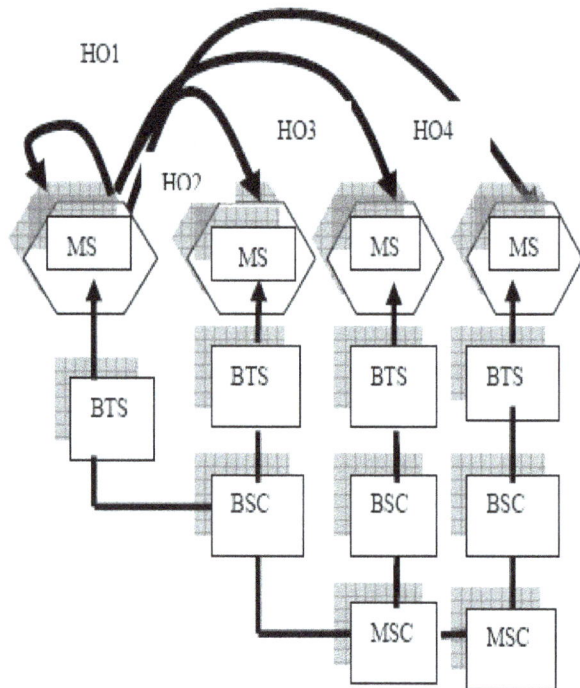

Figure 6.   Possible types of handover

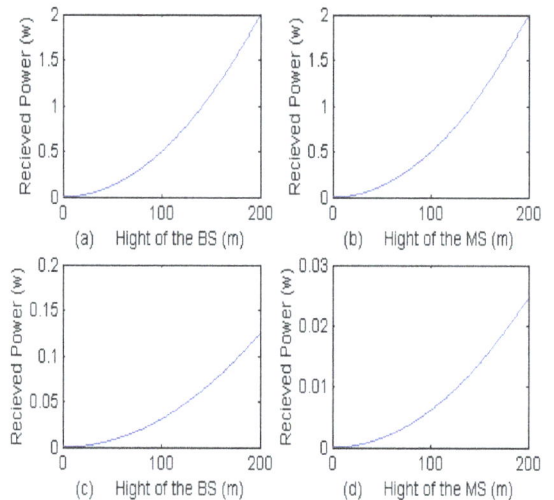

Figure 7.   The received power with respect to the antenna height

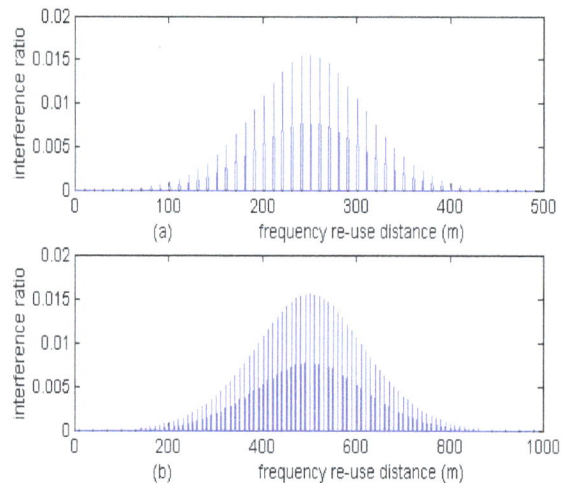

Figure 8.   The received power with respect to the re-use frequency distance

VIII.   SIMULATION

The Cellular systems require HO procedures to maintain the entire service area within an effective coverage range. The smaller cells require fast HO to recover calls within an adequate time and typical levels of the received signal, but this is different in large cells.

The behavior of the received signal level while a mobile device moves away from one BS to another seems to vary from one level to another. In this case, the hand over decision does not depend on the actual value of the received signal level, but on the average value.

The simulated results of the received power refers to the distance between the base station and the mobile station as demonstrated in Figure 7 (a) and (b) which are calculated at the same cell size. Graphs (c) and (d) in Figure 7 duplicate the cell size to determine its effects. Figure 8 illustrates the variation of carrier to interference ratio with respect to re-use frequency distance. These results are affected intensively by the topographic terrain.

IX.   CONCLUSION

The popularity of mobile phones and the number of mobile device users is continuously increasing, and at the same time mobile phone manufacturers are striving to introduce new feature-packed devices to hopefully attract potential new customers.

This work is limited and applied in a selected area that is situated in crowded noisy zones based on using GSM systems. This is due to the limitations that are already mentioned earlier in the introduction.

The presented HO algorithm is flexible and successful to select a proper significant path to take a correct decision. It also provides the ability to distinguish between real problems and false alarm problems. One of the important conclusions of the obtained results is that the pro-

proposed algorithm procedure maintains the ability to predict the adequate signal of the effective coverage cell.

More sophisticated handover mechanisms are needed for the transition from 2G networks to the more advanced mobile systems such as 3G networks, which are not available in all countries. The handover field offers a wide area for research and more challenges appear in the future when huge number of users access different types of wireless networks and different generations of mobile systems.

ACKNOWLEDGMENT

We would like to express our appreciations to Dr. Omar Daoud for his valuable assistance in proofreading this work.

REFERENCES

[1]   A. Jagoda, M. Viillepin, and De Viillepin, "Mobile Communications", John Wiley and Sons, 1993.

[2]   J.Dunlop and D.G. Smith, "Telecommunication Engineering", Prentice Hall, 3rd edition, 1992.

[3]   A. Alexiou, P.Kastarakis and V.N. Chand ristofilakis, "Interaction between GSM Handset Helical Antenna and User Head: Theoreti-

cal Analysis Experimental Results", The Environmentalist, 25, 215-221, 2005. (doi:10.1007/s10669-005-4286-6)

[4] L. Rassian, "A Permutation Code Evolutionary Strategy for Multi-Objective GSM Network Planning", J. Heuristics, 14,1-21, 2008. (doi:10.1007/s10732-007-9024-4)

[5] W. Scharnhorst et al., "Environmental Assessment of End-of-Life Treatment Options for A GSM 900 Antenna Rack", Int. J. LCA, 11, 6, pp.425-436, 2006. (doi:10.1065/lca2005.08.216)

[6] V.B. Manelis, I.V. Kaioukov, and A.V. Novikov, "Identification and Analysis of Interference Effects of GSM Base Stations", Radioelectronics and Communications Systems, vol.52, no.2, pp.55-62, 2009. (doi:10.3103/S0735272709020010)

[7] L. Zhang etal., "A 1.8 Tri Mode $\sum\Delta$ Modulator for GSM/WCDMA/WLAN Wireless Receiver", Analog Integr Circ Sig Processing, 49, pp.423-441, 2006.

[8] K. Al-Begain et al., "Analysis of GSM/GPRS Cell with Multiple Data Service Classes", Wireless Personal Communication, 25, pp41-57,2003. (doi:10.1023/A:1023603308841)

[9] J. Tigang and F. Pingzhi, "Channel De-Allocation Schemes for GSM/GPRS Networks", Journal of Electronics (China), no.6, vol.22, 2005.

[10] J. Mar and J. Huang, "The Complementary Use of 3G WCDMA and GSM/GPRS Cellular Radio Networks", Wireless Personal Communications, 43, pp.511-531, 2007 (doi:10.1007/s11277-007-9247-6)

[11] N. C. Lobely, "GSM to UMTS Architecture Evolution to Support Multimedia", B T Technol J, no.1, vol.19, January 2001.

[12] D. Obradovic and R. L. Scheiterer, "Troubleshooting in GSM Mobile Networks based on Domain Model and Sensory Information", ICANN LNCS, 3697, pp. 729-734, 2005

[13] M. E. Kounavis, and A. T. Campbell, "Design, Implementation and Evaluation of Programmable Handoff in Mobile Networks", Mobile Networks and Applications, Kluwer Academic Publishers, 2001.

[14] Pubudu N. Pathirana et al, "Mobility and Trajectory Prediction for Cellular Network with Mobile Base Station", 4th international symposium on mobile Ad hoc networking and computing, 1-3, June 2003.

[15] M. E. Kounavis et al, "Supporting Programmable Handoff in Mobile Networks", 6th International workshop on Mobile Multimedia Communications (MoMuC'99) San Diego, Ca, (November 1999).

[16] S. Seshan, et al, "Handoff in Cellular Network: the Daedalus Implementation and experience", Kluwer International Journal on Wireless Communication Systems, 1996.

[17] M. Dillenger and S. Buljore, "Reconfigurable Systems in Heterogeneous Environments", Software Defined Radio: Architectures, Systems and Functions, John Wiley and Sons, Ltd 2003.

[18] Payam Taaghol, "Optimization of WCDMA", Bechtel Telecommunication Technical Journal, Vol.2, No.1, 2004.

[19] S. Aust et al, "Policy Base Mobile IP Handoff Decision Using Generic Link Layer Information", Proceedings of the IEEE International Conference on Mobile and Wireless Communication Networks (MWCN 2003), Singapore, 27-29 October 2003.

[20] Gertie Alsenmyr et al, "Hand Over Between WCDMA and GSM "Ericson Review, No.1, 2003.

AUTHORS

Muzhir Al-Ani was born in Iraq in 1956. He received B. Sc. degree in Electrical Engineering from Sulaimania University, Iraq in 1979, Higher Diploma in Electronic and Communication Engineering from College of Engineering, University of Baghdad, Iraq in 1981, M. Sc. degree in Electronic and Communication Engineering from College of Engineering, University of Baghdad, Iraq in 1983. He received the Ph. D. degree in 1 July 1994 for his Thesis entitled Fast Algorithms of Digital Signal and Image Processing from Electronic Department of E.T.S.I.I. from University of Valladolid, Spain.

He joined in 7 July 1984 the Technical Institute of Al-Anbar as an Assistant Lecturer. He worked as Assistant of the Dean in 1985, the Head of the Electrical Department during the years 1985 to 1988.

He joined in 1994 the Electrical Engineering Department, College of Engineering, University of Al-Mustansiriya, Iraq, as a lecturer. He joined in 1996 Computer and Soft Engineering Department at the same College, working as the Head of Computer and Software Engineering Department during the years 1997 to 2001. In 5 May 1999 he was promoted Assistant professor (equivalent to Associate professor in Jordan) at the same Department. In 2 October 2001 he joined the Department of Computer Science and Information Systems in the University of Technology as the Head of the Department during the years 2001 to 2003.

He joined in September 2003 Electrical Engineering and Computer Department Applied Science University, Amman, Jordan as Associate professor.

He joined in September 2005 Management Information System Department Applied Science University, Amman, Jordan as Associate professor, then in September 2008 Computer Science Department at the same University.

His research interests include Digital Signal Processing, Parallel Processing, Digital Filters, Digital Image Processing, Image Compression, Computer Vision, Information Systems, Information Hiding and Steganography, Wireless Networks Mobile Communications and related work.

(e-mail: muzhir@gmail.com)

Wael Hassan Al-Sawalmeh received Ph.D. degree in 1998 for his thesis entitled Working Out and Investigating of checking method of Color Production with help of sensors Based on LEDs and Fiber Optics of University of Telecommunication, Saint- Petersburg State-Russia.

He received M.Sc. degree in Electrical Engineering from Electro technical Institute of communication Leningrad, Bonch-Bruyevith, Russia in 1993.

He joined in 1998 university of Omar Al-Mouktar in Libya as assistant professor. In September 2001 he joined the Higher Institute of Comprehensive Vocational in Al-Baida, Libya as a part time lecturer. In October 2001 he joined the Shahat Company for Computer Techniques in Al-Baida, Libya as a part time lecturer, in September 2002 he joined the Electrical Engineering and Computer Department Applied Science University, Amman, Jordan as Assistant professor. He joined in September 2004 Communication & Electronics Department Philadelphia University, Amman, Jordan as Assistant professor.

His research interests include TV, mobile communications and Digital Signal Processing.

(e-mail: waelalsawalme@hotmail.com)

26

A Review of the Navigation HCI Research During the 2000's

Teija Vainio

Tampere University of Technology, Tampere, Finland

Abstract—Two of the common problems associated with navigation research in the field of human computer interaction (HCI) systems and human factors, are the relatively narrow view that the display is able to provide for a large information space, and the impact that a diversity of contexts has on the users' divided attention. In recent years, much research has been focused on the development of navigation presentation techniques to address the first problem, and the development of capturing the context with multimodal interaction in order to address the second challenge. However, the growing number of new terminologies and techniques that has been developed has caused considerable confusion for HCI researchers, consequently making the comparison of these techniques and the generalisation of empirical results of experiments very difficult, if not impossible. This article provides a taxonomy of current navigation research, which describes clearly the navigation research on desktop and non-desktop environments; it also helps to identify research domains that afford and promote the direction of navigation research. This review reveals gaps where navigation research has identified challenges but has not yet explored them.

Index Terms—Human Factors, Navigation, review

I. INTRODUCTION

Over 500 years ago some of the most well-known feats of navigation were accomplished when Christopher Columbus sailed across the Atlantic Ocean. Even then, he knew that navigation tasks could be practiced beforehand in order to find an easier route. During Columbus's era, navigation was understood as moving from one place to another and finding one's way in a real physical environment. Hundreds of years later, navigation now also signifies something else for those of us who live with hypermedia applications, search information within large databases, listen to driving instructions on a SatNav device in our cars, and locate local sights as pedestrians using mobile tourist guides.

Navigation has also been the focus of a huge amount of scientific research over the past decades. For example, in 1948, Tolman (1948) argued that navigating is to some extent a personal ability or skill, and a human being creates a personal cognitive map in order to be able to navigate better in the world. Lynch (1960) explored to what extent these cognitive maps could support navigation in Chicago during the 1960s. In this type of navigation research, the focus is on w*ayfinding*, i.e., searching and finding a route from point A to point B. Thus, studies on wayfinding not only cover the wayfinding task in real environments, they are also the focus of navigation research in three-dimensional virtual environments; (see for example

Vinson, 1999, Cliburn et al., 2007). In virtual environments, users are exploring great quantities of information and one of the navigation challenges lies in how to support users in order to enable them to find what they are searching for. Three-dimensional virtual environments are also an example of a certain kind of large information space with interconnected and massive amounts of digital files or documents. Research on how to navigate in large information systems seems to focus on how to support document navigation for example, how to move with perspective-drag technique (Guiard et al., 2007)or scroll and zoom between documents; see Mehra et al. (2006). As stated, navigation research has an extensive and broad field of interesting starting points. However, in order to understand navigation research and to be able to take steps forward in the future, researchers should be able to place their own scientific contributions within the wider context of research into human computer interaction. This review of navigation research in the 2000s draws upon the bigger picture of scientific navigation research. The primary objective of this study is to capture an overview of navigation research being conducted in the human computer interaction research field for the period of 2000 to 2007.

Why cover this wider field of navigation research and not just focus on, for example, navigation research on mobile devices? The argument for taking the wider view is two-fold. First, it is assumed that navigation research shares some general characteristics whether it concerns mobile or desktop environments. One example of this similarity uses the concept of a landmark. A 'landmark' is defined by Lynch (1960) as an external point-reference, which a user (an observer) does not enter. Landmarks are physical objects, and they may be local or distant, or even mobile; they may include a storefront, a tower, or the sun. The impact of using landmarks to support navigation has been explored, for example, when studying auditory landmarks in mobile navigation (Baus et al., 2007), designing guidelines for landmarks to support navigation in virtual environments (Tory et al., 2004), developing dynamic landmark placement as a navigation aid in virtual worlds (Cliburn et al., 2007), and analyzing the navigability of Web applications for improving blind usability (Takagi, Saito, Fukuda and Asakawa, 2007). Therefore, it can be assumed that there might be other interesting similarities concerning navigation research across various environments and conducted with different devices. Second, the differences that, for example, desktop and non-desktop navigation research may possess could reveal something critical. For example, until recently multimodal interaction was the focus of navigation research mainly in desktop navigation; see e.g Wall and Brewster (2006) . Thus, Jöst et al (2005) explore the influence of social and situational contexts on multimodal interaction; they stated that multimodality can enhance the usability of navigation tasks for mobile pedestrians as well. Finally, even if navigation research in human computer interaction has been conducted in desktop environments for longer than in non-desktop environments, it is still interesting to examine how some of the key concepts used, that is, navigation, landmarks, and wayfinding, have counterparts in navigation in a real physical environment. When reviewing navigation research both within the desktop and non-desktop research, it assumed that the possible shift that occurs between these two areas can be acknowledged.

This paper is organized into five main sections. Following the Introduction, the Methodology section introduces the steps of the reviewing process. The third section focuses on the characteristics of the navigation research from the 2000s studied in this paper, categorized according to the technology and keywords focus of the research and key concepts. Section four describes the key results and possible gaps that should be taken into account in future research agendas in the field of navigation research. The fifth section discusses the implications of this description for the research of navigation in the field of human computer interaction.

II. METHODOLOGY

The main aim of this review is to describe the path for future research of navigation for any researcher in the field of human computer interaction. The research questions in this study are as follows:

1. What is the focus of navigation research in the field of human computer interaction?

2. What types of results has navigation research in the 2000s accomplished?

3. What are the main challenges and directions to fill in for future research?

The published navigation research articles were examined in the following stages:

To identifying the relevant articles, the first three steps were taken (see Swanson and Ramiller, 1993) 1): *selecting/searching articles for review* 2) *filtering relevant articles* and 3) *identifying content and structure*. In addition, in order to analyze the content, the following four steps were taken: 4) *evaluating the content and structure* 5) *grouping according to the objectives* 6) *clustering and validating* and finally 7) *triangulating* (see Patton, 2002).

The research method conducted for this study was content analysis. In previous studies reviewing HCI research and research methods, see e.g. (Kjeldskov and Graham, 2003); there are only a few, if any, reviews of navigation research. In this study, the articles published from 2000 to the end of 2007 were included from five HCI-related conferences and three journals. The selected articles were blind-reviewed and were published in:

1. SIGCHI conference on human factors in computing systems (CHI)

2. Nordic conference on human computer interaction (NordiCHI)

3. Human computer interaction with mobile devices and services (MobileHCI)

4. International conference Intelligent User Interfaces (IUI)

TABLE I.
CONFERENCE PROCEEDINGS AND JOURNALS STUDIED

Conference or journal	Title	Abstract	Keywords	Total
CHI 2000-2007	11	51	30	58
NordiCHI 2002-2007	1	5	4	6
MobileHCI 2004-2006	5	21	13	23
IUI 2000-2007	7	20	20	29
HUMAN-COMPUTER INTERACTION 2000-2007	5	4	1	5
TOCHI 2000-2007	2	13	7	14
PUC 2000-2007	4	11	4	14
Total	35	125	79	149

5. Human Computer Interaction

6. ACM Transactions on Computer Human Interaction (TOCHI)

7. Personal and Ubiquitous Computing (PUC)

All the articles have been peer-reviewed and the journals and conferences selected have an established reputation as the main publications for HCI research. Out of more than 200 articles consulted, 149 were kept for more detailed analysis.

The articles were selected using the following criteria:
1. navigation is mentioned in the title, or

2. in the abstract, or

3. as one of the keywords.

The total number of articles found is summarized in Table 1. Where overlap was found between the articles, that is, where navigation is mentioned in the title and in the abstract/keywords, or navigation is mentioned in the abstract and in the keywords, those articles were removed. The total number of selected articles was 149.

III. CHARACTERISTICS OF NAVIGATION RESEARCH

Navigation is the process of moving through an environment, either virtual or real. Thus, there are different kinds of navigation; it can, for example, be goal-directive or explorative (Darken and Sibert, 1996). Wayfinding is an essential part of navigation. The tasks of wayfinding can be categorized as naïve search, primed search, and exploration. Purposeful movement during navigation improves with increased spatial knowledge of the environment. Spatial knowledge can be described as three-level information: landmark knowledge, procedural knowledge and survey knowledge. (Darken and Sibert, 1996). Based on the articles reviewed for this study, navigation research can be roughly categorized into research on navigation *within* information systems, such as in databases, web pages, other digital documents and three-dimensional virtual environments; and on navigation *with* information systems, such as mobile guidance systems or wearable maps. Furthermore, navigation research can focus on 1) *navigation with information systems designed for desktop computers* or 2) *navigation with information systems designed for non-desktop computers* or 3) *navigation in real physical environments with non-desktop computers*. The two latter types of research include mobile computers, large displays, wearable computers and embedded technology. The objects, that is, navigators, are usually human beings but in recent research, studies on robots navigating in real physical environments (Amant and Christian, 2003) have been explored as well.

Navigation research can be divided into several categories according to the technology that is used when users are completing the navigation task. Based on the articles reviewed, navigation research during the 2000s can classified in the following ways:

1. Navigation in information systems with desktop computers

2. Navigation in 3D virtual environments with desktop computers and other type devices (e.g. haptic)

3. Navigation in information systems with mobile devices

4. Navigation in information systems with wearable devices

5. Navigation in information systems with embedded technology

Even with technology-based classification, which seems to be clear and well-restricted, there are some cases that are not easy to categorize according to the technology. For example, research on multi-device environments and social interaction are not so well suited to this type of technology-driven classification.

The primary motivations for navigation can be exploring or searching for information or completing a wayfinding task in all of these five cases, that is, when human beings are 1) navigating in information systems with desktop computers, 2) navigating in three-dimensional virtual environments, 3) navigating in information systems with mobile computers, 4) navigating with wearable computers, and 5) navigating with embedded technology. However, the wayfinding task in a real environment or context can focus only on mobile, wearable or embedded technology. Furthermore, these navigation studies with mobile, wearable and embedded technology can take place indoors or outdoors. The main difference between navigation with desktop and mobile, wearable and embedded computers is that the latter is designed to be carried out by pedestrians or vehicle drivers. In these cases, the navigation may not be the primary task of the users to the same extent that it could be when users are navigating with the desktop computers at their offices. In the first and second categories (see Table 2), there are similarities in research, for example, exploring orientation issues see (Tory et al., 2004). Moreover, in the last four categories there can also be similarities found in research, particularly if the focus is on wayfinding tasks. Wayfinding tasks are one of the key issues both when choosing the viewpoint in the research of three-dimensional virtual environments, and when providing instructions for finding the correct route for pedestrians or vehicle drivers.

In the research field of human computer-interaction, keyword classification systems reflect navigation research, which is currently more related to navigation in information systems. One of the most cited classifications is the ACM Computing Classification System (1998). According to this classification, navigation is part of the class H.5.4 Hypertext/Hypermedia, Navigation. However, if we analyze the empirical data of this review, the researchers themselves have taken a much broader view, describing their navigation research using the primary keywords of the ACM Classification system. They describe their research mainly using the ACM Classification

TABLE II.
SUMMARY OF ARTICLES CATEGORISED BY TECHNOLOGY

	Total share of reviewed articles
Navigation in information systems with desktop computers	53 %
Navigation in 3D virtual environments with desktop computers and other devices	11 %
Navigation in information systems with mobile devices	31%
Navigation in information systems with wearable devices	2 %
Navigation in information systems with embedded technology	5%

system with primary keyword (53 % of the total papers) into the class H.5.2 User Interfaces. In addition, classes such as K Computing Milieux, C.3 Special-Purpose and application-based systems, Real-time and embedded systems, J. Computer Applications, and D Software, are used to describe the navigation research. Analyzing the five different research focuses (see Table 2) in general terms by ACM, it seems that navigation research with desktop computers covers all the areas, whereas the non-desktop navigation research still primarily focuses on issues of design and human factors, and less on experimentation and theory development. This is the case particularly with navigation research of wearable and embedded technology. It seems that research related to design, human factors, and performance is covered in all three technology-based navigation research domains. Even if studies are categorized with the technology-driven classification presented in the Table 2, more detailed analysis with key concepts, navigation research methods, and outcomes of the studies is needed. With better understanding of the concepts, methods used and the results of previous navigation research, more detailed explanation of prior research within the research stream is found and therefore arguments for future research trends and directions can be based on more solid ground.

A. Navigation in information systems with desktop computers

According to our data, current navigation research in the field of human computer interaction primarily concentrates on examining how users navigate within large information systems. This trend is easy to recognize by the amount of articles reviewed for this study, particularly from the CHI conference and the two HCI journals, TOCHI and Human Computer Studies. The total share of the articles reviewed in this study is 53%. This figure does not include navigation research into three-dimensional virtual environments. The main focus of navigation research (see Table 4) covers a wide range of issues from user-related issues to social navigation and multimodal interaction. It is clear that this type of navigation research has the longer research tradition compared with non-desktop navigation research, and therefore desktop navigation research covers a wider range of research topics. The key topics of user-related issues of research includes, for example, gender and age issues in addition to user adaptability, whereas the rest of research focuses on the impact of navigating with other users (social navigation), user interface design, navigation techniques (panning, scrolling etc.), searching and browsing, filtering and recommending, and multimodal interaction. However, research into developing usability evaluation methods is also the current centre of attention of this type of navigation research.

The primary research methods for navigation research on desktop navigation are quantitative and experimental research methods. The data-gathering methods and data-analyzing methods are most often based on statistical analysis. Only a few, (see Wall and Brewster, 2006), if any, research investigations were conducted using qualitative methods. Main findings covers a wide range of navigation techniques, theoretical developments, design guidelines for user interfaces, usability evaluation methods, new systems for recommend users, and issues related to social navigation.

As a summary, navigation HCI research with desktop computers during the 2000s seems to explore how to develop techniques to navigate, how to present and search/browse information, what kind of differences there are between women and men navigators and users with different ages, in addition to the impact of other navigators. Moreover, how to evaluate usability of navigation is also an important research topic. As a result, system development, design guidelines and method in addition to theoretical development can be presented.

B. Navigation in three-dimensional virtual environments

A total of 11% of the articles reviewed concerned navigation research in three-dimensional virtual environments. The key focus of the research (see Table 5) is covering user-related issues, social navigation and viewpoint and orientation issues. However, evaluation methods, multimodal interaction and user interface issues are lacking this type of research based on the reviewed articles. The main research methods are quantitative, experimental methods, for example, task-based controlled laboratory tests.

The main findings of this type of navigation research includes, for example, the development of new navigation techniques, e.g., (Tan et al, 2001), choosing orientation (Tory et al., 2004), or viewpoint Tatemura (2000). Furthermore, user interface-related studies resulted in an intelligent navigation interface with a personalisable assisting mechanism (Li and Hsu, 2004), and more explicit or exaggerated representations of actions (Hindmarsh et al., 2000).

To summarize, navigation research of three-dimensional virtual environments during the 2000s covered how users orientate and choose a viewpoint, what kind of navigation techniques users have, how to present the information of an avatar, the impact of other navigators and the gender of a user. The navigation research related to virtual environments is resulting techniques of navigation, choosing viewpoint and orientation and mechanisms of representations.

C. Navigation in information systems with mobile technology

In relation to research on navigation in information systems with non-desktop computers, and particularly with mobile (or handheld devices), the total share is 31% of the all reviewed articles.

TABLE III.
KEY CONCEPTS OF NAVIGATION IN INFORMATION SYSTEMS WITH DESKTOP COMPUTERS

Research focus	Concepts
Navigation techniques	panning, hop, scrolling, fisheye, click action, crossover effect, dynamic/ static peephole
Searching and browsing	faceted search, visual data browsing, interacting with search-engines
Social navigation	social affordance, socially translucent systems, social search, social markers
User-related issues	gender and age issues, user adaptability, web pages visited
Multimodal interaction	tactile interaction, speech recognition, multi-point interaction
Usability evaluation	evaluation of 3D techniques, cognitive walk-through, easy prototyping, interpretability, automatic checking tools
Information visualization	scalability of information visualisation
Filtering and recommending	content-based and collaborative filtering, personalization in e-commerce, resource adaptability,
User interface design	overview display, levels of detail, off-screen target selection, halo interface, reconnaissance agents, zoomable user interfaces with and without an overview
Other issues	task-based taxonomy, goal-directed behaviours, Hyperbolic Tree, user's success rates, page re-visitation, entry points in to video, effects of context information (structural and temporal), contextualising navigation, automatically build associations in different media

TABLE IV.
KEY CONCEPTS OF NAVIGATION IN 3D VIRTUAL ENVIRONMENTS

Research focus	Concepts
Navigation techniques	a personalisable assisting mechanism, tracking system, Speed-coupled Flying with Orbiting
Orientation	performance differences for 3D orientation, spatial orientation tasks
Viewpoint	Semi-distorted views to support peripheral awareness; transition between local and global view, views blocked by obstacles
Social navigation	sensitive to the actions of others
Navigation task	spatial orientation task
User related issues	gender bias, gender-specific navigation,
Information vizualisation	explicit or exaggerated representations of actions provided by avatars

It is argued here that the main differences in navigation research between research on navigation in information systems and navigation with mobile devices in real environments is the impact of context. Tamminen et al.(2004) have stated that in mobile contexts, users' internal factors are different and external factors are dynamic and unpredictable. Furthermore, they argue that in relation to a mobile context, users solve navigation problems with social solutions, both national and within the mobile context, and there are temporal tensions, such as acceleration, declaration, hurrying, normal and waiting. Regarding the user interface issues, modality selection and interruption management are two of the main issues (Tamminen et al., 2004) in mobile navigation. According to the data of this study, the research focus of navigation research with mobile devices is on searching and exploring, multimodal interaction, user-related issues, user interface and navigation techniques (see Table 6). The key issues in user-related themes are limited to the impact of age and the urban environment.

Main findings of non-desktop navigation research are related to interaction design, for example, Holland et al., (2002) and adaptability, e.g., Baus et al., (2002).

To summarize, navigation research with mobile technology during the 2000s explores navigation techniques, how to search and explore information, the possible impact of age and spatially-aware displays on navigation, in addition to how to design user interfaces and tailor the presentation information. In multimodal interaction, speech seems to be emphasized as the most popular modality.

D. Navigation in information systems with wearable and with embedded technology

In the research field of HCI, navigation with wearable computers is not yet in the mainstream of the current research. Of the reviewed articles, the total share was less than 2% of all the articles. The main focus of the research concentrates on interaction and user interface topics. The key concepts of research (see Table 7) involves primarily how simple interaction with wearable computers can be conducted. Other research interests include studies on positioning and measuring.

The main findings include issues such as presenting a platform or a prototype for wearable computing, or a map application, see for example, Raghunath and Narayanaswami (2002), Rantanen et al., (2002). Navigation HCI research during the 2000s is at an early phase. The amount of published articles are as yet very few and the topics concentrated on issues such as basic interaction and the design of user interfaces in this type of navigation research.

The embedded technology in this study is defined as including large displays, outdoors or indoors, ubiquitous and pervasive systems. The total share of this type of navigation research in this study is 5%. The main focus of this type of research seems to focus on navigation techniques and displays (see Table 8). In other words, issues such as the adaptability of displays and tangible displays, besides the design of sizes and visual angles of displays, are highlighted in the navigation research.

The main findings relate to the design issues of the displays, such as size [24], adaptability and choosing the visual angle, or comparing physical and virtual navigation, for example, (Ball et al., 2007).

E. Special case: Navigation in a real physical environment with mobile/wearable/embedded technology

Of the reviewed articles in this study, navigation with mobile devices includes special cases, for example, where the users' actions involves navigation tasks in a real environment with a mobile system particularly designed for support navigation and wayfinding in a real environment. It is argued here that in most cases of this type of navigation, the role of navigation in the information system is the secondary task, whereas the primary task is the navigation (by walking, driving a car, or a bike) in a real physical

TABLE V.
KEY CONCEPTS OF NAVIGATION IN INFORMATION SYSTEMS WITH MOBILE COMPUTERS

Research focus	Concepts
Navigation techniques	static and dynamic peephole navigation
Searching and exploring	navigating web pages, effective browsing, tailoring information
Multimodal interaction	voice response
User-related issues	age, urban people
User interface	inconsistent UIs, speech-based UI, an expressive representation for location
Context	spatially aware displays
Other	improving productivity and efficiency, personalised service, hybrid system, decision-theoretic handheld system

TABLE VI.
KEY CONCEPTS OF NAVIGATION IN INFORMATION SYSTEMS WITH WEARABLE COMPUTERS

Research focus	Concepts
Interaction	Automatic emergency message
User interface	Navigation aids, a map application, a visualisation method
Other	Communication, positioning. measuring the human and the environment, a wearable computing platform on a wrist

TABLE VII.
KEY CONCEPTS OF THE ARTICLES CONCERNING NAVIGATION IN INFORMATION SYSTEMS WITH EMBEDDED COMPUTERS

Research focus	Concepts
Design factors	display size, visual angle constant
Multimodal interaction	tangible displays
User interface	adaptive displays
Navigation technique	Physical vs. virtual navigation, spatial orientation, directional signs
Other	user performance

environment. This is not the case when users are navigating, for example, with desktop computers in virtual environments. The possible misunderstanding and uncertainty with navigation concepts between these research fields may be apparent. Therefore, it is suggested here that we should clarify whether the research is about virtual navigation (in an information system with desktop/mobile/embedded technology) or (real) navigation (in a real environment with technology). In most cases, this may be obvious, but above all, when the research into virtual navigation is seeking and developing novel methods and concepts based on the principles of real navigation, for example, when using landmarks and utilizing navigation strategies, the potential for misinterpretation is real.

When exploring navigation HCI research that focuses on supporting navigation tasks with mobile technology, Table 9 describes how that navigation research focuses on issues of context (indoor, outdoor, in-vehicle, pedestrian). The user-related key concepts seem to focus on the impact of visual ability and the impact of indoor or outdoor navigation, as well as whether the navigators are acting as a driver or a pedestrian.

TABLE VIII.
KEY CONCEPTS OF THE ARTICLES

Research focus	Concepts
Wayfinding	Wayfinding strategies, route guidance, situated navigation, navigation task
Indoor	Hybrid system
Outdoor	Wayfinding, context awareness
Pedestrian navigation	Navigation assistance, perceptible landmarks, uncertain predictions
User interface	Wearable adjustable 3D version of the map application, the maintenance of forward/ up correspondence, spatially aware display, adaptive route directions
In-vehicle navigation	Driving performance and listening synthetic and natural speech, speech-based UI, mismatch between the complexity of maps and the attention demands of driving
Embedded technology	Scalability of information visualisation, target acquisition, tangible user interface that facilitates retrieval of historical stories in a tourist spot. Orientation, movement, relative positions of physical, mental rotation tasks
User-related issues	Visually-impaired people
Interaction	One-handed thumb
Other	Eyes-free navigation, environmental factors

To summarize, navigation HCI research to support real navigation tasks is for the most part divided into indoor or outdoor navigation and/or into pedestrian navigation and in-vehicle navigation.

IV. CONCLUSIONS

The main results of navigation research between 2000 and 2007 tend to contribute mainly by providing novel systems, guidelines ,e.g., Tan et al. (2006), for designers' user interface design, e.g., Raghunath and Narayanaswami (2002), McGrenere et al. (2007), systems or prototypes of systems, e.g. Holland et al. (2002), Cheverst et al. (2002), Yee (2003), Rantanen et al. (2002), and novel evaluation methods e.g., Takagi et al., (2007). It is quite interesting how research on navigation, whether it is focusing on navigation within an information system, or wayfinding in a real physical environment, possesses some similarities. On a general level, such research focuses on navigation strategies, and navigation techniques connect different kinds of navigation research (see Table 10). However, there are clear differences as well. For example, research on social issues in mobile navigation is lacking. Most of the research presents a cross-sectional view about the navigation issues. There are not so many studies focusing on the theory development of navigation or longitudinal studies indicating the impact of using navigation systems in a real or virtual environment. Fewer studies are about the possible transformation of user actions when they are navigating or using navigation systems.

One of the almost totally missing research topics of navigation HCI concerns navigation with multi-device environments. Nevertheless, most users are already using multiple devices in their everyday life. This review covers the field of navigation research during the 2000s so far. Based on the analyzed data, there are plenty of challenges for researchers and several directions for their future work. A general proposal for future work includes sugges-

TABLE IX.
TAXONOMY FOR THE RESEARCH FOCUS

	Research focus
Navigation in information systems with desktop computers	Evaluation methods, navigation techniques, multimodal interaction, social navigation
Navigation in 3D virtual environments with desktop computers and other devices	Social navigation, performance differences, orientation, viewpoint
Navigation in information systems with mobile devices	Searching, exploring, user interface, pedestrian navigation, in-vehicle navigation
Navigation in information systems with wearable devices	Navigation aids, communication, user interface
Navigation in information systems with embedded technology	Display size, adaptive displays, user performance

tions for conducting more qualitative research alongside quantitative research. For example, exploring the context and its impact on navigation with qualitative research methods, whether with desktop or mobile computers, may reveal critical factors about navigation. Furthermore, due to the fact that we are living and navigating in *multi-device* environments, much more research should be carried out on exploring what kinds of possible impact changing devices during a navigation task can cause. In additions, *longitudial studies* of navigation are needed to better understand how navigation is possible when changing the technology that users are provided with.

For non-desktop navigation research, there are gaps to fill if we compare navigation research into desktop navigation. First, it is strongly emphasized here that describing who the users are and how the sample of users for the tests and evaluations were chosen could be mentioned more precisely. People do have different kind of navigation strategies or techniques, see for example Czrewinski et al., (2002) and Guiard et al. (2007), and therefore it is more than crucial that these *user-related issues* should be explored. Navigation in a variety of contexts (indoor, outdoor, pedestrian, in-vehicle) has been explored, but it does not describe for example what kind of impact gender or age has on mobile navigation. Therefore, possibilities how to adapt novel navigation systems according to user's profile should be explored. Second, what is missing almost totally from non-desktop navigation research, is research on evaluation methods. For non-desktop navigation research, as distinct from the desktop environment, the fact that navigation cannot be the primary task of a user as often it is in a desktop navigation, should be noted in the research as well. In other words, what are the main challenges when people are navigating in real environments? Furthermore, what kind of challenges people face if navigation guidance is not presented on the screen of mobile device but it is, for example, projected on different surfaces.

Finally, based on this review, due to the possible misunderstanding and uncertainty with navigation concepts between the research fields of navigation in information systems and navigation with information systems, it is suggested here that we should clarify whether the research is about *virtual navigation* or *navigation* in a real environment.

REFERENCES

[1] Amant, R., St. Christian, D. B. (2003). Environment modification in a simulated human-robot interaction task: experimentation and analysis. *IUI'03 Proc. of the 8th international conference on intelligent user interfaces.*

[2] Ball, R., North, C., and Bowman, D. A. (2007). Move to improve: promoting physical navigation to increase user performance with large displays. *CHI '07 Proc of the SIGCHI Conference on Human Factors in Computing Systems.* CHI '07. ACM, New York, NY, 191-200.

[3] Baus J., Krüger, A., Wahlster, W. (2002). A resource-adaptive mobile navigation system. *IUI '02: Proc of the 7th international conference on intelligent user interfaces.*

[4] Baus, J., Wasinger R., Aslan I., Krüger, A., Maier, A., Schwartz, T. (2007). Auditory perceptible landmarks in mobile navigation. *IUI'07 Proc of the 12th international conference on Intelligent user interfaces.*

[5] Cheverst, K., Davies, N., Mitchell, K., Friday, A., and Efstratiou, C. (2000). Developing a context-aware electronic tourist guide: some issues and experiences. *CHI '00 Proc of the SIGCHI Conference on Human Factors in Computing Systems.* CHI '00. ACM, New York, NY, 17-24.

[6] Cliburn, D., Winlock, T., Rilea, S., Van Donsel, M. (2007). Dynamic landmark placement as a navigation aid in virtual worlds. VRST '07: *Proc of the 2007 ACM symposium on virtual reality software and technology.*

[7] Czerwinski, M., Tan, D. S., and Robertson, G. G. (2002). Women take a wider view. *CHI '02 Proc of the SIGCHI Conference on Human Factors in Computing Systems: Changing Our World, Changing Ourselves.* ACM, New York, NY, 195-202

[8] Darken, R., Sibert, J. (1996). Navigating large virtual spaces. *Int Journal of Human-Computer Interaction, January-March 1996, 8* (1), 49-72. doi:10.1080/10447319609526140

[9] Guiard, Y., Du, Y., and Chapuis, O. (2007). Quantifying degree of goal directedness in document navigation: application to the evaluation of the perspective-drag technique. *CHI '07 Proc of the SIGCHI Conference on Human Factors in Computing Systems..* ACM, New York, NY, 327-336.

[10] Hindmarsh, J., Fraser, M., Heath, C., Benford, S., and Greenhalgh, C. (2000). Object-focused interaction in collaborative virtual environments. *ACM Transactions in Computing-Human Interaction, 7* (4), 477-509. doi:10.1145/365058.365088

[11] Holland, S., Morse, D. R., Gedenryd, H. (2002). AudioGPS: Spatial Audio Navigation with a Minimal Attention Interface, *Personal and Ubiquitous Computing, 6* (4). doi:10.1007/s007790200025

[12] Jöst, M., Häußler, J., Merdes M.,, Malaka R. (2005), Multimodal interaction for pedestrians: an evaluation study. *IUI '05: Proc of the 10th international conference on intelligent user interfaces.*

[13] Kjeldskov, J. & Graham, C. (2003). A Review of MobileHCI Research Methods. In *Lecture Notes in Computer Science: Human-Computer Interaction with Mobile Devices.* 5th International Symposium, Mobile HCI 2003, pp.317-335. Berlin, Heidelberg: Springer-Verlag.

[14] Li.,T-Y, Hsu, S-W., (2004). An intelligent 3D user interface adapting to user control behaviors. *IUI '04: Proc of the 9th international conference on intelligent user interfaces.*

[15] Lynch, K. (1960). *Image of the City.* Cambridge: The Technology Press & Harvard University Press.

[16] Mehra, S., Werkhoven, P., and Worring, M. (2006). Navigating on handheld displays: Dynamic versus static peephole navigation. *ACM Transactions in Computing-Human Interaction 13* (4) , 448-457.

[17] McGrenere, J., Baecker, R. and Booth, K., S. (2007): A field evaluation of an adaptable two-interface design for feature-rich software. In *ACM Transactions on Computer-Human Interaction,* 14 (1), 3

[18] Patton, M., Q. (2002). Qualitative Evaluation and Research Methods. Thousands Paks. CA: Sage Publication.

[19] Raghunath, M. T., Narayanaswami, C., (2002), User Interfaces for Applications on a Wrist Watch. *Personal and Ubiquitous Computing, 6* (1). doi:10.1007/s007790200002

[20] Rantanen, J., Impiö, J., Karinsalo, T., Malmivaara, M., Reho, A., Tasanen, M., and Vanhala, J. (2002). Smart Clothing Prototype for the Arctic Environment, *Personal and Ubiquitous Computing, 6* (1). doi:10.1007/s007790200001

[21] Swanson E.B. and N.C. Ramiller, (1993). Information systems research thematics: Submissions to a new journal, 1987-92. *Information Systems Research 4,* (4) 299-330. doi:10.1287/isre.4.4.299

[22] Takagi, H., Saito, S., Fukuda, K., and Asakawa, C. (2007). Analysis of navigability of Web applications for improving blind usability. *ACM Transactions on Computer-Human Interaction 14* (3) 13. doi:10.1145/1279700.1279703

[23] Tamminen, S., Oulasvirta, A., Toiskallio, K., Kankainen, A. (2004). Understanding mobile contexts, May 2004 *Personal and Ubiquitous Computing, 8* (2). doi:10.1007/s00779-004-0263-1

[24] Tan, D. S., Robertson, G. G., and Czerwinski, M. (2001). Exploring 3D navigation: combining speed-coupled flying with orbiting. *CHI '01 Proc of the SIGCHI Conference on Human Factors in Computing Systems..* ACM, New York, NY, 418-425.

[25] Tan, D. S., Gergle D., Scupelli P., Pausch, R., (2006). Physically large displays improve performance on spatial tasks. *ACM Transactions on Computer-Human Interaction, 13* (1). doi:10.1145/1143518.1143521

[26] Tatemura, J., (2000), Virtual reviewers for collaborative exploration of movie reviews. *IUI '00: Proc of the 5th international conference on Intelligent user interfaces*

[27] Tolman, E. (1948). Cognitive maps in rats and men. *Psychological Review, 55,* 189-208. doi:10.1037/h0061626

[28] Tory, M., Moller, T., Atkins, S., Kirkpatrick, A., E., (2004). Combining 2D and 3D views for orientation and relative position tasks. *CHI '04: Proc of the SIGCHI conference on Human Factors in Computing Systems.*

[29] Vinson, N., G., (1999). Design guidelines for landmarks to support navigation in virtual environments. *CHI '99: Proc of the SIGCHI conference on Human Factors in Computing Systems: the CHI is the limit.*

[30] Wall, S. and Brewster, S. (2006). Feeling what you hear: tactile feedback for navigation of audio graphs. *CHI '06 Proc of the SIGCHI Conference on Human Factors in Computing Systems.* R. Grinter, T. Rodden, P. Aoki, E. Cutrell, R. Jeffries, and G. Olson, Eds. CHI '06. ACM, New York, NY, 1123-1132.

[31] Yee, K. (2003). Peephole displays: pen interaction on spatially aware handheld computers. *CHI '03 Proc of the SIGCHI Conference on Human Factors in Computing Systems.* CHI '03. ACM, New York, NY, 1-8.

ACKNOWLEDGMENT

The author would like to acknowledge the partners of TOPI project at the Tampere University of Technology and at the University of Tampere

AUTHOR

Teija Vainio. Author is a researcher at the Human Centered Technology Unit, Tampere University of Technology in Finland (e-mail: teija.vainio@tut.fi).

27

Extending LIP to Provide an Adaptive Mobile Learning

Mona Laroussi[1,2], Alain Derycke[2]

[1] Riadi Laboratory /ENSI , Tunis, Tunisia

[2] LIFL Laboratory - Lille 1 University Lille France

Abstract—In this paper, we discuss the issue of adaptativity in mobile learning. To adapt a system, we need a learner model. We discuss the limits of learner model in the environment of mobility. We present LIP (Learner Information Package) and we propose an extension of LIP (C-LIP) in order to take into consideration new dimensions induced by mobility. To test this extension, we conceive and develop a prototype of an editor aiming at assisting the teacher in the implementation and in the update of C-Lip.

Index Terms—adaptativity, context, mobile learning, learner context, LIP

I. INTRODUCTION

We are currently working on solutions to provide adaptativity in an environment of mobile learning. we particularly investigate how to integrate the advantages of mobile learning with strategies and techniques successfully employed in web-based educational systems, especially methods and techniques developed for adaptive educational hypermedia systems. Research in adaptive educational hypermedia has ascertained several techniques for navigational level and content level adaptation [11].

In adaptive educational hypermedia, the focus is on the learner. In fact, adaptativity implies the integration of a learner model in the system and uses this model to adapt navigation, content and interaction.

In this regard, the past eight years have seen a rapid growth in research, development and deployment of mobile technologies to support learning. Although research in this area began with the seminal work of Kay and colleagues at Xerox PARC [10] it is only recently that both technology and educational needs have converged. The new technology includes multimedia-equipped mobile phones, personal digital assistants (PDAs) and pen tablet computers; the new emphasis in education is on supporting the learner, in collaboration with peers and teachers, through a lifetime of education, both within and outside the classroom [11].

However, the majority of mobile learning systems do not take into account the heterogeneous needs of learners by providing them with the same learning material and process to learners without taking into account their various contexts. A solution to this problem can be best achieved through the use of learner models used in the traditional learning system or in hypermedia system.

The combination of learner model standards with current and emerging mobile technologies offer better information presentation that take into account the characteristics of the learner performing the search; thus, achieving personalized adaptive learning.

This paper is concerned with extending the standard related to the learner model to support context of mobile learning. The paper is structured as follows: Section 1 deals with the terminology necessary to understand the paper. Section 2 gives an overview of the main issues and requirements related to learner models. Section 3 introduces the most important standards dealing with learner modeling with an emphasis on LIP, Limits of LIP in a mobile learning context and possible extension. Section 4 presents the extension proposed in learner model and so Extended LIP editor (C-LIP). Finally, Section 5 gives a summary of the paper and outlines perspectives for the future.

II. BACKGROUND

This section presents the definitions and standards associated with this work. The mobile learning is described in section A. Section B presents the context and the different elements composing it includes.

A. Mobile Learning

Mobile learning: is learning through mobile computational devices: Palms, Windows CE machines, even digital cell phone.

The vision of mobile computing is that of portable computation with rich interactivity, total connectivity, and powerful processing. This small device is always networked, allowing easy input through pens and/or speech or even a keyboard when necessary (though it may be something completely different like a chord keyboard), and the ability to see high resolution images and hear quality sound. It may be that the image is overlaid on the world through glasses that act like a Heads Up Display.

Mobile learning can be considered from two viewpoints. The first one is a technically oriented perspective regarding traditional behaviorist educational paradigms as given and tries to represent or support them with mobile technologies. A main concern from this perspective is how to create, enrich, distribute and display learning material on mobile devices; the main benefits are to personalize the way of learning (where you want, when you want, what you want, as fast as you want, how you want; etc.) [12].

The second one, learning is not only the simple use of mobile devices for pedagogical purposes; even if this mobility favors the distant learner, but it gives a broader definition of mobility. By this second definition we think about the continuous connectivity (anytime and anywhere), the distributed task between wired and wireless devices, intense interaction between learner and environment.

Mobile learning is becoming widely accepted. There is a need to reduce time away from the job to deliver in a more flexible and highly adaptive way

Some students live in distant or rural areas with poor transportation systems thus it is important to reduce the costs of delivering training to large numbers of people.

people need to access instructional content 24 hours a day, seven days a week, or just in time while on their jobs. In addition in mobile learning, we can collaborate and cooperate to solve a special task. We can also access to learning material in a free way – any time in any place and with any device – but this freedom impose new constraints linked to the whole environment (indoor and outdoor) thus two notions appear context and context-aware.

B. Context

Based on previous definitions [4][7][15][1][13][16] and previous work [14]. In mobile learning environment, we have defined context as a set of element that we consider appropriate to favourite interaction between the user and the application.

Given the diversity of context information, it is useful to attempt to categorize it to make it easier to apprehend in a systematic manner. To this aim, we introduce a simple classification of context information, based on categories of contextual information.

We introduce two essential categories of context information—individual context and shared context [14].

Individual context includes information relevant to the interaction between the learner and mobile learning applications.

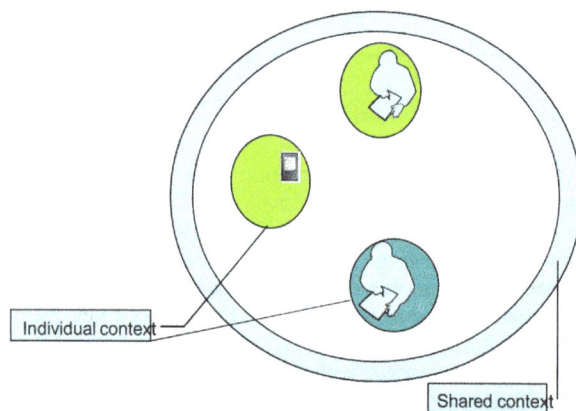

Figure 1.   Shared and individual context in mobile learning

Shared context includes information relevant to collaborative group work or learners sharing common interests. (figure 1).

III. LEARNER MODEL

A. Definition

A learner model represents relevant learner characteristics, like preferences, knowledge, competencies, tasks, or objectives. The majority of educational adaptive hypermedia systems use an *overlay model* of user knowledge [2]. The key principle of the

overlay model is that for each domain model concept, an individual user knowledge model stores some data that is an estimation of the user knowledge level on this concept. A weighted overlay model of user knowledge can be represented as a set of pairs "concept-value", one pair for each domain concept. Some systems store multiple evidences about the user level of knowledge separately. Another alternative to model the user knowledge is provided by a *historic model* that keeps some information about user visits to individual pages. Some systems use this model as a secondary source of adaptation.

The learner's goals can be modeled as a set of concepts (competencies) that can be represented similarly to the overlay model. Additionally to these dynamic dimensions the learner model includes also a more static one – *user preferences*. The most relevant ones are preferred cognitive and learning styles, as well as the language.

The following standards relate to user modelling:

IEEE Public And Private Information – specifies both the syntax and semantics of a 'Learner Model,' which will characterize a learner and his or her knowledge/abilities.

IMS Learner Information Package – learner information data exchange between systems that support the Internet learning environment.

B. Learner model in adaptive mobile learning

We can define an adaptive mobile learning as a mobile system which supports both learner adaptation and device adaptation for constructing an adaptive learning content.

We can distinguish two approaches for adaptation, learner adaptation is to present learning contents based on the profiles and preferences of individual learners. Device adaptation is the process of automatically transforming the source content to an adaptive content according to the specifications of mobile devices (screen size, resolution, alimentation, etc.).

To provide adaptive content, we need information about learner and his environment. Information about learner is stored on a learner model that should be maintained.

C. . Standards for Learner Model

The two most important standards for learner modeling are IEEE LTSC Personal and Private Information Standard (PAPI) [22], and IMS Learner Information Package (LIP) [21]. Both standards deal with several related categories of information about a learner, some of them are used in this work. Characteristics of the main standards of learner models are presented next.

The IMS LIP standard contains several categories of data about a user. The identification category presents demographic and biographic data about a learner. The goal category presents learner targets, career expectation and other objectives. The *QCL* category is used for the identification of qualifications, certifications, and licenses from recognized authorities. The activity category contains learner-related activity in any state of completion. The interest category maintains any information describing learner hobbies and recreational activities. The relationship category maintains relationships between core data elements. The competency category serves as slot for skills, experience and knowledge acquired. The *accessibility* category points toward general accessibility to learner information by means of language capabilities, disabilities, eligibility, and learning preferences. The *transcript* category presents a summary of academic achievements. The *affiliation* category presents information about membership in professional organizations. The *security key* is used for setting learner passwords.

The PAPI standard distinguishes *personal*, *relations*, *security*, *preference*, *performance*, and *portfolio* learner information. The *personal* category contains information about names, contacts and addresses of a learner. *Relations* serve as a category for relationships of a specific learner to other persons (e.g. classmate).

Security aims at providing access rights. *Preference* indicates the types of devices and objects which the learner technological support is able to recognize. *Performance* contains information about measured performance of a learner through learning material. *Portfolios access* the previous experiences of a user.

Many tools exist enabling conversion between two standards.

There are other proposals for a standardization of the learner data, but they not enter the objective of this paper. We can note two standards AICC3 and SCORM. The standard SCORM offers a data model for managing all the learning productions. This model comprises a set of fields in order to allow a standardized exchange of data between a runable training unit and the platforms.

The learner's follow-up was one of the principal concerns of AICC. In this model, the data exchanges between the learning system and a given training module are done via files. This approach allows the division of data between several modules constituting the training.

IV. LIP (IMS)

A. Definition

Learner Information is a collection of information about a Learner (individual or group learners) or a Producer of learning content (creators, providers or vendors). The IMS Learner Information Package (IMS LIP) specification addresses the interoperability of internet-based Learner Information server may exchange data with Learner Information systems with other systems that support the Internet learning environment. The target of the specification is to define a set of packages that can be used to import data into and extract data from an IMS compliant Learner Information server, a Learner Delivery system or with other Learner Information servers. It is the responsibility of the Learner Information server to allow the owner of the learner information to define what part of the learner information can be shared with other systems. The core structures of the IMS LIP are based upon: accessibilities; activities; affiliations; competencies; goals; identifications; interests; qualifications, certifications and licences; relationship; security keys; and transcripts.

B. Concept

The Learner Information Packaging Requirement Specification [21] introduced the base learner information system architecture. The underlying process components (circles) and data structures (thin rectangles) and the actors (stick-people) are shown in Figure 2.

The key components of the learner information system are:

- Local learner information system - local server(s) that are directly accessible by the corresponding user community;
- Remote learner information system - a reflection of the distributed nature of a learner information server, i.e. different parts of the 'learner information' could be stored on several servers;
- Other systems - others systems that may be interconnected to the learner information servers e.g. e-mail. The interfaces to these systems are beyond the scope of this specification;
- Data structures;
- Learner info - the actual learner information data itself;
- Access - the access rights to the learner information data i.e. who can see what;
- Messaging - the messaging protocol used to implement the actual profile interchanges; and
- Actors - the different roles of the users accessing a profile server. The different actors shown in Figure 2 are not an exhaustive list.

C. Learner Data Structure

The Learner information is separated into eleven main categories. These structures have been identified as the primary data structures that are required to support learner information. This composite approach means that only the required information needs to be packaged and stored.

- Identification: Biographic and demographic data relevant to learning;
- Goal: Learning, career and other objectives and aspirations;
- Qualifications, Certifications and Licenses (qcl): Qualifications, certifications and licenses granted by recognized authorities;
- Activity: Any learning-related activity in any state of completion. Could be self-reported. Includes formal and informal education, training, work experience, and military or civic service;
- Transcript: A record that is used to provide an institutionally-based summary of academic achievement. The structure of this record can take many forms;
- Interest: Information describing hobbies and recreational activities;
- Competency: Skills, knowledge, and abilities acquired in the cognitive, affective, and/or psychomotor domains;
- Affiliation: Membership of professional organizations, etc. Membership of groups is covered by the IMS Enterprise specification;
- Accessibility: General accessibility to the learner information as defined through language capabilities, disabilities, eligibilities and learning preferences including cognitive preferences (e.g. issues of learning style), physical preferences (e.g. a preference for large print), and technological preferences (e.g. a preference for a particular computer platform);
- Security key: The set of passwords and security keys assigned to the learner for;

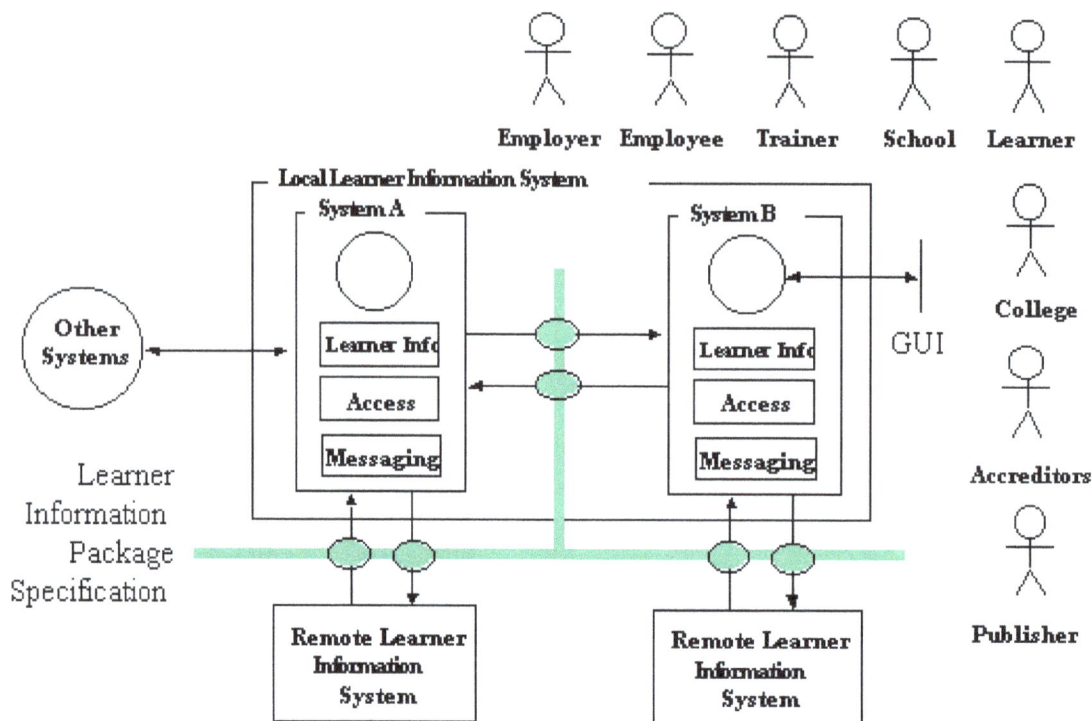

Figure 2.   Learner information system component representation

- transactions with learner information systems and services; and
- Relationship: The set of relationships between the core components. The core structures do not have within them identifiers that link to the core structures. Instead all of these relationships are captured in a single core structure thereby making the inks simpler to identify and manage.

V.   LIMITS OF LIP IN MOBILE LEARNING ENVIRONMENT

The LIP Standard was initially designed to describe the learner without considering a mobile learning environment.

However, the availability of wireless technologies and the popularity of handheld devices have opened up new accessible opportunities for education: mobile learning. Hence, the LIP core specification becomes unsuitable to satisfy the new constraints imposed by this new learning environment. Indeed, the design features such as the volatile memories or the interfaces (screen, stylet..) raise strong constraints. They impose a reflection on the fond and the form of the pedagogical subjects to present. Thereafter, we describe two different scenarios in order to explain LIP Limits.

A.   Scenario 1

Alice is a master student at university in Paris. The number of hours studied in her master is very tiny.

Thus Alice decided to seek a work in a company in Marseille. She plans to follow her studies via her mobile phone when she cannot attend the courses.

Alice faces a problem when she knows that it was proved that the profile and registration preferences must be done respecting LIP specification.

The standard does not specify preferences about her device nor the context of the environment in which she will study.

This can have bad consequences on the content and the presentation of her learning activities.

B.   Scenario 2

Alain is a service boss in a factory. He spends most of his time far from his work place because of his various commitments. The team reporting to him is generally faced with problems when using new machines. To resolve problems, the worker on the machine must contact him.

In order to facilitate the resolution of these problems, the factory finds the following solution: provide each worker with a PDA to contact the expert via Internet.

To contact the expert, any employee (learner) must be identified, and information taken by sensors integrated in the PDA (the ambient temperature, frequencies of noise which it releases...) must be provided to the expert who offers the precise solution according to case description, and the worker concerned.

This information provided to the expert must respect LIP specification. Very quickly the company was unable to carry out its objective because of the limits of the LIP core. The components of the standard did not allow a rich description of the work context of learner.

C.   Synthesis

From these two scenarios, we can see that the current structure of LIP does not support the mobile learning. Indeed the current version of LIP does not allow a geographical representation of learner and does not allow adapting the contents to the mobile devices by taking account the various contextual elements described in section 2.B.

Our objective is to define a new LIP core component that allows adaptive mobile learning (and the definition of he various parts needed for mobile learning).

VI. FROM LEARNER MODEL TO CONTEXT MODEL

Most of previous and current works on adaptive e-learning focus on user model [11][6] which takes into account only internal environment of the user: personal information, interest centers, preferences… etc. but the rush of the wireless and mobile technologies has created a move from e-learning to m-learning and P-learning (pervasive learning).

An efficient mobile learning system has to be sensitive not only to the user model information but also to the whole context that characterizes the interactions between users, applications and the surrounding environment. For that, we attempted to extend the learner model into more abstract learner context model that includes the user model and gathers all contextual elements relevant to pervasive and mobile learning systems.

Many of the traditional learner model attributes remain relevant (knowledge, preferences, misconceptions, etc.), but there are also issues that do not usually apply in PC environments. For example, how may location of the user affect an interaction? How desktop and mobile PCs might be integrated to allow the user to interact with whichever device is most convenient at the time (cf Figure 3).

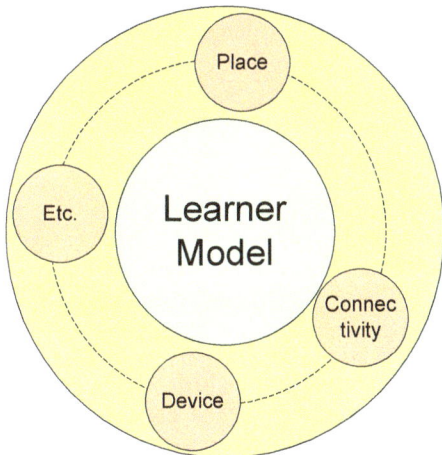

Figure 3. New dimensions for learner model induced by mobility

VII. EXTENDED LIP: C-LIP

In a situation of mobile learning, besides the traditional elements which are in relation with pedagogy, logistics and the model of learning, new elements appear: these elements can be summarized by the context as have been presented above.

We thus thought of adding an element context to LIP. But the element context already exists. We therefore took into account this element to show the limits of

these components and to modify them with regard to our contextual elements.

A. The context in LIP

The introduction of the context into the LIP was carried out by IMS which performs updates on the existing standard.

Changes were brought on its core. Thus, a new version was proposed under the name ACCLIP "Accessibility for Learner Information Package ".

The ACCLIP is an extension of the LIP specifications v1.0 IMS [21]. This new specification adds components which define accessibility preferences.

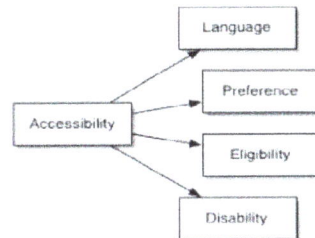

Figure 4.   previous sub-elements of «Accessibility » in LIP.

These new elements added in the core are designed to be compatible with all the work carried out previously on the LIP with regard to the intimacy, the access and the integrity of information. They provide means to describe the preferences of a learner when he/she wants to access an elearning situation.

LIP "Learner Information Package" has been modified in the information model specification. At the beginning, the component **"accessibility"** was set up to make sure that the products and technologies are able to support handicapped people.

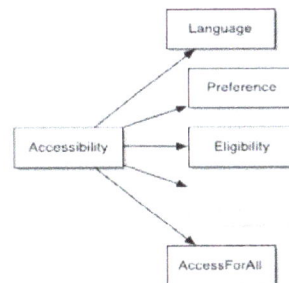

Figure 5.   new sub-elements of « Accessibility » in LIP

The specification of this component was modified. A new element named "AccessForAll" was defined under the element "Accessibility". The "Disability" element was removed and another element named "Accommodation" was added under "Eligibility".

Figures 4 and 5 illustrate the various components modifications. This new presentation allowed the addition of the context under the new component "AccessForAll".

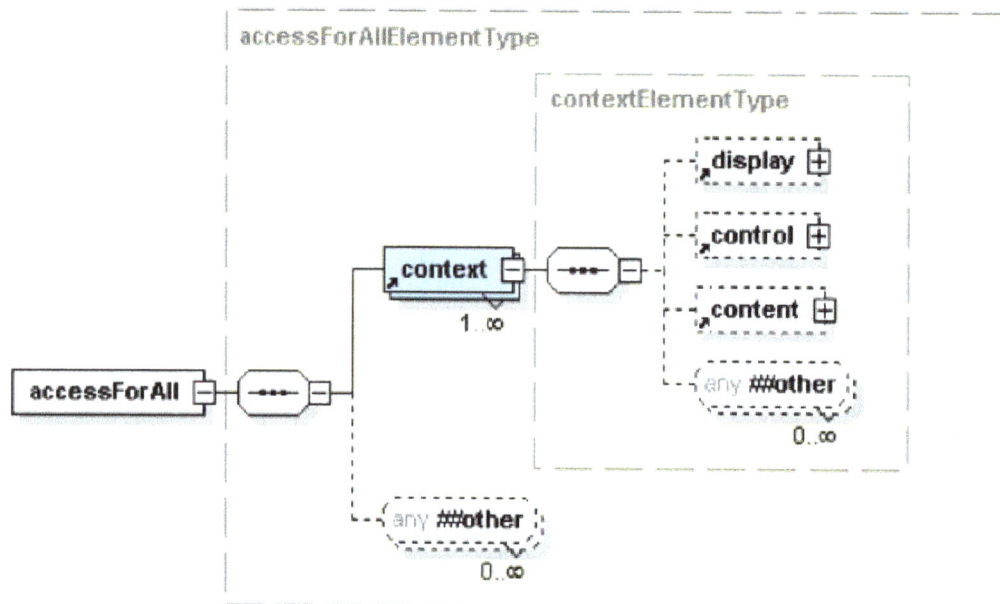

Figure 6. composition of the access for all elements

B. ACCLIP definition of context:

The new specification brought on the structure of the LIP on the level of the "Accessibility" component has led to the addition of a new element "AccessForAll" under which the context is defined. The figure 6 shows the composition of this element. The sub-elements of the context are of three Types:

1) Display:

It specifies the preference of display technology i.e how the user interface and the contents should be presented. This component has a paramount role in the display technique of the user interface. It describes in a detailed way the various parts which make an interface.

This element also defined the manner of display of the text, its color, and the sights contents. It includes also components which make it possible to define visual alert, the methods used during the sonorities integration.

2) Control:

It specifies the technologies which envisage alternative manners to command a device. This component will make it possible, for example, to control all the short cuts used by the keyboard, and the improvements in the keyboard. It defines components related to the mouse such as the speed of pointing, and sound alerts.

3) Content:

It specifies the preferences of the contents, indicating all desired transformations: determining the language used to describe the various components of the interface, and the way in which the additive contents will be presented.

C. Limits of LIP context

The part of the context, defined in the standard, provides only the least possible amount of information about the environment. It does not include the parts of the learner entourage. It is limited to define some components

related to display, keyboard, mouse and the sound. These components are not sufficient when we aim to adapt content to a mobile learner. Mobile learner implies constraints related to device such as small screens, limited autonomy, limited scale of colors, limited size of programs etc. Constraints are also related to environments: connectivity, noise, luminosity etc.

Since, we are interested in the content adaptation to learner in a mobile learning situation and we project to use LIP as a standard for the modeling learner attributes, we propose to enrich ACCLIP.

D. The core of C-LIP

We have demonstrated that the element context defined in LIP isn't sufficient to model mobile learner.

Figure 7 explains the element context added to the core of LIP. The extension is mainly represented by the context which is characterized by its nature, extension and origin and in order to manage Contextual elements, we must differentiate between them by giving them different features. Table 1 illustrates these features: nature, acquisition type, acquisition mode, relevance, evolution, adaptation and frequency of updating. The extended LIP is labelled C-LIP for context LIP. Elements of context depend on scenario and use of mobile device. Figure 7 shows the new core of LIP. The core takes into account the context with elements predefined in LIP. Elements in the added context are more detailed in order to give an idea about all features. Technically, theses features are coded as metadata in an XML file.

Once the core of C-LIP is defined, we thought of facilitating the task of the teacher by offering him/her a tool allowing data acquisition for C-LIP. The editor is based on the core of LIP and offers the traditional fields of the LIP and the new fields induced by mobility.

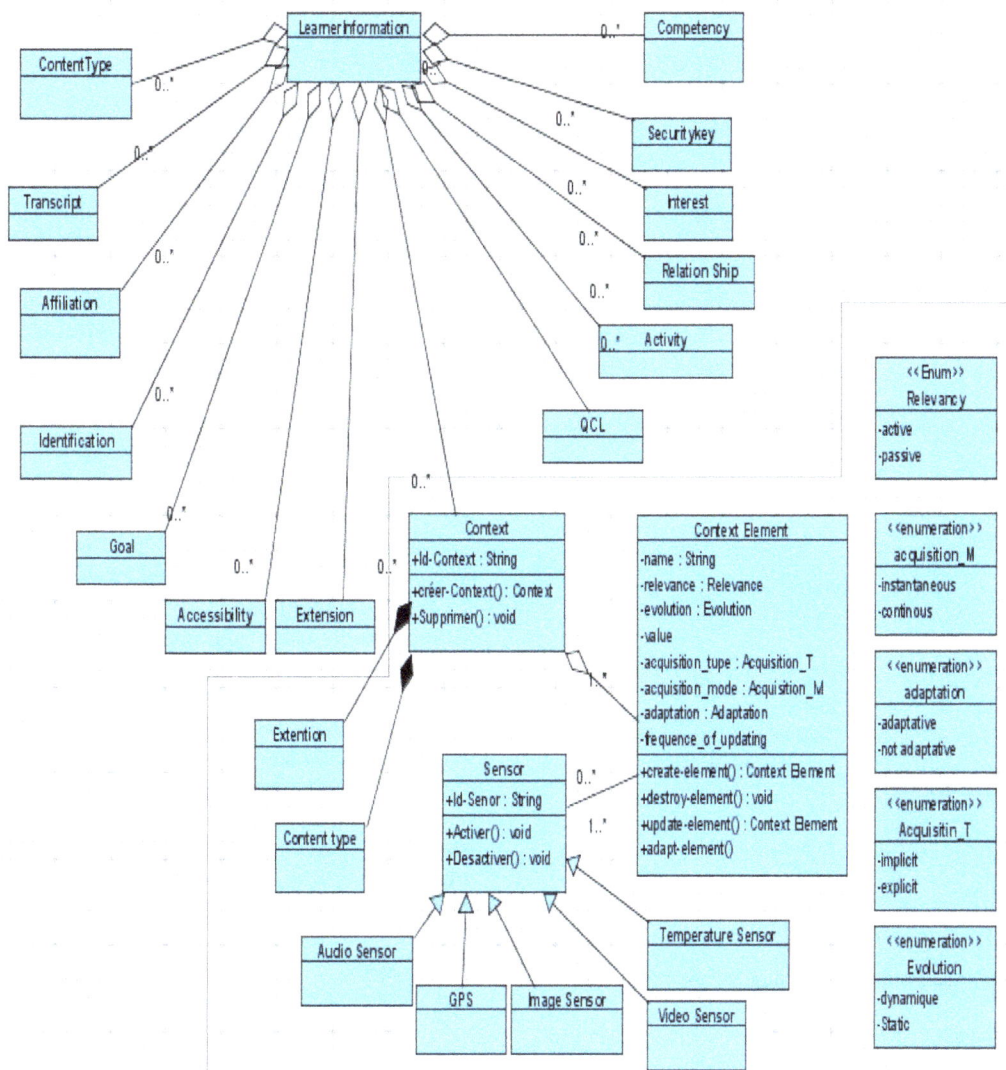

Figure 7.   The added elements in the core of LIP

TABLE I.
CONTEXTUAL ELEMENT FEATURES

Contextual element features	Possible values
Nature	Natural: temperature Artificial: the sound of the stereo channel
Acquisition type	Explicit acquisition: contextual element is directly acquired. Automatic acquisition: contextual element is sensed automatically (e.g., by sensors). Manual acquisition: contextual elements are given by learner. Implicit acquisition: contextual elements are inferred from others stored contextual element.
Acquisition mode	Instantaneous: contextual element is acquired only once at the beginning of the interaction (e.g., date). Continuous: contextual element is acquired continuously during an M-learning session (e.g., noise level).
Relevance	Active: contextual element relevant to the interaction between learner and the system (e.g., if learning type is a Visio-conference, noise level is an active element) Passive: isn't relevant to a given interaction between learner and the system (e.g., if the learner's task consists of reading a text, the name of learner is a passive element).
Evolution	Dynamic: contextual element change during the interaction (e.g., noise level). Static: contextual element does not change during interaction (e.g., season).
Adaptation	Adaptable. Not adaptable.
Frequency of updating	This feature ensures the newness of contextual elements.

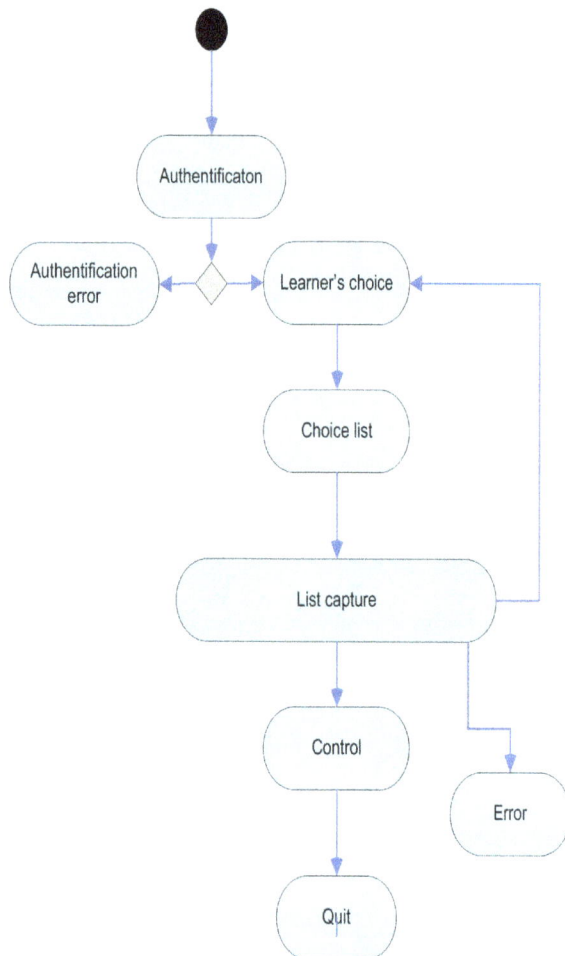

Figure 8.   Workflow of activity diagram in C-LIP

VIII. C-LIP EDITOR

A.   Functional architecture

We have designed and implemented C-LIP editor. This editor enables a semi-automatic incremental data acquisition of C-LIP fields.

Figure 8 explains the workflow function of C-LIP editor. After authentication the teacher selects student and completes the missing elements of C-LIP.

This incremental editor doesn't overload the teacher. All fields are optional and data are stored between sessions and can be restructured incremented or suppressed. The teacher makes his/her choice between different elements composing LIP and those elements appear as angles in the following interfaces (figure 10).

This editor gives the opportunity to the teacher to personalize his/her course using data saved in C-LIP. This personalization can either display the name or forename of the student only or include details about learning styles, profiles and/or mobile devices used in a particular context.

Each element of the interface implies a new interface with elements composing it.

B.   Technical details

The data are stored in XML file (Figure 9). XML file enables the reuse and the interoperability of the

information even after extension. For each group, teacher disposes of file regrouping all students of a group.

C.   Tests and evaluations

The C-LIP tool was tested in classroom study involving students and teachers. Each student received a handheld Pocket PC that could be used in all classes, and taken home, and thus could access the Internet via the school's wireless network.

In order to better understand the use of C-LIP, two aspects are examined. First, from the students' point of view we evaluate whether the use of this software enhances the quality of content and second from the teachers' point of view.

Does the use of C-lip bother the teacher?

How can the teacher use results of C-lip to adapt courses?

The primary data analyzed for this experimentation was collected from students and teachers.

At the beginning, some teachers found the form filling and the use of the interface of C-LIP very time-consuming. But they found this context modelling to adapt course material very useful.

Some students had difficulty understanding or using interface elements but they enjoyed finding personalized interface.

This small evaluation highlights several issues that should be considered for the next version such as the number of pertinent questions about the real role of student and teacher etc.

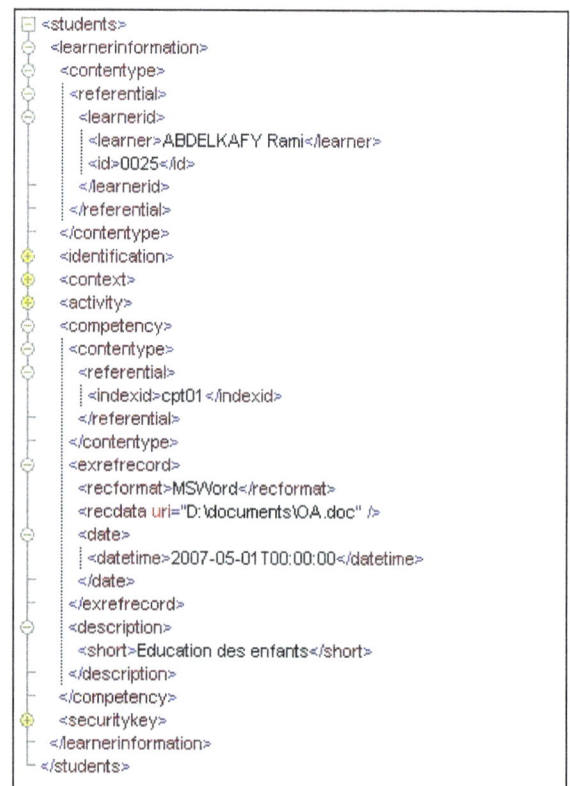

Figure 9.   XML storage

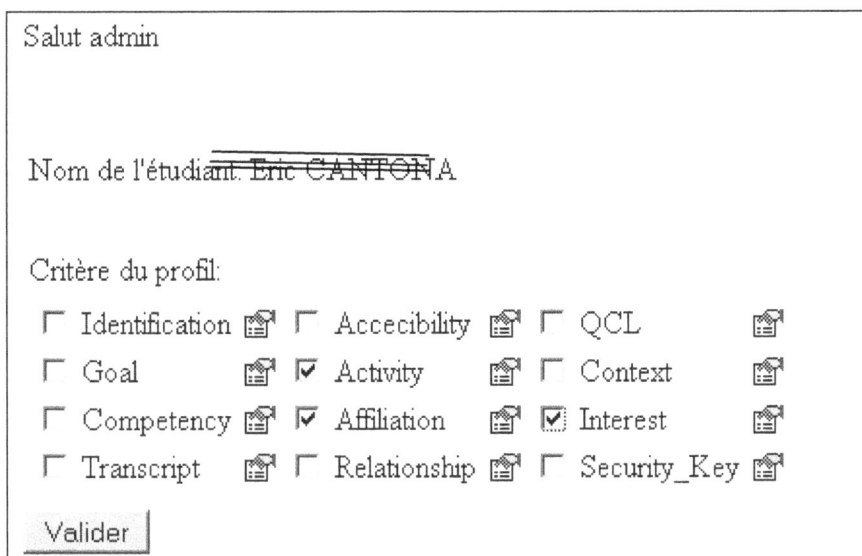

Figure 10. elements choice interface in LIP editor

IX. CONCLUSION AND FUTURE WORK

The focus of this paper was adaptive mobile learning by the use of an extended LIP. The investigation of existing learner modeling standards revealed that the IMS LIP specification is not sufficient to model mobile learner.

Thus, we proposed an extension of the LIP core in order to introduce new contextual elements induced by mobility.

Adding context in the learner model enabled us to model new elements introduced by mobility. The editor C-LIP facilitates the task in the sense that the teacher is not obliged to handle directly an XML. The result of its modelling can be read by any LMS.

The inconvenience is that all the fields are captured by the teacher and are not automatically deduced from various work of learner. We are now working on analysing learning session traces to aid learner model.

We plan to integrate C-LIP environment in web service learning platform SOLEIL [18] using SCORM as learning resource standard.

We mention that, in the same project SOLEIL, we have proposed an adaptation of SCORM to support learner's learning styles and mobility [3].

We project to create a web service based on C-LIP editor providing interoperable and distributed mobile learner contextual LIP.

REFERENCES

[1] Anind K. DEY, Understanding and Using Context, Personal and Ubiquitous Computing, Volume 5, Issue 1, February 2001, Pages: 4 – 7.

[2] P. Brusilovsky,. and C. Peylo, (2003) Adaptive and intelligent Web-based educational systems. In P. Brusilovsky and C. Peylo (eds.), International Journal of Artificial Intelligence in Education 13 (2-4), Special Issue on Adaptive and Intelligent Web-based Educational Systems, 159-172.

[3] R. Drira, M. Laroussi, A. Derycke., «SCORM dans l'apprentissage mobile». Atelier Apprentissage mobile EIAH'07, Lausanne Juin 2007.

[4] J. Ensing, Software architecture for the support of context aware applications, Preliminary study, February 2002.

[5] V. Jo, Jones Ubiquitous Learning Environment: an Adaptive Teching System using Ubiquitous Technology. In R Atkinson, Mcbeath, Joans-Dwyer (eds) Beyond the comfort zone:proceeding of the ASCILITE conference, Perth, Australia, 5-8 December 2004, pp 468-474.

[6] S. Iksal, Declarative specification and semantic composition for adaptive virtual documents. PHD thesis, December 2002.

[7] M. Kaenampornpan,., O'Neill, E., An Integrated Context Model: Bringing Activity to Context, in "Workshop on Advanced Context Modelling, Reasoning and Management", UbiComp 2004, Nottingham, UK, 2004.

[8] Kinshuk & T Goh. (2003). Mobile Adaptation with Multiple Representation Approach as Educational Pedagogy. In Uhr W., Esswein W. & Schoop E. (Eds.),

[9] ,Keil-Slawick, R. Hampel, T. Ebman, B. Re-Conceptualising Learning Environments: A framework for Pervasive eLearning. Proceedings of 3rd Int'l conf. on Pervasive Computing and Communication (PERCOM), IEEE press, 2005, 6 p.

[10] Kay, A., & Goldberg, A. (1977, March Conlpurer; 10(3), 31-42. (Reprinted in A. Goldberg (Ed.). (1988). A history of tations (pp. 254-263). Reading,MA: Addison-Wesley.)

[11] M. Laroussi, Conception et réalisation d'un système hypermédia adaptatif didactique : Le système CAMELEON, PhD Thesis, Ecole National des Sciences Informatiques, Tunis, Mars 2001.

[12] M. Laroussi, A. Derycke, New e-learning services based on mobile and ubiquitous computing, CALIE'04, Grenoble France, february2004.

[13] Z. Maamar, S. Mostéfaoui, Q. Mahmoud, Context for Personalized Web Services, Proceedings of the 38th Hawaii International Conference on System Sciences – 2005.

[14] J. Malek., M Laroussi., A Deryicke., "How to adapt context to mobile and collaborative learning", Multi-channel Adaptive Context-sensitive systems Workshop (MAC'06), University of Glasgow, 15th May 2006.

[15] R. Rupnik, M. Krisper, M. Bajec, A new application model for mobile technologies, International Journal of Information Technology and Management (IJITM), Vol. 3, No. 2/3/4, 2004

[16] .Salber, A.Dey and G. Abowd, Designing and Building Context-Aware Applications, 2001.

[17] K Sushil. Sharma and Fred L. Kitchens, Web Services Architecture for M-Learning, Electronic Journal on e-Learning Volume 2 Issue 1,P 203-216, February 2004;

[18] I. Tirellil, M. Hajjouni, M. Laroussi; "Mobile and collaborative platform based on web services". ML'05 Malte Juin 05.

[19] L. Uden,. 'Activity theory for designing mobile learning', Int. J. Mobile Learning and Organisation, Vol. 1, No. 1, pp.81–102. (doi:10.1504/IJMLO.2007.011190)

[20] Learning   Tecnology   standards   committee: http://ltsc.ieee.org

[21] IMS Consortium: http://www.imsproject.org.

[22] PAPI (2000). Public and Private Information; IEEE 2000, Draft Standard for Learning Technology - Public and Private Information (PAPI) for Learner, IEEE P1484.2/D6, 2000. http://ltsc.ieee.org/

AUTHORS

M. Laroussi (e-mail: mona.laroussi@univ-lille1.fr) is currently Assistant Professor at the Department of Computer Science in National Institute of applied sciences and Technology (Tunisia) and an associated researcher in Noce Laboratory   (Lille   France)   http://noce.univ-lille1.fr/cms/

A. Derycke. (e-mail: alain.derycke@univ-lille1.fr), is currently Professor in University of science and technology of Lille1. He is also the director of Noce Laboratory (Lille France) http://noce.univ-lille1.fr/cms/

28

A Framework for Educational Collaborative Activities Based on Mobile Devices

A Support to the Instructional Design

Cruz-Flores, René and López-Morteo, Gabriel

Engineering Institute, Autonomous University of Baja California, Baja California, México

Abstract—In this paper we present a framework (CA-Mobile Framework) for developing educational activities for collaborative learning based on mobile devices. The framework has two main components, the Reference Document component and an API for developing mobile software. This framework is an alternative approach for developing m-learning activities with an integral approach that includes educative and technological aspects through a process model. The framework components, including a process model for applying the framework are described. An evaluation of the value of the Reference Document component through the design of an educational activity was conducted with a group of eight teachers, without previous experience on the design of educational activities based on mobile devices, divided in small groups working independently on the design of the same activity. The design products of each group were compared with a control design developed by authors. The results shown that all groups were capable of replicate the control design using the CA-Mobile Framework

Index Terms—API library, collaborative learning, m-learning, framework, instructional design, process model.

I. INTRODUCTION

With the increase in the use of mobile learning (m-learning) as an alternative for enforcing and providing learning processes, many educators and researchers are interested in taking advantage of its unique attributes and features such as ready-at-hand, ad-hoc wireless networks, multimedia support, moderate and massive storage and eventually, on-the-fly communication and collaboration create several kinds of learning environments that cannot be easily replaced using other types of technology like the desktop PC (Attewell, 2005; Vahey and Crawford, 2002). This has generated the "anywhere and anytime" concept that refers to the capacity that these mobile devices have to access several sources of information or connect with other devices (mobile or not) through wireless connections in practically any place and when needed, even while moving (Rieger and Gay 1997; Wagner 2005; Shen 2008).

Instructional designers have taken these technological characteristics into account to design educational activities based specifically on the use of mobile devices. Due to the diversity of applications and uses of m-learning, different types of these applications have been developed and presented in several reported works, including text based applications like quizzes or class notes (Leung and Chan 2003; White and Byrd 2000), graphical and game based applications (Facer et al. 2004;Milos et al. 2009), video and camera based applications (Yoon et al. 2008;Chen et al. 2004) and recently reality augmented based applications (Spira, D. 2009; Mark, P. 2009;Alapetite, A. 2010).

Moreover, the use of mobile devices has generated growth in several collaborative processes supported by mobile devices. Cell phones, Smartphones and PDAs create new "on-the-go" services (Metcalf and Marco, 2006), allowing people to communicate, negotiate, socialize and learn at the same time in a cooperative and collaborative way. As a consequence, a new model was created under the name of Mobile Computer-Supported Collaborative Learning (Zurita and Nussbaum, 2004; Roschelle et al., 2005). This model begins with the premise of creating collaborative learning situations using mobile devices as the main mediator element.

Therefore, the technology used for supporting formal educational activities could allow active collaboration through real-time chats, shared screens and boards, support for team creation, awareness of participation, and control time of activities such as those described in projects by Benford et al. (2004), Hamid and Fung (2007) and Jarkievich et al. (2008). This kind of complex interactions require a control of interactions between participants, data and technology, time control and session work integrity, in a synchronous or asynchronous way.

These features are commonly implemented in many collaborative m-learning projects. However, these were designed following one specific approach, technical or educative, instead of an integral approach that uses all the mobile technology features in a didactical way according to a previous instructional design, in order to impact directly on learners. By following a non-integral approach, the result of the activity can have a low impact on learning. For example, a rich technological implementation without an adequate instructional design can distract from the educational objective. While a purely educative implementation can underuse unique characteristics of mobile devices.

Another disadvantage of not using an integral design is that it complicates reusability of didactical and technical resources. This can result in a replication of the development process when development is done for a unique or special case, thus making it hard to use these resources in another activity. Also the design process cannot be documented and organized properly.

To achieve an integral design of activities, it is necessary to take into account both educational and technical aspects throughout the complete development process including concepts of motivation (Cheng and Yeh 2009), learning experience and objectives (Parson et al. 2007), new models focused on just-in-time, just-in-place models (Metcalf, D. 2006) and different learning activity levels (Wang et al. 2004). In the same way, collaborative activities have special requirements which will be considered from instructional design to software development. For example, a previous experience with collaborative tools (Huang et al. 2009), Peer-assisted learning procedures (Tsuei 2009) and social aspects such as positive interdependence, face-to-face interaction, individual accountability/personal responsibility and teamwork skills (Smith et al. 2005).

Although these aspects are not always clear at the beginning of the design, they are an important part of the activity and it is highly recommended to describe them in greatest detail. This is important to facilitate the implementation of all functions associated with these features, both in software and didactic scripts. Also, we consider that these characteristics impact in some degree on the success or failure of the activity in a real educational scenario.

Taking these aspects into account, we developed a framework for developing m-learning activities with an integral approach that includes educative and technological aspects. The framework contains two main components; a Reference Document component, and a specialized Application Programming Interface library (API) for the development of mobile software, including a process model for applying the framework. The propose of this paper is to show the results of an evaluation of the Reference Document component (the first component of CA-Mobile Framework) realized by a group of eight teachers, in order to know their assessment of the usefulness and clarity of this component, and show evidence that the Reference Document component provides a structured and organized route to create the instructional design of activities based on mobile devices.

II.   Considerations for Designing Collaborative M-Learning Activities

The creation of a learning environment mediated by computers, where students can work together to reach shared goals, is a complex task from a technological and educational viewpoint. From a technological approach, the learning activity can use one or several interaction mechanisms as of direct communication channels (verbal, text, audio or video), shared places (boards, screens and documents), awareness of collaboration (presence and status rosters, definition and description of roles and status of the common goals), and control and monitoring of course tasks and student's learning paths. Moreover, from the educational perspective, the different skills of students as well as the environmental conditions where the activity will be developed are factors that increase its complexity. All educational activities that include the use of technology as an active actor require a differentiated educational design compared to traditional face-to-face educational activities that involve only teacher and students (Patten et al., 2006). In this sense, there exist several factors that the instructional designer should take into consideration in activity design; factors such as type of technology, physical space, use conditions, setup time, as well as training of participants for the use of technology. These, are just

some of the factors that lead the design and should be considered when developing the activity in order to obtain successful results. They can be seen as non functional requirements according to Avellis et al. (2005) and Mostakhdemin-Hosseini (2009).

In the same way that the characteristics of the desktop PC impose certain criteria and constraints for activity design, the use of mobile devices imposes its own criteria and constraints associated with its mobility and technological restrictions. Unlike the desktop PC, mobile devices have small screens, moderate storage capacity and reduced keyboards. These constraints are in some way compensated with their capacity for portability, mobility, support for multimedia, and ease with which to create ad-hoc networks between two or more devices for example, using Bluetooth, WiFi o IrDA (Cruz-Flores and López-Morteo, 2008). All of these characteristics should be considered when activities based on mobile devices are designed, even if the activities require the use and interaction of mobile devices and desktop PCs at the same time.

Thus, from an instructional design point of view, and considering the professional profile of the activity designer (experience like teacher and experience designing activities supported by computers), it seems to be necessary to have within his capabilities a huge knowledge about the potential and unique features of mobile devices, in order to determine the role of technology on the activity execution. As an aid to identify both humans and technical needs by the instructional designer, we present a classification of the needed skills, also known as pre-conditions that this professional has to develop in order to create reliable instructional designs based on the use of mobile devices. The categories of the pre-conditions for the design of these educational activities can be classified in three categories;

- Human. Considers all skills and knowledge needed to execute tasks described in the activity (e.g. experience using cell phones).

- Technical. Includes all the processes associated with technology like installation, configuration, and setup of the systems that participate in indirect tasks (supporting tasks), but are not part of the main educational activity (e.g. deployment of mobile software into devices).

- Organizational. Considers all tasks for the organization of participants before the main educational activity. This category also contains the teacher's guide for the development of the activity, as well as the administration of the computational resources (mobile devices and/or PC) needed for the execution of the activity (e.g. grouping by teams or by peers).

If these elements are considered during activity design, then the contexts for the activity can be clearly defined, because they are pre-conditions that, if not specified and considered, can significantly and negatively affect activity execution. We believe that this situation can happen because the pre-conditions are not optional and serve to create the environment where the activity will take place, and compliance with it will be necessary even if they are not considered explicitly in the design. These activities are intended to achieve their pedagogical objective through the execution of focused group tasks.

For this work, the actions like start of interaction, type of interaction, type of communication, duration of interaction, and definition of roles could be used by any collaborative educational activity that uses mobile devices. All these considerations served to organize and develop the components in the framework according to their educative or technical functionality.

III. CA-MOBILE FRAMEWORK FOR DESIGNERS AND DEVELOPERS

The interest on using mobile devices in the educational environment is not new and several works have been reported by specialized literature in the last 10 years. However, the design process for developing software for mobile devices with an educational approach has not had a substantial growth in the use of tools and programming languages. The creation of educational activities based on these devices is very dispersed. As we mentioned previously, some works are oriented mainly towards the educational context, while some others privilege the technological aspects. We believe that the aspiration of achieving the educational objective supported by technology can be done by merging features from these two fields while taking into account that the primary need is to support the learning process using mobile devices.

We propose an integral framework (CA-Mobile Framework) to support the process of designing collaborative educational activities based on mobile devices including educative aspects and the development of software to support these activities.

The CA-Mobile Framework (see Fig. 1), defines all the mechanisms necessary to support the development process of the educational activities that we are interested in. The framework artifacts are a set of documents for the specification of didactical requirements, functional requirements, and a specification/adaptation of an instructional script of activities which was adapted to consider the use of mobile devices. Moreover, the framework also provides a set of application programming interfaces (CA-Mobile API), to create and implement features in software to mediate the activity. This API contains reusable software components to implement the following functionalities: communication, turn control, textual message exchange, textual and graphical resource management, and user interface management.

The CA-Mobile API was developed for the Java programming language and includes two versions of the API, the SE edition for desktop applications and the ME edition for mobile devices with JME support. The two versions were required because the CA-Mobile Framework was designed to include mobile devices with optional interaction with desktop PCs (e.j. a chat application between cellular phones and PCs.) in the same activity. The API versions share many features like Bluetooth communication or floor control. However, others are exclusive of the ME edition (e.j. Camera control or canvas graphics control). These components also, allow controlling the execution of the process stages and the products resulting from each one. All the components of the framework are grouped by functionality and are detailed in section five of this paper.

Due to the holistic approach of the framework with respect to the educative and technical aspects, the CA-Mobile Framework can be analyzed from two viewpoints; educational approach and technological approach, which are described below.

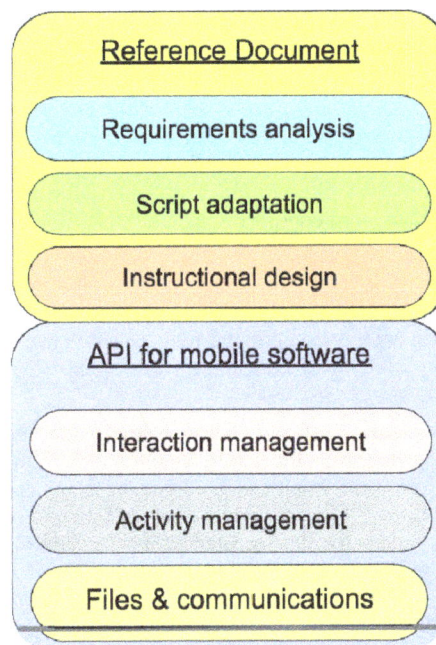

Figure 1. CA-Mobile Framework components overview grouped by educative and technical functionality.

IV. EDUCATIONAL COMPONENTS

One predominant factor in instructional design is the educational foundation of activities, thus the CA-Mobile Framework contains several components for describing and specifying these pedagogical features grouped as follows:

A. Instructional Design

This component acts as the starting point of the development process. It contains all elements (documents, guides and reference videos) for supporting the description of activities in didactical terms. External associated requirement are included too.

- **Activity description**. - Provides a structured way to describe educational activities, and serves as a basis for the beginning of the development of the process activities.
- **Description of external actions associated with the educational activity**. – Describes an additional process necessary to accomplish the objective of the activity, such as forming student groups, task explanation, setup of mobile devices and desktop PC, as well as software configuration.

B. Script adaptation

The Framework supports the process of adapting the original script to several forms. It also includes reference guides regarding the use of these forms. Both elements: forms and guides are part of the Reference Document.

- **Didactical script adaptation.** - Modifies and adapts the original script, to include the use of mobile devices and/or PC Desktop, specifying the moment where these devices will be used, and the actions that will be supported by these devices.
- **Specification of human, technical, and organizational considerations**. - This specifies the preconditions that must be fulfilled before the activity

takes off. This component is associated with the description of external actions that are in turn associated with the educational activity. It is also related to the actions that are necessary for the execution or the repetition of the activity, such as user training, connection between devices. Tasks related to previous coordination like grouping students by team or special classroom conditions are described here.

- **Forms use guide.**- Provides a reference to the forms and templates. It is used by other components of this reference document to describe in detail the fields within each document and their possible uses, as well as an analogy of concepts in the cases where it applies.

- **Specification of associated roles inside the activity.**- Determines the actions and responsibilities of all the roles and resources involved in the execution of an activity, like the student, the teacher, mobile devices and the software, describing the characteristics and skills of each one.

C. *Instructional Design*

- **Describes the stages of an activity through use cases**. - Describes through use cases, all the detailed actions to be executed by each role in a specific stage of the activity. These descriptions will be the basis for the final script of the activity.

- **Specification of functional requirements.**- Extracts from use cases, those requirements that must be covered by mobile and desktop software, excluding human coordination and negotiation tasks (non functional requirements).

- **Association of requirements vs. API components.**- Makes an association between each requirement described in the specification of functional requirements and the selected API components in order to know which software component will be used to implement the required functionality.

Each one of these components is used by the instructional designer to fill out or complement a structured document in the framework, either an activity description, use cases or requirements document.

V. Technological Components

Once the requirements are specified, the next task is to create software to support the activity that follows the requirements associated with technology. Several software components were added as part of CA-Mobile Framework to support the development of mobile software using the Java programming language and JME libraries. The components simplify the implementation of these libraries and the resulting software can be executed in any device with JME support. These components are grouped in an API and are described below.

A. *Interaction management*

This part of the API includes components to build user interfaces such as capture forms, graphics resources manager, and camera control.

- **User interface control.**- Implements the components to create menus and capture forms, as well as persistent look-and-feel in several mobile devices and desktop PC.

- **Video and camera control.**- Implements the necessary components to handle photo and video camera of mobile devices.

- **Images control.**- Implements the components to handle graphics contents on the display like a single image or with textual description.

B. *Activity management*

This part of the API contains components to begin the interaction between participants, handle work sessions, as well as manage turn taking and shared resources.

- **Peer administration.**- Implements the mechanisms to manage the start of interaction between peers through mobile devices and desktop PCs.

- **Session control.**- Implements the mechanisms to start a collaborative work session between two or more peers through mobile devices and a desktop PC.

- **Turn control.**- Implements the components for the management of turns between participants in a collaborative work session.

- **Floor control.**- Implements the components for the management of shared resources, such as screens and message logs.

C. *Files and communications*

This includes components to establish wireless communication, handle messages exchanged between peers, and file exchange management.

- **Communications.**- Implements the mechanisms to establish wireless communication between mobile devices and desktop PCs.

- **File management.**- Implements the components for the management of files exchanged between mobile devices and a desktop PC.

- **Message control.**- Implements the mechanisms to manage the exchange of messages between applications and users, message handling, and visualization of textual messages.

These components, the reference document and the API for mobile software, work together in the development process of educational activities. However, these components by themselves do not define how and when the components will be used in the development process. For this reason, a process model was created to support the use of the forms, guides and video references, as well as API components of CA-Mobile Framework. This model presents a workflow of the tasks involved during development process. All details of the model and the relationship between components of the framework are described in the next section.

VI. Process Model of CA-Mobile Framework

The process model created to guide the use of the framework components is divided into five stages, which define the roles included in the development process, the artifacts, and the products at each stage. The actors that perform the roles need some degree of experience using mobile devices and are classified as instructional designer, teacher or content expert, and programmer. The responsibilities and tasks of these roles are described in Table 1.

TABLE I.
ROLES IN THE PROCESS MODEL

Role	Responsibilities
Instructional designer	Is responsible for the specification of the activity in terms of structure, learning style, educational objective, specifying the activity requirements in terms of CA-Mobile Framework and adapting the didactical script to include the use of newly developed mobile software.
Teacher or content expert	Is responsible for suggesting and choosing the contents that will be included in the activity, participation in the definition of the educational objective, and participation in the final product evaluation in terms of the original didactical requirements.
Programmer	Specifies, in conjunction with the instructional designer, the requirements of the software in functional terms, chooses the API components that will be used to satisfy the requirements, and builds the software for mobile devices and, if it is the case, for Desktop PC. Participates in adapting the didactical script as well as in final product evaluation.

In addition to the roles, the process model (see Figure 2) proposes six main stages to develop the educational activity. In the same way, the model specifies each of the artifacts (documents and tools) used as inputs and outputs of each stage.

A. Stage 0: Instructional design

This stage describes, in a general way, the activity in didactical terms including objectives, skills fostered, organization requirements (previous to activity execution, including any technical requirements), estimated time, and preliminary and general script for the activity. The result of this stage is a form (part of CA-Mobile Framework) where the activity is specified including typical data (name, time, audience) and a preliminary didactical script specifying the use of mobile devices inside the activity.

B. Stage 1: Characterization and specification of activity requirements

At the beginning of this stage, the first requirements specification of the activity based on a preliminary script is written, associating concepts of the instructional field to functionalities that will be performed by both the participants of activity and mobile technology. At the end of this stage, a document is created where all task and organization requirements are specified in coordination terms, as well as features that will be mediated by mobile devices and the desktop PC. All tasks defined in the instructional design stage are described using a special use case format, which is part of CA-Mobile Framework. Although some tasks won't be supported by mobile devices such as human coordination or verbal negotiations, all must be described using these formats.

C. Stage 2: Association of requirements vs. API components

Once all tasks are described, the functional requirements are extracted from the forms (only the ones supported by mobile devices or desktop PC). A selection is made of the API components that will be used to build mobile software to cover these functional requirements of the activity, taking into account that the functionality of

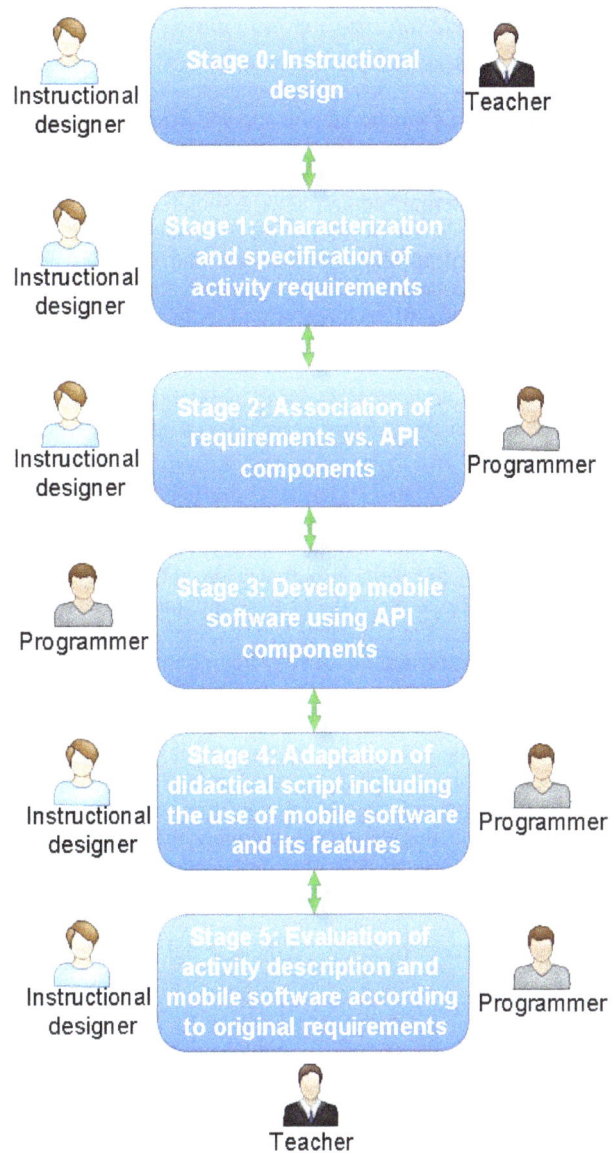

Figure 2. The six stages used in the process model of CA-Mobile Framework including the roles involved in each one.

the selected components corresponds to one or more requirements. For this, the Requirement and components specification document, part of CA-Mobile Framework, is used. This document presents the relation between a functional requirement and the selected API component that covers it.

D. Stage 3: Development of mobile software using API components

The mobile software to support the activity is developed in this stage covering all functional requirements using pre-selected API components specified in the requirements and components specification. Another external tools might be used like a specific API, e.g. for a specific rich graphical interface, or extended API for integrating a Desktop PC using Bluetooth communication, e.g. a Bluetooth library for PC. A functional prototype of software is the result of this stage. In some cases, the result is one version of software for mobile devices and a different version for Desktop PC.

E. Stage 4: Adaptation of didactical script including the use of mobile software and its features

In this stage an adaptation of didactical script is performed, taking previous scripts extracted from use cases specified in the characterization and specification of activity requirements stage. This adaptation must include the use of mobile software as part of tasks that require it and defines the external requirements associated with the use of mobile devices or Desktop PCs, for example user training or pre-configuration of software and devices. Once finished, the adapted didactical scripts are extended in order for all tasks of each activity to include details about the use of software and times for each task.

F. Stage 5: Evaluation of activity description and mobile software according to original requirements

At the end of the process, the didactical scripts and software are evaluated in terms of the functional requirements specified. Necessary changes are made both in the didactical script and mobile software, maintaining the consistency of the relation between the different parts of the activity. Both the didactical script and software are provided to test the complete activity including performing the necessary actions of resource control and coordination of participants.

In this way, the process model proposes a structured way to develop educational activities maintaining at all times a control by stages, generated products of each one, roles and responsibilities. Moreover, the model provides a reference about the stages that integrate the complete development process of activities based on mobile devices, from an integral approach of educational and technological aspects. The products generated could be used to design a new activity or to extend the original.

VII. REFERENCE DOCUMENT EVALUATION

In this paper we present a qualitative evaluation of the Reference Document component focusing on the instructional designer's perspective, in order to know the assessment about the usefulness and clarity of forms, guides and training videos in the design process. The CA-Mobile API is being evaluated and validated in another study with a technical approach to found evidence of usability, completeness and simplification that the API provides for the development of mobile software.

This evaluation was carried out through an experiment with education professionals, in order to test the forms, guides and reference videos of the Reference Document with one exercise; at the end of the exercise they answered a questionnaire about the use of these artifacts. The main propose of this evaluation was to capture the perception of teachers of the usefulness of the Reference Document of the CA-Mobile Framework for design of educational collaborative activities based on mobile devices from an educational perspective.

The evaluation was performed with a group of eight teachers; all of them are postgraduate students of a science education program, of which 75% have a college education in science education and the remaining 25% in mathematics and engineering.

Usefulness was evaluated by an analysis of coincidence of description activity forms, comparing each form with a pre-designed form. Later a questionnaire about the artifacts (forms and guides) was applied to all participants.

Thus, the evaluation was divided in two parts. In the first part, teachers were exposed to one experimental scenario presented in a video, which described an activity based on mobile devices and their associated process. We also showed all the activities related to the setup of the mobile devices and desktop PC. In the second part, participants had to create the preliminary instructional design of the activity and all associated process of the observed scenario, material organization, devices configuration, and task coordination using the forms and guides of the Reference Document component of the framework.

The video included the tasks preformed before the execution of the activity, in order to identify those processes related to setting up and deploying software in mobile devices, as well as forming teams including a record of the mobile devices that would be used. All these processes were performed on the teacher's platform. Subsequently an example of how students perform the activity inside and outside the classroom was presented. After that, the teachers answered a questionnaire about the usefulness of artifacts used (see figure 3).

A. Description of the educational activity

The activity used in the scenarios with participants was "Recognizing geometric shapes" which was developed and adapted to use mobile devices by Cruz-Flores and López-Morteo (2009). The main objectives of this activity are to develop a better understanding of shapes as a whole, to identify the parts of a shape or identify full shapes inside a physical environment, to understand how shapes differ and how they are alike using appropriate vocabulary during class discussion, similar to activities presented by CETConnect (2009), Instructor WEB(2009) and Learn, N. (2004), which proposes the same educational objective, but do not consider the use of mobile devices.

This activity was developed using the CA-Mobile Framework and applying the process model, where the instructional designers specify the educational purpose of the activity and how the use of mobile devices will provide support for this activity. After that, the programmers developed the software in two versions: Mobile and PC. In this case, both versions of software perform complementary tasks in the same activity. At the end, the soft-

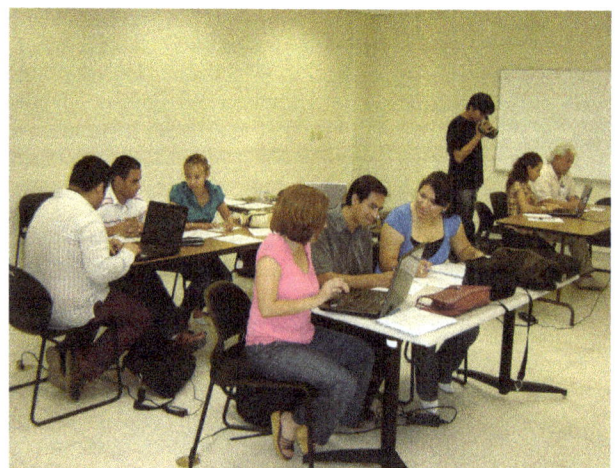

Figure 3. Teachers during Reference Document component evaluation: Working in teams to describe the educational activity.

ware and didactical scripts were evaluated and validated by teachers, instructional designers and programmers in functional terms. This activity was used as an example of how to introduce participants of this evaluation into the design of educational activities based on mobile devices. A description of the "Recognizing geometric shapes" activity is detailed below.

Four team members (mobiles peers) will explore the physical environment near the school in search of any building or parts of a building which have at least one of the basic geometric shapes: circle, triangle, square and rectangle. With the mobile device, they will take pictures of the buildings they found and send them to the other team members (local peers) who are working on the Desktop PC. These peers will receive the photos and they will classify them according to their geometric shape. Once identified, local peers will build a brochure with the photos, and the name of the geometric shape that it represents.

Team members can also communicate with each other through an integrated chat between mobile device and PC. It is important to communicate in case another photo is needed, or to communicate that a geometric shape is missing. At the end of the activity, all brochures made by each team are presented and discussed in class

The descriptive information that teachers should describe was composed of; Activity name, Date, Author, Version, Description, Educational/didactical objective, Skills fostered, Audience, Requirements, Estimated time, Development, Roles, Deliverable product and Required material by team.

B. Evaluation

At the end of the exercise, the three teams of instructional designers delivered their own detailed description of the activity. The description made by each team was compared against a pre-defined master description in order to determine the degree of coincidence between the descriptions of the teams and the original version, finding similarities and differences in all core fields. These coincidences allow us to see whether the fields and structure of the forms and the guides are enough to clearly describe each part of the activity.

The results of the three teams were evaluated with regards to interpretation of the text in each field, and not in terms of words and concepts used. The classification of range of values for this analysis of coincidence is the following:

- High.- Level of coincidence with the original from 65% to 100%.
- Medium.- Level of coincidence with the original from 30% to 65%.
- Low.- Level of coincidence with the original from 0% to 30%.

It is important to highlight that the levels of coincidence were analyzed document by document in order to interpret written ideas in the artifacts by each team. Using these parameters, the result of the analysis of all the descriptions of each team is detailed in Table II.

These results show that, in spite of the variations, the three teams made a description of the activity that was similar to the original. The case with a smaller degree of coincidence was team one (with two teachers) in the field: estimated time and development. Based on the observa-

TABLE II.
ANALYSIS OF COINCIDENCE. HIGH (0%-30%), MEDIUM (30%-65%) AND LOW (65%-100%)

Field	Team 1 (two people)	Team 2 (three people)	Team 3 (three people)
Description	Medium	High	High
Educational/didactical objective	High	High	Medium
Skills fostered	High	Medium	High
Audience	High	High	High
Requirements	High	High	High
Estimated time	Low	Medium	High
Development	Low	High	High
Roles	Medium	High	High
Deliverable product	High	High	High
Material by team	High	High	High

TABLE III.
RESULTS FOR EACH ITEM IN THE QUESTIONNAIRE USING MEDIAN

Item	Def. yes (1)	Prob. yes (2)	Undecided (3)	Prob. no (4)	Def. no (5)	Median
1	7	1	0	0	0	1
2	7	1	0	0	0	1
3	8	0	0	0	0	1
4	5	3	0	0	0	1
5	7	1	0	0	0	1
6	8	0	0	0	0	1
7	3	4	1	0	0	2
8	1	7	0	0	0	2
9	7	1	0	0	0	1
10	7	1	0	0	0	1

tions of the experiment, we believe that these differences could be caused by the viewpoint of one participant who was a teacher with many years of instructional experience who considered that the estimated time should be higher than proposed in the original design and likewise considered that the development of the activity should include aspects related to the delivery of assigned equipment and how the teacher obtains the final evidence of the activity.

At the end of the first part of the experiment (activity description), a questionnaire with ten closed questions and two open questions was applied to participants in order to obtain a general opinion regarding the artifacts. Through the use of this questionnaire we expected to observe the assessment of the participants regarding their experience using the Reference Document artifacts. At the end, the results were analyzed using a median and the results are presented in table III. The items of questionnaire are:

1. Could you use all artifacts (forms, guides and videos) of the Reference Document?
2. Is the purpose of each document clear?
3. Are artifacts of reference and training (forms, guides and videos) useful?
4. Are the names of those forms appropriate and refer to their use?
5. Do the field names correspond to their content?
6. Is the vocabulary used in the artifacts adequate?

7. Do you consider the guides that accompany the forms useful?

8. Is the structure of the forms appropriate?

9. Do the forms help to specify the use of mobile technology as part of the activity?

10. Is the Reference Document useful for defining and describing activities using mobile technology?

11. Describe any suggestion of change, adaptation and / or extension of the artifacts presented.

12. What is your general assessment of the Reference Document of the CA-Mobile Framework?

The responses for the first item indicate that participants used all artifacts (Definitely yes=87.5%, Probably yes=12.5%, Undecided=0.0%, Probably No=0.0% and Definitely No=0.0%). This results show that most participants used all artifacts at least once during the description activity process (Median=1).

The second item is related with the clarity of the objective of each artifact (Definitely yes=87.5%, Probably yes=12.5%, Undecided=0.0%, Probably No=0.0% and Definitely No=0.0%). This means that participants understand the objective of the artifacts used during the design process (Median=1).

According to the third element (Median=1), the assessment about the usefulness of the artifacts shows that all participants found these resources useful (Definitely yes=100.0%, Probably yes = 0.0%, Undecided = 0.0%, Probably No=0.0% and Definitely No=0.0%).

The fourth item is related to the names of each artifact and whether it refers to its use (Definitely yes=62.5%, Probably yes=37.5%, Undecided=0.0%, Probably No=0.0% and Definitely No=0.0%). This indicates that the names assigned to the artifacts are adequate, although they could be clearer. In this case, some participants suggested name changes that would be more representative of the content of each document (Median=1).

Regarding the fifth element that concerns the congruence between the fields in each form and their content (Definitely yes=87.5%, Probably yes=12.5%, Undecided=0.0%, Probably No=0.0% and Definitely No=0.0%) the results show that the majority of participants consider the names and the described content congruent (Median=1).

The sixth item refers to vocabulary used to define both the fields and their descriptions in the guides, in order to know the assessment of participants regarding the words used and their meaning in the pedagogical field (Definitely yes=100.0%, Probably yes = 0.0%, Undecided = 0.0%, Probably No=0.0% and Definitely No=0.0%). This mean that the vocabulary used is representative for participants (Median=1).

The seventh item refers to usefulness of the guides that accompany the forms (Definitely yes=37.5%, Probably yes = 50.0%, Undecided = 12.5%, Probably No=0.0% and Definitely No=0.0%). These results show that the guides were used as support to fill the forms. This item shows a different median value (Median=2), because the participants found all the guides useful, whoever they thought that these guides were not essential. This opinion may be caused by experience level of participants with regard to instructional design. In this evaluation all participants are

professional educators and were selected for their skills to describe educational activities.

The eighth element measures the organization and structure of the forms (Definitely yes=12.5%, Probably yes = 87.5%, Undecided = 0.0%, Probably No=0.0% and Definitely No=0.0%), the responses show that teachers though the contents of the forms are suitable, although their structure and organization was not the most appropriate (Median=2).

The ninth item determines whether the artifacts used facilitate the use of mobile devices and desktop PC as part of the tasks of an activity (Definitely yes=87.5%, Probably yes=12.5%, Undecided=0.0%, Probably No=0.0% and Definitely No=0.0%). This means that most of the participants found the contents inside artifacts are directed to describe and include the use of mobile devices like a part of the activity (Median=1).

The tenth item questions the usefulness, in general terms, of the Document Reference to design educational activities based on mobile devices from the instructional perspective (Definitely yes=87.5%, Probably yes=12.5%, Undecided=0.0%, Probably No=0.0% and Definitely No=0.0%). This means that teachers believe that the artifacts are useful to describe this type of activity (Median=1).

Finally, the last two questions gave a textual assessment about the Reference Document, which show that participants found these artifacts useful as support tools to develop the instructional design of activities based on mobile devices. Also, teachers made general recommendations regarding the structure of the contents, and the vocabulary that might be useful for the fields.

Once all responses from the participants were analyzed, it was possible to summarize the following points:

- Instructional designers consider that the Reference Document is useful as a support tool to design educational activities that include use of mobile devices as a core part of the activity.

- The contents inside the forms cover all the information needed about didactical and pedagogical aspects of the activity to generate the didactical scripts and to support the software development process.

- The guides and reference videos are a good complement to the forms of specification of the activity.

VIII. Conclusion

When mobile devices are used to support educational activities, it is possible to create an interesting and attractive learning environment for students. However, the characteristics that this technology imposes for its use in the classroom, imply that the development process of activities based on mobile devices must take into account several features from pedagogical and technological approaches, such as the definition of the educative objective, the creation of all coordination tasks, the specification of the role of mobile devices in the interactions between participants, and the definition of a didactical script to include the use of mobile devices and desktop PCs as an important part of the activity in order to reach the educative goal proposed.

Group activities require a special design because they involve a collaborative learning model, which is different from an individual learning model. These differences imply that mobile technology act as a dynamic actor to mediate the interactions between participants, manage resources and products, permit the negotiation, coordination and socialization process through mobile devices.

The CA-Mobile Framework presented in this paper offers several elements to develop educational activities based on mobile devices, through its two main components: the Document Reference and the API for mobile software. These components provide support at several levels of the development process. This framework includes a process model to coordinate the application of its different elements in a structured way, allowing the definition of user roles for those involved in the development of the educational activities based on the use of mobile devices. In addition, the process model helps to maintain control of the execution of the stages, indicating the order of execution and the inputs and outputs required during the process.

Because of the nature of the framework, it is possible to evaluate it from two main approaches: pedagogical and technological. This paper showed an evaluation from a pedagogical approach where the usefulness of the documental reference of CA-Mobile Framework was evaluated and the results gave evidence of that it is possible to describe entire educational activities based on mobile devices and make an instructional design using the artifacts provided taking in account the following elements: advantages of mobility and portability of devices, use of wireless communication to share real time information and resources, and tasks performed beforehand to deploy and install software.

The results show that using the guides, forms and videos of the Reference Document allowed the replication of the procedure for describing and defining activities, obtaining similar results to the original design, the assessment of the participants in the experiment gave evidence that the Reference Document component provides a structured and organized route to create the instructional design required by process model in stages 0, 1 and 2. These stages provide the foundations for the entire development process. The table 4 shows that the activity description of each team has a high coincidence grade about the original description. Team number one reported a 60% of high coincidence, 20% of medium coincidence and 20% of low coincidence. Team number two reported a 80% of high coincidence and 20% of medium coincidence and the last team reported a 90% of high coincidence and 10% of medium coincidence. The three teams described all elements of the activity using the Reference Document component's artifacts as reported in the questionnaire. The results allowed us to determine that the Reference Document of CA-Mobile Framework provides support to instructional designers to create a consistent design of collaborative educational activities based on mobile devices through the use of the forms, guides and video references of Reference Document. The forms description of activity, use cases and description of requirements filled out by instructional designers, will be used by mobile programmers to develop the software for the activity based on the specified requirements.

IX. FUTURE WORK

In order to have an integral evaluation about CA-Mobile Framework, we require an evaluation of the API

component for developing mobile software. This evaluation will be complementary to the evaluation presented in this paper. For that, an evaluation of CA-Mobile API in terms of usability and completeness is being performed. Taking a technological approach, the API will be evaluated as a software component applying metrics like the cognitive dimension framework (Clarke and Becker 2003). To perform an evaluation from a technological approach, it is necessary to create a complete educational activity based on mobile devices. This will allow the evaluation of the complete development process from the educational and technological approach, including the last stages of the process model, which can only be evaluated with a complete final product.

REFERENCES

[1] Alapetite, A. Dynami 2D-barcodes for multi-device Web session migration including mobile phones Personal and Ubiquitous Computing, 2010, 14

[2] Attewell, J. (2005). Mobile technologies and learning: A technology update and m-learning project summary.

[3] Avellis, G., Finkelstein, A., and CSATA, V. T. (2005). Innovative use of mobile learning for occupational stress: Evaluation of non functional requirements and architectures. Wireless and Mobile Technologies in Education.

[4] Benford, S., Rowland, D., Flintham, M., Hull, R., and Reid, J. (2004). Savannah: Designing a location-based game simulating lion behaviour. International Conference on Advances in Computer Entertainment Technology.

[5] CETConnect. (2009). Recognizing Geometric Shapes, Vertices: geometry meets art. From http://www.cetconnect.org/Vertices/lesson_recognizing_geometric_shapes.aspx

[6] Chen, Y.; Kao, T.; Yu, G. & Sheu, J. A mobile butterfly-watching learning system for supporting independent learning Wireless and Mobile Technologies in Education, 2004

[7] Cheng, Y.-C. & Yeh, H.-T. From concepts of motivation to its application in instructional design: Reconsidering motivation from an instructional design perspective British Journal of Educational Technology, 2009, 40, 597-605 doi:10.1111/j.1467-8535.2008.00857.x

[8] Clarke, S. and Becker, C. (2003) Using the cognitive dimensions framework to evaluate the usability of a class library. In Joint Conf. EASE & PPIG, Petre & D. Budgen (Eds), pages 359-366, 2003.

[9] Cruz-Flores, R. and López-Morteo, G. (2008). A model for collaborative learning ob jects based on mobile devices. ENC '08. Mexican International Conference on Computer Science, 2008, page 7.

[10] Cruz-Flores, R. and López-Morteo, G. (2009). Actividades educativas colaborativas a través de dispositivos móviles empleando ob jetos educativos móviles (emo): Un ejemplo de implementación. Proceedings of Conferencia Conjunta Iberoamericana sobre Tecnologías para el Aprendizaje 2009.

[11] Hamid, S. A. and Fung, L. (2007). Learn programming by using mobile edutainment game approach. Proceedings of the The First IEEE International Workshop on Digital Game and Intelligent Toy Enhanced Learning, pages 170–172.

[12] Huang, Y.-M., Jeng, Y.-L., and Huang, T.-C. (2009). An educational mobile blogging system for supporting collaborative learning. Journal of Educational Technology & Society, 12(2):163-175.

[13] Jarkievich, P., Frankhammar, M., and Fernaeus, Y. (2008). In the hands of children: exploring the use of mobile phone functionality in casual play settings. MobileHCI '08: Proceedings of the 10th international conference on Human computer interaction with mobile devices and services.

[14] Learn, N. (2004). Exploring geometric shapes, Mathematics North Carolina Curriculum Alignment. Learn, NC. From http://www.learnnc.org/lp/pages/3821

[15] Metcalf, D. S. and Marco, J. M. D. (2006). mLearning: Mobile Learning and Performance in the Palm of Your Hand. HRD Press, Inc., Amherst, Massachusetts.

[16] Meng, Z.; Chu, J. and Zhang, L. (2004) Collaborative learning system based on Wireless Mobile equipments. Electrical and Computer Engineering, 2004. Canadian Conference on, 2004, 1, 481-484

[17] Mostakhdemin-Hosseini, Ali. (2009) Usability Considerations of Mobile Learning Applications. International Journal of Interactive Mobile Technologies, 2009, 3, 29-31

[18] Milos, M., Miroslav, M., Miroslav, L. and Dusan, S. (2009) Mobile educational game: adventure anywhere MobileHCI '09: Proceedings of the 11th International Conference on Human-Computer Interaction with Mobile Devices and Services, 2009

[19] Patten, B., Sánchez, I. A., and Tangney, B. (2006). Designing collaborative, constructionist and contextual applications for handheld devices. Computers & Education. 46:294-308. Elsevier. doi:10.1016/j.compedu.2005.11.011

[20] Power, M. (2009)Augmented Reality - A Game Changer in Mobile Learning JISC CETIS, 2009.

[21] Roschelle, J., Rosas, R., and Nussbaum, M. (2005). Towards a design framework for mobile computer-supported collaborative learning. Proceedings of conference on Computer support for collaborative learning: learning 2005: the next 10 years.

[22] Rieger, R. and Gay, G. (1997). Using mobile computing to enhance field study. CSCL '97: Proceedings of the 2nd international conference on Computer support for collaborative learning, International Society of the Learning Sciences, 1997, 218-226.

[23] Smith, K. A., Sheppard, S. D., Johnson, D. W., and Johnson, R. T. (2005). Pedagogies of engagement: Classroom-based practices. Journal of Engineering Education, pages 1-15.

[24] Spira, D. (2009). Google Goggles: Augmented Reality for Mobile Learning Danspira.com, 2009.

[25] Shen, R.; Wang, M. and Pan, X. (2008) Increasing interactivity in blended classrooms through a cutting-edge mobile learning system British Journal of Educational Technology, 2008, 39, 1073-1086. doi:10.1111/j.1467-8535.2007.00778.x

[26] Tsuei, M. Development of a peer-assisted learning strategy in computer-supported collaborative learning environments for elementary school students British Journal of Educational Technology, 2009, 1-19

[27] Leung, C. and Chan, Y. (2003) Mobile Learning: A New Paradigm in Electronic Learning 3rd IEEE International Conference on Advanced Learning Technologies (ICALT'03), 2003

[28] Vahey, P. and Crawford, V. (2002). Palm education pioneers program: Final evaluation report. Technical report. From http://palmgrants.sri.com/PEP_Final_Report.pdf.

[29] Wagner, E. D. Enabling Mobile Learning. EDUCAUSE REVIEW, 2005, 40, 40-52

[30] White, F. and Byrd, L. (2000) The award for excellence in teaching, learning and technology SIGUCCS '00: Proceedings of the 28th annual ACM SIGUCCS conference on User services: Building the future, 2000

[31] WEB, I. (2009). Identifying Geometric Shapes and Parts, Geometric Shapes and Parts Lesson. Instructor WEB. From http://www.instructorweb.com/lesson/parts.asp

[32] Zurita, G. and Nussbaum, M. (2004). Mcscl: Mobile computer supported collaborative learning. Computers & Education, 42:289–314. Elsevier. doi:10.1016/j.compedu.2003.08.005

[33] Yoon, Y.; Ahn, Y.; Lee, G.; Hong, S. & Kim, M. Context-aware photo selection for promoting photo consumption on a mobile phone MobileHCI '08: Proceedings of the 10th international conference on Human computer interaction with mobile devices and services, 2008

[34] Parsons, D., Ryu, H., & Cranshaw, M. (2007). A design requirements framework for mobile learning environments. Journal of Computers , 2 (4), 1-8. doi:10.4304/jcp.2.4.1-8

AUTHORS

Cruz-Flores, Rene. Ph.D. Student. Computer Science, Instituto de Ingeniería, Universitada Autónoma de Baja California, Mexicali, BC. México. (e-mail: renecruz@uabc.mx).

López-Morteo, Gabriel , Professor. Computer Science Department, Instituto de Ingeniería, Universitada Autónoma de Baja California, Mexicali, BC. México. (e-mail: galopez@iing.mxl.uabc.mx).

29

Towards for analyzing alternatives of Interaction Design Based on Verbal Decision Analysis of User Experience

Marília Mendes [1], Ana Lisse Carvalho[1], Elizabeth Furtado[1, 2], Placido Rogerio Pinheiro[1, 2]

[1] University of Fortaleza (UNIFOR, MIA), Fortaleza, Brazil
[2] University of State of Ceará (UECE), Fortaleza, Brazil

Abstract—In domains (as digital TV, smart home, and tangible interfaces) that represent a new paradigm of interactivity, the decision of the most appropriate interaction design solution is a challenge. HCI researchers have promoted in their works the validation of design alternative solutions with users before producing the final solution. User experience with technology is a subject that has also gained ground in these works in order to analyze the appropriate solution(s). Following this concept, a study was accomplished under the objective of finding a better interaction solution for an application of mobile TV. Three executable applications of mobile TV prototypes were built. A Verbal Decision Analysis model was applied on the investigations for the favorite characteristics in each prototype based on the user's experience and their intentions of use. This model led a performance of a qualitative analysis which objectified the design of a new prototype.

Index Terms—Human Computer Interaction, Operational Research, Mobile Digital Television, Verbal Decision Analysis, Interaction Design.

I. INTRODUCTION

There are some facts which motivated us to have an assumption that applications for Digital TeleVision (DTV) and for mobile devices will be very useful in the next years to users access a huge of services (internet, Learning, Health and Government). First, some nations are preparing their broadcast structure. In UK, now 70% houses have DTV, 100% in 2012 [4]. Many network operators (such as in Europe, USA, Japan, Korea and Canada) have started to broadcast TV on handhelds. Second, in emergent countries, a great part of the population has a TV in their homes and uses cell phones [5], but only fews have computers and access to the Internet. Therefore designing usable interactive applications for TV (called iTV applications) is a need to motivate people to use them. Since the DTV represents a new paradigm to interaction design, designers don't have enough knowledge concerning a new domain or technology [19]. There are several practices that designers can apply in order to reduce this difficult. There are many HCI works that give suggestions about the way as professionals should apply user-centered practices when designing and analyzing iTV applications ([12][11]).

Recent researches in HCI are moving from a perspective based on predicting users' behavior (as cognitive features) to a perspective based on observing and understanding the users' behavior (as their experiences [10] and emotions [6]). During workshops [20], we could realize that in organizations, as Microsoft, anthropologists and ethnographers are working with usability experts and project managers in order to obtain data related to the following questions: what do the people want to experience? What should the experience feel like for them? Positive and negative comments are obtained from users when describing their experience using software, for instance. Taking into account the users' preferences is an action that has also gained ground in these works when designers are analyzing the better fitting solution(s). Participants of a project must accomplish changes in functionalities from users' contributions for improving the final version of the system. However, the traditional processes of evaluation are quite strict and not flexible to the emergence of new project alternatives and new ways of considering these alternatives. For example, it is typical to find the following scenario: designers evaluate two or three interface solutions applying usability tests, and choose one to implement. At traditional means, usability tests are applied to all alternatives. This work goes beyond the evaluation of this traditional view, by allowing designers to focus on only some criteria of the presented alternatives and by giving them the possibility to think about a new option.

In our approach, the evaluation process is conducted in three steps: first, the designers project high fidelity prototypes with characteristics that want to assess; second, designers carry out the usability tests and analyze the interaction users-TV content shown through each alternative of user's interfaces under the light of user's experience criteria (as users' preferences, their familiarity with technology) and; third, they organize their subjective questions by applying verbal decision analysis in order to define the best characteristics selected by the users during the test process. As a result of this procedure a new alternative of Prototype for Mobile Television Applications could be produced.

We have chosen to apply the Verbal Decision Analysis strategy on the purpose of organizing the usability tests results with sophisticated interactive applications. The reason was that applying to problems which have qualitative nature and difficulty to be formalized, called unstructured [7], may help designers to understand and organize their subjective questions.

The main challenge of this paper is to demonstrate the odds of the construction of a new prototype (a design alternative) starting with the users' opinions, collected through usability tests and classified through a computational method like multicriteria model.

We do this, by integrating two different areas (HCI and OR - Operational Research) when we describe an approach for evaluating the Interaction design in a subjective perspective of OR. This approach is co-evolutionary because the evaluation process can restart (as many times as the designers want) being all the design alternatives used again by the same or similar sets of users by feeding the design of a final version.

In this paper, we first present the definition of the prototypes of a Mobile Digital Television Application we developed and implemented. Then we discuss about Verbal Decision Analysis. In the fourth Section, we show a new interaction project. Lastly, some conclusions and indications for future works are provided.

II. DEFINITION OF THE PROTOTYPES OF A MOBILE DIGITAL TELEVISION APPLICATION

We had an assumption that a mobile TV application should have the same interaction purpose that an application for DTV. When designing and evaluating a mobile TV application for a specific project being executed, our first impulse was to develop for mobile TV with the same functionalities already developed for DTV, being sure to make the appropriate mapping of visual information between the devices. However we knew new services could be developed to support the users when interacting on movement. Despite that our scenario was the following: the end-users who were involved in this project did not have any experience lived with mobile TV, and we did not have any mobile TV application available to show them as example. Then we decided to apply the following strategy to do the elicitation and validation of new users' requirements. First at all, we looked for existing works in mobile TV to know their main functionalities. Then we analyzed each functionality previously implemented for DTV to verify if it should still be considered and to define how to consider it. Then we implemented one simple visual prototype in order to provide the users a generic idea about its possible usefulness (such as to see movie, to interact with it, to communicate with others, etc).

During a meeting session with the four stakeholders (such as two usability experts, one designer, and one programmer) eight volunteer end-users were presented the prototype in a PDA Palm OS. Each user received one Palm and manipulated the prototype for a period of 20 minutes, performing some non predefined scenarios. They talked one with other when they were not able to do any task. After this period, we asked them about the problems they had experimented as well as the needs they had with respect to their current cell phones and their expectations of mobile services. Some of the most relevant users' comments obtained in terms of interaction design were: 1) when we open a TV we see immediately a channel, when interacting with this mobile application, we need to look for it in the menu options, 2) it was unlikely that mobile TV would be watched in a place where it was not safe, 3) we could only the see the TV services of our interest.

At a certain moment we noticed users familiarized with the iTV applications inspected to interact of the same way, with the same look and feel, even though the RC could not be used. This fact made us to investigate our assumption that the users' experience with technology influences in their preferences for an interaction solution. For this, we defined and implemented different candidate solutions of navigation across the screens, which will be shown as follow.

We used different navigation patterns to design the executable prototypes of this mobile application.

Some features that differentiate applications for DTV and for mobile devices were considered when designing the solutions. These features are: DTVs are used to access TV services at home requiring a (Set-Top-Box) STB to store the applications that implement these services. The existing STBs have capability very limited. Mobile devices are used to access TV services anywhere, and the iTV applications are download directly in the device. In mobile devices, input typically is made using a stylus or finger on a touch sensitive display. In DTV, it is made using the Remote Control (RC). Although applications for DTV have less limitation to place TV information on their screen than applications for mobile devices do, the RC does not allow the direct manipulation on the interaction elements.

As this experiment focused on the users' experience related to the navigation and selection tasks, these differences (mainly the size of screen of the devices and their input styles) were considered in three mobile prototypes. Each prototype presented a different navigation pattern that usually includes icons, links and control of pages to allow the user to perform the navigation and selection tasks. The users access TV services by categories represented by icons or links. For each category, users can navigate through the options to choose a service by selecting the desired service in order to view it (use it then close it).

The three mobile prototypes were the following:

- In the first solution, the designers kept the same style guide of the iTV application (see Figure 1). The most important decisions concerning this pattern were [14]: (1) the navigation arrows, which continued on the bottom and the top with the numeration of the pages in order to inform users that there are more categories of services available (see Figure 1 left side); (2) a navigation arrow, which was included to return to the previous page (see Figure 1 right side); and (3) the navigation bar, which was preserved but without the colored options, because the interaction is made of through the stylus.

- In the second solution, the designers looked for the consistence with other PDA applications and the main change was the following: the navigation arrows to navigate among the categories were substituted by a scroll bar, which is typical in the other PDA applications, and no control to divide the pages was done (see Figure 2, left and right side).

- In the third solution, the style guide was a little bit similar to desktop applications. The designer decided to represent all the TV services as menu options displayed in toolbar located on the top of the screen. In the middle of the screen, the image of option selected can be showed (see Figure 3, left and right side).

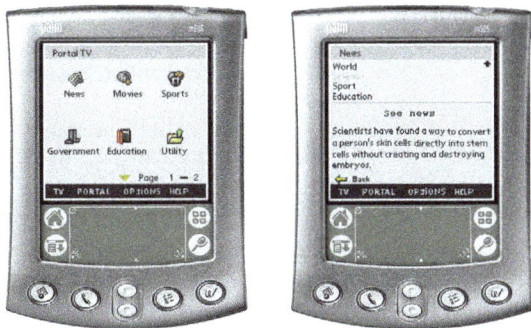

Figure 1 - Prototype 1. Similar to TVD applications

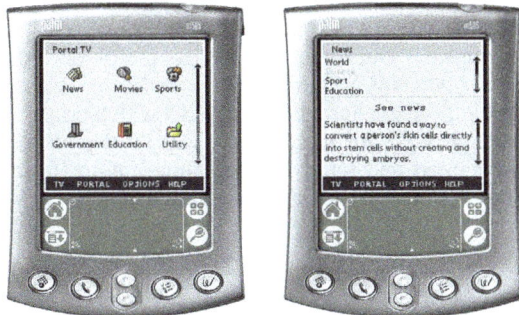

Figure 2 - Prototype 2. Similar to Palm applications

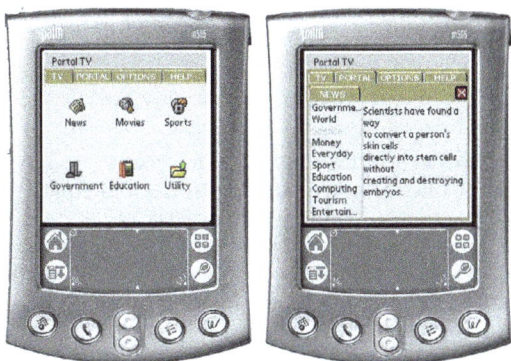

Figure 3 - Prototype 3. Similar to Desktop applications

The scenarios and the questions applied were defined from some hypotheses, that were previously elaborated taking into account the interaction user-applications and the context of use (as the environment, the TV content shown through the user interfaces, the users experience, their emotions, etc.).

The hypotheses were the following:

• Hypothesis 1: The evidence of the of the interface functions facilitates the use and influences in the effort spent by the user to localize himself/herself in the application. We consider here the design decisions that resulted in functions are easily found by the users;

• Hypothesis 2: The user experience with applications which have similar ways of navigation will influence the choice of an interface. Aspects such as facility of use and, accuracy of an interface and user familiarity are included in this hypothesis;

• Hypothesis 3: The locomotion of the user while manipulating the device will influence in the choice of the interface. This hypothesis refers to design decisions that result in less precision to navigate between the op-

tions and screens facilitate the navigation of a person while manipulating the application;

• Hypothesis 4: The involvement with the content influences in the user's choice, so that, if the content is interesting, it may be decisive for the user to choose the interface. This hypothesis refers to the holistic evaluation view: when the user uses an interface that has a content which attracts him, s/he will prefer this interface;

• Hypothesis 5: The emotion felt by the user when using the interface exercises a considerable influence in the choice. Aspect such as the user feeling states as pleasure in interactive experiences is considered in this hypothesis.

The hypotheses helped to identify what criteria could bring implication for choosing one specific solution or for choosing just some attributes belonging to the existing solutions. The investigation of this implication was made from the results of the tests [8] that are entered as data for the multi-criteria model used in this paper. Next we present a summary of the ZAPROS III method in order to get a better understanding on its implementation. After we will present how this investigation takes part.

III. VERBAL DECISION ANALYSIS

The ZAPROS III method belongs to the Verbal Decision Analysis – VDA framework. It combines a group of methods that are essentially based on a verbal description of decision making problems. It was developed with the aim of ranking given multicriteria alternatives, which makes it different from other verbal decision making, such as ORCLASS [15], and PACOM mainly due to its applicability. The Verbal Decision Analysis supports the decision making process by verbal representation of the problem [16]. It can be applied to problems with the following characteristics [17]: the decision rule must be developed before the definition of alternatives; there are a large number of alternatives; criteria evaluations can only be established by human; the graduations of quality inherent to the criteria are verbal definitions that represent the subjective values of the decision maker.

The decision maker is the key element of multicriteria problems and all necessary attention should be given in order to have well formed rules and consistent and correctly evaluated alternatives. Thus, the categorization of preference will be adequately obtained according to the principle of good ordering established by Zorn's lemma [18].

The application of the method ZAPROS III to modeling problem can be applied following this three-step procedure:

• Elicitation of Criteria and their Values: Once the problem is defined, the criteria related to the decision making problem are elicited. Quality Variations (QV) of criteria are established through interviews and conversations with specialists in the area and decision makers;

• Organization of the Ordinal Scales of Preference: An ordinal scale of preference for quality variations for two criteria is established based on pair wise comparisons. The preference between these two criteria is chosen according to the decision maker and the obtained scales of preferences are denominated Joint Scale of Quality Variation (JSQV) for two criteria;

• Comparisons of Alternatives: The ranking of the alternatives is constructed by comparisons between pairs of alternatives.

By going through these phases, a problem modeled on the ZAPROS III method results in a display of alternatives [9]. The display gives us a quantitative notion of the order of preference, in an absolute form (in relation to all possible alternatives) and also a relative form (in relation to a restricted group of alternatives). The following is the modeled case study.

IV. EVALUATION OF ALTERNATIVES BY APPLYING THE ZAPROS III METHOD

The specialists urged to analyze the aspects that had the greatest influence in the choice of a determined interface project. Then they established verbally some criteria for the implemented prototypes of mobile TVD applications as soon as the possible values for each criterion.

In table 1, the values are shown for the criteria directed to the aspects on which the definition of the influence among the standards is based. For instance, for functions evidence criterion there are three possible values related to the difficulty of the user for identifying the system functionalities.

Table 1 - Criteria and associated values

Representation	Criteria	Values
A	Functions Evidence	A1. There was no difficulty on identifying the system functionalities; A2. There were some difficulties on the system functionalities' identification; A3. It was hard to identify the system functionalities
B	User's familiarity with a determined technology	B1. No acquaintance was required with similar applications of a determined technology; B2. It was required little user's acquaintance with applications of a given technology; B3. The manipulation of the prototype was fairly easy when the user was proverbial with similar applications.
C	User's locomotion while manipulating the device	C1. The user was not hindered in any way when manipulating the prototype while moving; C2. The user was occasionally confused when manipulating the prototype while moving; C3. The spatial orientation of the application was hindered when the user was moving.
D	Content Influence	D1.There was no influence of content on choosing the interface; D2.The content exerted some influence on choosing the interface; D3.The content was decisive on choosing the interface.
E	User's feeling states	E1. The user felt fine (safe, modern, comfortable, etc.) when using the interface; E2. The user felt indifferent when using the interface; E3. The user felt bad (uncomfortable, unsafe, frustrated) when using the interface;

The order of preference among the criteria values was established from the results of the tests (observation, questionnaires). For example, it was observed that when the users were moving and trying to execute a task in a determined prototype, they complained that it was difficult to move and manipulate the device at the same time. After the tests, the responses to the questionnaires were gathered and evaluated. Questions like "What prototype did you prefer? And Why?" indicated the order of preference among the project alternatives and also which criteria values were decisive for the choice. The JSQV was gradually elaborated and validated with information from the tests, and resulted in the following sentence:

$$A1\, \pi\; A2\; \pi\; B1\; \pi\; B2\; \pi\; C1\; \pi\; E1\; \pi\; D1\; \pi\; E2\; \pi\; D2\; \pi\; B3\; \pi\; D3$$
$$\pi\; C2\; \pi\; C3\; \pi\; A3\; \pi E3$$

In a simplified way we read the sentence above from A1 to E3 in this way: A1 (No difficulty was found on identifying the system functionalities) is the preferable criteria value, and E3 (The user felt bad (uncomfortable, unsafe, frustrated) when using the interface) is the least desirable because it was not perceived.

The next step of the method was to carry out the comparison of the alternative standards. Each alternative was studied in order to define which criteria value the materialized prototypes. The usability tests also supplied important information on how the users described the alternative standards (for example, the majority of users said that access to content using prototype 3 (three) was quite easy. Three alternatives the most preferred of all were established from this information (presented by preference order):

• Prototype 1 - A2 B1 C2 D1 E2 (Alternative 1);
• Prototype 2 - A2 B3 C1 D1 E1 (Alternative 2);
• Prototype 3 - A2 B1 C1 D1 E2 (Alternative 3).

In a simplify way, the majority of users said for the second prototype that the interaction with the application was not damaged while moving (C1) and that they had fun using it (E1). The applied model was useful for the choice of the most favorite prototypes (details of this multicriteria model application can be obtained in previous works of the authors of this paper [8], [1], [2] and [13]). However it is not enough to choose a single option among the presented three, instead we want to identify the best characteristics (represented by the criteria values) existing in the analyzed prototypes. Next we aim at showing how it is possible to elaborate a new prototype based on the identified characteristics.

V. ELABORATION OF A NEW INTERACTION PROJECT

This section we will have three sub-sections. In the first one, we show how the design was done collaboratively with the designers, in the second one, we will analyze the values of criteria of each prototype that will be useful to develop a new proposal (project). In the third one, we will design this new project. For the creation of the new project it was considered JSQV and the selection of the best characteristic of each prototype with base on the values of criteria presented by them. To sketch the new interface, a tool was used called SketchiXML. SketchiXML is a multi-platform multi-agent interactive application that enables designers and end users to sketch user's interfaces with different levels of details and support for different contexts of use [3].

A. Collaborative Design

We conducted a collaborative design session aiming to understand the views of designers on the outcome of the case study, to design a new solution for interaction. For this, we considered the scale JSQV and the selection of preferred features of each prototype based on the values of criteria presented. In this sub-section we explain the methodology applied during the collaborative design.

The collaborative design session lasted an hour and involved a group composed of the designer and analyst (specialist in ZAPROS) who participated in this case study, as well as three other volunteer designers.

The main designer led the session in a lab environment, using a data show, presenting, through explanatory slides, a summary of the case study conducted with the proposed strategy. The results of the strategy (ordering of alternatives and criteria values) were reported and interpreted by the designer and analyst, aiming to explain to three designers invited the meaning of the ranks in the preferences of users. This analysis will be explained in more detail in the next sub-section B.

All the participants discussed the characteristics of each alternative. The three designers (Figure 4) found the results of the strategy very interesting and useful to know the preferences of users, extract the best features and improve the weaknesses identified. Through these discussions, a new prototype has been developed collaboratively (Figure 5). The result of the design Collaborative has been consolidated in the design of interaction using the tool SketchiXML which will be explained in sub-section C.

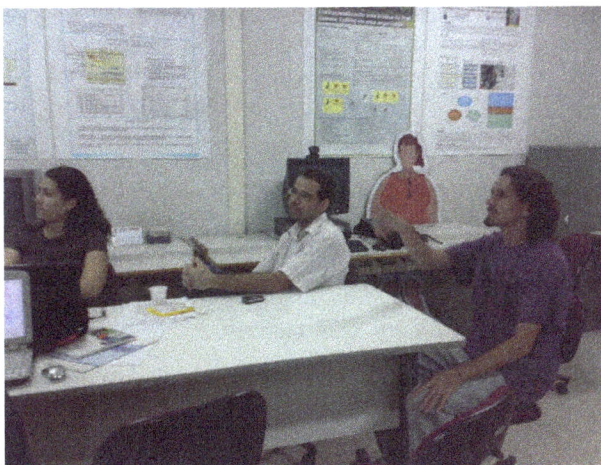

Figure 4 - The three designers to make decisions based on the results of case study

Figure 5 - Collaboration during the design of the new alternative

B. Analysis of the values of criteria of each prototype

In Figure 6, we can visualize in prominence the values of better rank in the scale JSQV of each prototype. They will be the base for analysis and subsequent elaboration of the new prototype.

Prototype 1 - A2 (B1) C2 D1 E2 (Alternative 1);

Prototype 2 - A2 B3 (C1) D1 (E1) (Alternative 2);

Prototype 3 - A2 (B1)(C1) D1 E2 (Alternative 3).

Figure 6 - Selection of the values in agreement with the scale

When a value is repeated in the three prototypes, as the value A2 (that means, the users had some difficulty on identifying the system functionalities), we make use of the same value for the three prototypes.

In the second column, we obtained two different values: B1 and B3. In agreement with the rank, B1 is preferred in relation to B3, for that reason we will consider the value B1 that was identified in the prototypes 1 and 3 with the characteristic: no acquaintance is required with similar applications of a determined technology (B1), meaning that it is not necessary to have experiences with a given technology for a good use of the two prototypes.

In the third column, we obtained two different values: C1 and C2. In agreement with the rank, C1 is preferred in relation to C2, for that reason we will consider the value C1 that was identified in the prototypes 2 and 3 with the characteristic: the user was not hindered in any way when manipulating the prototype while moving (C1), therefore we can consider in the new prototype, similar navigation ways of the prototypes 2 and 3, like scrolling and tabs. In the fourth column, the value D1 is repeated for the three prototypes. The value says that there was not influence of content on choosing the interface (D1). In the last column, we obtained two different values: E2 and E1. In agreement with the rank, the value E1 is preferred in relation to E2, for that reason we will consider the value E1 that was identified in the prototype 2 with the characteristic: the user felt fine (safe, modern, comfortable, etc.) when using the interface, therefore we should consider some of the characteristics of the prototype 2 that turned its comfortable use for the tested users.

C. The new interaction project

The new prototype (see figure 7) was projected based on analyzes explained in the previous item.

The prototype 1 contains criterion value B1 that is with rank 3 in the final scale, this means that it is very important that the new prototype does not demand any user experience with technology. Hence, we selected the arrow with explanatory label for symbol of the navigation: lateral (returning the previous screen) and of lower or of top (with passage of pages down and upward, respectively).

The third prototype presents besides the value B1, the value C1, because it makes the manipulation of the prototype easier, when the user is in movement. Therefore, it was necessary also to select some characteristic of this prototype. The selected characteristic was the one that possesses a larger area of contact through the tabs, then a junction of B1 was accomplished with C1, resulting in an arrow of such adult for navigation. Another selected characteristic of the prototype 3 was the located menu in the superior part of the screen.

The second prototype contains the value E1 as differential, and also contains the characteristic that was taken as advantage in the new prototype: the scrolling bar that was used for navigation interns through the widgets combox for list of title of the news and listbox for the content of the selected news.

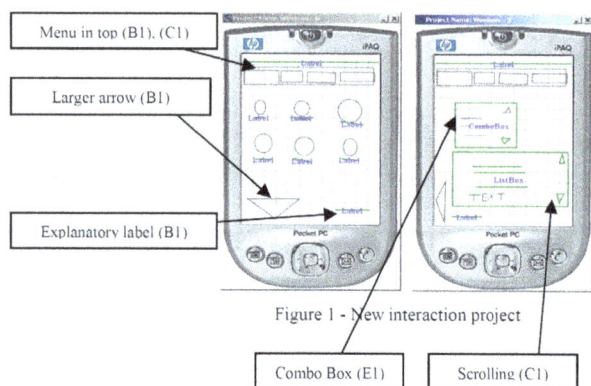

Figure 1 - New interaction project

Figure 7 - New interaction project

We expect this proposal to be more adequate according to the following reasons: the use of bigger arrows makes possible the change of screens especially when the users are moving; the use of explanatory labels help the users about the context of the application; the menu located at the superior part of the screen was favored by the user because it makes easy the visualization of the options.

VI. CONCLUSIONS AND FUTURE WORKS

From this conclusion we described some points that can be classified in HCI and OR areas: In HCI, the points refer to both what to design and evaluate and in OR, they refer to analysis of these design and evaluate through the ZAPROS III multicriteria method.

In HCI area, the points are the followings:

- Designers can produce applications more innovative and adapted to the users' cultural issues if they understand better the technological scenarios lived by the users. Theirs should also provide users with TV content design artifacts rather than just focus on the user

interfaces of the iTV applications. Designers should express their ideas, by showing users several entertainment situations they could have by anticipating and proactively perceiving their emotions (feeling states, involvement with the content).

- Users should understand better what they will have at the end of the project. It is common to hear the following phrase: the users don't know what they want. Making the users to explore a possible solution by themselves simulating situations that occur in the real world can be important to designers to feel the users' acceptance.

- Designers should go beyond the usability when doing usability tests with users. Traditionally, the usability of a developed system has been evaluated to assure both its effectiveness (such as the number of successful task completions) and efficiency (such as the time required to complete an interactive task). Recently, these assumptions have been revisited and broadened to embed the concepts of the affective quality theories. The affective aspect (such as users' feeling states, pleasure and their involvement with the content) is particularly relevant in the context of iTV, since entertainment depends strongly on how synergetic the involvement of the viewer with the TV content.

In OR area, the point is:

- The multicrieria method was useful in the elaboration of another design alternative, because it showed a structured model focused on users' preferences and by defining a rank with the best distinct characteristics by the own users in usability tests. This new e final prototype contemplated these best criteria. In our approach, it is not imperative to build a new design alternative or a final prototype. From a more general point of view, some useful lessons about the utility and the usability of similar systems (e.g. user interface guidelines) can be learned and further applied in future projects.

The idea of using sketchiXML for visualization of the new prototype was considered useful because it is a fast way to obtain an initial sketch. In addition, this tool generates the interface specifications in a standard format, as the UsiXML. This format is widely interpreted by interface generation tools that are freely available for use. These tools allow the automatic generation of the final user interface. In our approach, executable prototypes need to be easily obtained and quickly tested by the users in order to support its evolutionary characteristic.

As future works and next steps, we intend to validate this new prototype with the users by performing a second test in order to verify if the prototype is really in agreement with the best suggested characteristics. Using the ZAPROS III multicriteria method, we will include the new developed prototype as another alternative in order to be sure that it is really the best of the fourth alternatives. This iterative feature characterizes the co-evolutionary process of the project

REFERENCES

[1] A. Carvalho, M. Mendes, P. Pinheiro, E. Furtado, Analysis of the Interaction Design for Mobile TV Applications based on Multi-Criteria, International Conference on Research and Practical Issues of Enterprise Information Systems (CONFENIS 2007), October 14-17, Beijing, China.

[2] A. Carvalho; M. Mendes; E. Furtado; P. Pinheiro. Avaliação de Projetos de Interação para Aplicações de TVD Móvel utilizando Multicritério.. In: XXXIX Symposium of the Brazilian Operational Research Society (XXXIX SBPO), 2007, Fortaleza. XXXIX Symposium of the Brazilian Operational Research Society, 2007.

[3] A. Coyette, J. Vanderdonckt. A Sketching Tool for Designing Anyuser, Anyplatform, Anywhere User Interfaces, Proc. of Interact'2005.

[4] A. Jaokar, T. Fish, Mobile Web 2.0. FutureText. 2006.

[5] Agência Nacional de Telecomunicações. http://www.anatel.gov.br/Tools/frame.asp?link=/biblioteca/releases/2005/release_15_09_2005(2).pdf

[6] D. Norman, Emotional Design. Basic Books. 2004.

[7] H. Simon, A.Newell, Heuristic Problem Solving: The Next Advance in Operations Research, Oper. Res., vol. 6, pp. 4-10, (1958). (doi:10.1287/opre.6.1.1)

[8] I. Tamanini; M. Mendes; A. Carvalho; E. Furtado; P. Pinheiro. A model for mobile television applications based on verbal decision analysis. In: Tarek M. Sobh. (Org.). International Joint Conferences on Computer, Information, and Systems Sciences, and Engineering: 2008, to appear.

[9] J. Figueira, S. Greco, M. Ehrgott,(Eds.), Multiple Criteria Decision Analysis: State of the Art Surveys Series: International Series in Operations Research & Management Science, Vol. 78, XXXVI, 1045 p, (2005).

[10] J. McCarthy, P. Wright, Technology as experience. MIT.2004.

[11] K. Chorianopoulos, D. Spinellis, User Interface Evaluation of interactive TV: a media studies perspective. Univ. Access Inf Soc. 5: 209-218. 2006. (doi:10.1007/s10209-006-0032-1)

[12] L. Eronen, User Centered Design of New and Novel Products: case digital television. Thesis. Publications in Telecommunications software and multimedia. 2004.

[13] M. Mendes, E. Furtado, Mapeamento De Um Portal De Acesso De Televisão Digital Em Dispositivos Móveis. IHC'2006.

[14] O. Larichev, Cognitive validity in design of decision-aiding techniques, Journal of Multi-Criteria Decision Analysis, 1(3): 127–138, (1992). (doi:10.1002/mcda.4020010303)

[15] O. Larichev, H. Moshkovich, Verbal Decision Analysis For Unstructured Problems, Boston: Kluwer Academic Publishers, (1997).

[16] O. Larichev, Ranking Multicriteria Alternatives: The Method ZAPROS III, European Journal of Operational Research, Vol. 131, (2001). (doi:10.1016/S0377-2217(00)00096-5)

[17] P. R. Halmos, Naive Set Theory, Springer, 116 p., (1974).

[18] S. Dow, T. Saponas, Y. Li, J. Landay, External Representations in Ubiquitous Computing Design and the Implications of Design Tools. Proc. DIS'2006. Pp 241-250. 2006.

[19] T. Winograd, T., CS147: Introduction to Human-Computer Interaction, 2006. Stanford, CA. http://cs147.stanford.edu.

AUTHORS

Marília Soares Mendes has a degree in Computer Science from University of Fortaleza, Ce, Brazil (2006). Master in Applied Informatics from University of Fortaleza, Ce, Brazil (2009). Currently, she is a doctoral student at the Federal University of Ceará, Brazil. She has experience in computer science, with emphasis on research in human computer interaction (HCI), usability and software engineering, working mainly in digital television and mobile devices. (e-mail: mariliamendes@gmail.com).

Ana Lisse Carvalho has a degree in Computer Science from University of Ceará, Brazil (2004), Master in Applied Informatics from University of Fortaleza, Ce, Brazil (2008). Currently, she is a Civil Servant of Federal Office for Data Processing. She has experience in the area of computer science, especially operational research and software engineering. (e-mail: ana.lisse@edu.unifor.br).

Elizabeth Furtado is a Researcher in human computer interaction (HCI) and a Visiting Scholar (a sabbatical study) at HCI laboratory in Stanford University, in 2006. She is a consultant of software quality process. She did her PhD in Computer Science in 1997 in France. She is Adviser of the graduate and undergraduate students in the Master program of the MIA, University of Fortaleza, Ce, Brazil since 1989, in usability of collaborative learning systems; definition of tools for working with guidelines and patterns in order to generate user interfaces for interactive television (iTV) and definition of scenarios to evaluate iTV applications. (e-mail: elizabet@unifor.br).

Plácido Rogério Pinheiro has a degree in Mathematics from Federal University of Ceará, Brazil in (1979). He is an Electrician Engineer from the University of Fortaleza, Ce, Brazil (1983), with Master degree in Mathematics at the Federal University of Ceará (1990) and PhD in Systems Engineering and Computation, Federal University of Rio de Janeiro, Ce, Brazil (1995). He is currently Professor at the University of Fortaleza. He has experience in the modelling of industrial processes using mathematical programming and multicriteria. His training allows publishing in the area of mathematics, with emphasis on discrete mathematics and combinatorics, working primarily in mathematical programming and multicriteria. (e-mail: placido@unifor.br).

30

Adoption of Mobile Learning Among Distance Education Students in Universiti Sains Malaysia

Issham Ismail, Rozhan M. Idrus, Azidah Abu Ziden and Munirah Rosli

Universiti Sains Malaysia, Penang, Malaysia

Abstract—This study was carried out in order to investigate whether mobile learning using Short Message Service (SMS) was a method of learning adopted by the students enrolled in the School of Distance Education, Universiti Sains Malaysia. As adult learners who are in vocation, time and isolation are the bane of self-study. Since all the students own a mobile device that can receive SMS, educational messages can be sent directly to their devices. This experimental study explored the impact of learner's characteristic, learning design and learning environment to their adoptability. This study utilised two models of data analysis, the Statistical Package for Social Science (SPSS) Version 12.0 and the Rasch model analysis for measurement. The analysis was conducted on a sample of 105 students based on gender, age, ethnicity, programme of study and mobile device ownership. The students were from four different courses which are Bachelors of Science, Bachelor of Arts, Bachelor of Social Science and Bachelor of Management. The questionnaire-answer session were administered by the respective course managers in their tutorial sessions during the annual residential intensive course in the main campus of the Universiti Sains Malaysia. The result indicated that mobile learning has helped them to pace their studies. By using mobile learning, learners easily get any information that they need at anytime anywhere. Learners would also like to take another mobile learning assisted course if the courses are relevant to their learning needs. Furthermore, the SMS educational content received through their hand phone are easily remembered.

Index Terms—Mobile learning, distance education, SMS,

I. INTRODUCTION

Emerging technologies are leading to the development of several new opportunities which enhance and guide the learning process to higher level compared to unimaginable conditions in previous years. The use of these technologies turns out to be well aligned with strategic educational goals such as improving student retention and achievement, supporting differentiation of learning needs, and reaching learners who would not otherwise have the opportunity to participate in education [1]. Rapid developments in handsets, networks, and mobile applications can make educational implementations using mobile phones can be a risky one [2]. A great deal of effort has also been devoted in understanding how mobile technologies are related to both traditional and innovative ways of teaching and learning, showing the applicability of mobile learning across a wide spectrum of activity [3][4] as well as highlighting the most important emerging issues [5]. Studies done by Malaysian Communications and Multimedia Commission [6] found that in Quarter 2, 2009, the pene-

tration rate for cellular phone in Malaysia is 100.8 %. Penetration rate over 100% occurs because of multiple subscriptions. The mobile phone is multipurpose device. It is not only used to transmitting voice communication but also can provide a number of other functions and services. The one of that is a short messages service. Previous studies have examined ways in which everyday life activities influence mobile phone use and to a certain extent SMS usage [7] [8].

For mobile users as well in all mobile applications, SMS messaging is found to be the most useful and convenient way of technology. SMS is inexpensive, supported by almost all phones as an unlimited offering, familiar to students, and rapidly gaining worldwide acceptance. SMS is a low-threshold application used widely by students to quickly send concise, text-based messages at any time. Text messaging, also known as the short message service or SMS, is changing the communications landscape on college campuses [9]. According to the International Association of the Wireless Telecommunications Industry, the number of SMS messages sent in the United States each month now exceeds 48 billion, up from just 10 billion per month in 2005 [10]. The aim of this research is to investigate the problematic of designing mobile learning among the students enrolled in the School of Distance Education, Universiti Sains Malaysia. The responses are viewed from three factors, via, learner characteristic, learning design and learning environment.

II. MOBILE LEARNING

Mobile Learning commonly referred to as, M-Learning, is a form of e-learning that specifically employs wireless communications devices to deliver content and learning support [11]. Mobile learning represents exciting new frontiers in education and pedagogy [12]. With the features of "wearable" computing and multimedia content delivery via mobile technologies, mobile learning becomes feasible and offers new benefits to instructors and learners [13]. M-learning is the exciting art of using mobile technologies to enhance the learning experience. It refers to the use of mobile and pocket IT devices, such as PDAs, mobile phones, Pocket PCs, laptops and the Internet in teaching and learning process. It helps people to learn and gain information just from their pocket devices.

III. LEARNING TRANSFER

From the educational psychologist's view, learning is defined as the relatively permanent change in behavior [14]. Learning transfer is the application of knowledge, attitudes and skill that are learned from one situation to another learning situation. It is because the learning context is often different from the context of application as the goal of training is not accomplished unless transfer occurs.

Successful transfer of learning requires that training content be relevant to the task, that the learner must be motivated and that the learner must learn the training content. There remains considerable controversy about how transfer of learning should be conceptualized and explained, whether it relates to learning or whether it exits at all [15].

IV. LEARNING DESIGN

A 'learning design' is defined as the description of the teaching-learning process that takes place in a unit of learning (e.g., a course, a lesson or any other design learning event) [16]. Due to the rapid emergence of wireless communications technology and mobile devices, the use of handheld technology in education has increasingly been the object of study in recent years. The key principle in learning design is that it represents the learning activities and it supports activities that are performance by different individuals (learners, teachers) in the context of a unit of learning. Due to their small size and familiarity, mobile phones in the classroom can be unobtrusive [17], requires no technology training, and are not intimidating to most users. All students can ask questions and comment (simultaneously if needed) without interrupting the in-class activities; interaction can continue after class [18].

V. LEARNING ENVIRONMENT

The presence of interactivity in the classroom is reported to yield benefits in relation to the promotion of more active learning environments, the building of learning communities, the provision of greater feedback for lecturers, and it also contributes towards student motivation [19] [20]. It is possible to study any where and any time with the development of wireless mobile network and the improve of the mobile communications equipment, within educational environments, students frequently move venues, [21]. The learner's mobility creates an ever-changing environment for learning:

"....the mobile technology, while essential, is only one of the different types of technology and interaction employed. The learning experiences cross spatial, temporal and/or conceptual borders and involve interactions with fixed technologies as well as mobile devices. Weaving the interactions with mobile technology into the fabric of pedagogical interaction that develops around them becomes the focus of attention". [22]

The learner's location positively affects the learning contents and method as well in constructive mobile learning.

VI. RESEARCH METHODOLOGY

This study was conducted in February-April 2009 by sending course content using SMS to students who have registered with the SMS learning programme. The survey was conducted for two week before the final examinations for the semester of 2008/2009. A total of 105 questionnaires were distributed to student from four different courses which are Bachelors of Science, Bachelor of Arts, Bachelor of Social Science and Bachelor of Management. The data was collected using simple random sample through secondary data based on the online databases and past researchers studies. The questionnaires were administered by the respective course managers in their tutorial sessions during the annual residential intensive course in the main campus of the Universiti Sains Malaysia.

A. Instrument

The questionnaire consisted of 6 parts which are demographic data, learning transfer, system elementary, SMS-learning services, technology acceptance and effectiveness of SMS-learning. For demographic data, it focused more on the respondent's demographic information and personal background such as gender, age, ethnic group, the courses types, year of study, monthly income, marital status, current CGPA and others. Learning transfer was viewed from the perspective of information on the learner characteristics, learning design and learning environment. System elementary investigates the accessibility, while SMS-learning services captured more on facilities, satisfaction of M-learning services and teaching and learning style. Technology acceptance investigated the perceived ease of use, security and privacy, perceived usefulness, amount of information, perceived enjoyment, social influence and new technology usability. The final component was the effectiveness of SMS-learning captured from the course package, usability and students responsiveness. All questions were measured using a 5-point Likert scale, which 1 stands for 'strongly disagree, 2 'was for 'disagree', 3 was for neutral, 4 was for agree and 5 was for 'strongly agree' except for questions on demographic. This study utilised two models of data analysis, the Statistical Package for Social Science (SPSS) Version 12.0 and the Rasch model for measurement.

B. Analysis and Finding

Statistical Package for Social Science (SPSS) Version 12.0 was used to analyze the data. The data was run by an analysis of variance (ANOVA) test.

TABLE I.
DEMOGRAPHY OF THE RESPONDENTS

		Frequency
Gender	Male	31
	Female	74
Age (years)	20-29	44
	30-39	46
	40-49	12
	50 and above	3
Ethnicity	Malay	60
	Chinese	11
	Indian	27
	Other	7
Programme	B. Science	2
	B. Arts	1
	B. Social Science	2
	B. Management	98
Mobile Device Ownership	Mobile Phone	96
	Both	6
	PDA/Pocket PC/Palmtop	3

TABLE II.
STATISTICS FOR THE PROBLEMATIC OF DESIGNING MOBILE LEARNING

Item	Statement	Infit MNSQ	Outfit MNSQ
Design 18	I can easily remember the term that I received on my mobile phone	1.38	1.41
Learner 6	This course by mobile learning experience was fun	1.13	1.28
Environment 28	I would like to see the SMS learning to be used in next semester as well	1.12	1.04
Design 17	I found the SMS learning enjoyable	1.10	1.04
Design 14	Mobile learning is convenient for communication with other course students.	1.25	1.24
Learner 7	I would take another mobile learning assisted course if relevant to my learning needs.	.86	.83
Design 15	The daily SMS messages assisted in my studies greatly.	.89	.89
Design 12	Using mobile learning, it is easy for me to access course content.	.85	.85
Learner 8	Mobile learning increases the quality of my distance education course.	.74	.73
Learner 9	Mobile learning has helped me pace my studies in my distance education course.	.68	.64
Mean		.99	1.00
S.D.		.21	.24

CRONBACH ALPHA (KR-20) PERSON RAW SCORE
RELIABILITY = .93

A total of three scenarios represents (10) ten items were analyzed in Learning Transfer. One scenario represents four items were analyzed in 'learning characteristics', second scenario represents five items were analyzed in 'learning design' and third scenario represents one items were analyzed in 'learning environment'.

A specification for the person center, or mean, to be located at zero was entered into the Winstep (version 3.57.1) code. For comparison purpose, an analysis was conducted and examined with and without this specification [23]. To determine the measure of stability and accuracy, the review begins with fit statistics to assess whether the assumption of uni dimensionality holds empirically (Linacre, 2004). OUTFIT mean-square fit statistics (MNSQs) are equivalent to a chi-square statistics value greater than 2.0 indicate unexplained randomness throughout the data [24].

Table 1 illustrate that all items included in the learning transfer measure fits the expectations of the Rasch model. The data shows that infits MNSQ and outfit MNSQ are not more than 2. From here we can conclude that the infits MNSQ is .99. This is almost perfect because it nears to 1, while mean for outfitMNSQ is 1 which is a perfect result. Standard division (S. D.) for both also good because more that 2.

Reliability is the degree to which measures are free from error and therefore yield consistent results. Sekaran stated that, the closer the reliability coefficient (Cronbach's Alpha) to 1.0 the better it is and those values over 0.80 are consider as good [25]. Values in 0.70 are acceptable while below than 0.60 considered as poor. In the reliability analysis, the alpha value that is closer the reliability coefficient to 1.00 is the better. Related to the table, reliability for this analysis is 0.93, which is good because it is closer to 1.0 and shows that respondents answered all questions consistently. From this study also we can see that six items were deleted from a total of ten (10) items, meaning that only 10 items that were include in range 0.7 until 1.4 logit.

The item and person map in Figure 1 displays a hierarchy of design, learner and enviroment preferences as rated by the participants and indicates the participant's willingness to endorse the items is generally very high and the item endorsability is quite easy, as noted by the mean, m, of items and person. The characteristic which participants rated as most preferable in a counselor was "a good listener". The characteristic which participants rated as the least preferable in a counselor was "a good listener". The characteristics which participants rated as the least preferable in a counselor were "sympathetic", "validates my thoughts", "uses humor" and "comfortable talking about issues of diversity" [23].

Design 9, stated that mobile learning has helped me pace my studies in my distance education course. The results show that mobile learning has assisted the learners in their study. This concept is also similar to other place like Open Universiti Malaysia. They use SMS as their supplementary learning tool. According to Nurhizam, SMS service which will be provided to learners are as follows; multiple choice questions with feedback, pre post self- test, quizzes and assignment notification, crucial assignment reminders, access to examinations and test

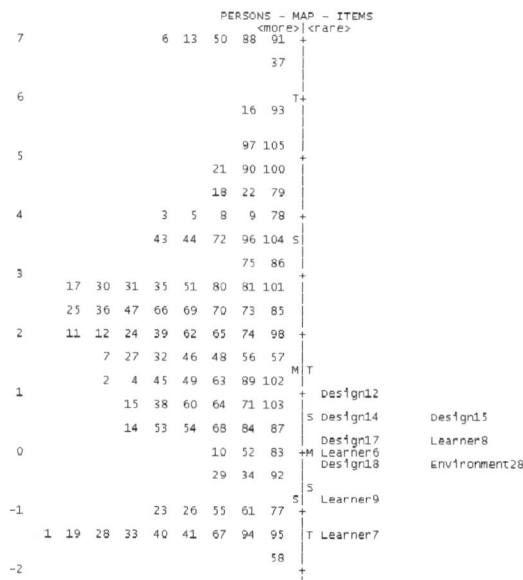

Figure 1. Hierarchy Map of Person and Items

marks, fact of the week, web links, reading materials lists and courses registration [26]. Mobile learning is the methods that were use to make their study more interesting and the learners can easily get any information that they need. Now, this is important to improve the mobile learning system that provides satisfactions to the students and having them enjoying and benefit from the use f this system.

Due to that, in design 7, the learners would take another mobile learning assisted course if relevant to my learning needs. This is show that the learners were required more mobile learning course in their study. They are interested in pursuing in this method of learning if there is any opportunity offered. It is the future of mobile learning.

Unfortunately, in design 12, there is a constraint in mobile learning. The bigger constraint in using mobile learning existed as the learners finds it difficult to access the course. This finding is consistent with our system. It is because our system does not provides the system which enable learners to access the course. However, according to study done by Hassan, W.Z. & Sulaiman, W.A., the results showed that the enjoyment with regards to the use of SMS is related to behavioral intention of SMS [27]. This indicates that, the learners still enjoys using SMS in their study although the system is not easy to access.

While design 14 suggested that mobile learning is convenient for communication with other course students. However, the results shows that the respondent disagree about the item. Consistent with our research, currently we did not provide the system which can communicate with other course students. Currently, mobile learning that has been offered in this case is a one way communication, whereby students are not allowed to communicate with the server. Therefore, learners need to participate with the combination of e-learning method. M-learning, is a form of e-learning that specifically employs wireless communications devices to deliver content and learning support [11]. Most existing typical e-learning systems are tailored toward PC-based web access and are not customized to be used through mobile devices (Woukeu et al., 2005; Goh and Kinshuk, 2006). Thus, by using this combination, the learners can get more information about their study easily and make their study more interesting.

For design 17 on the other hand, it shows that on average, respondents stated that SMS is less exciting. Currently, related with our research, we only send facts to our learners. For the future plan, we plan to insert more fun elements that will increase learner effectiveness like MMS, GPRS and many more.

VII. CONCLUSION

The result that we can derive from this study is that SMS is determined by the perception of usefulness. The terms that are received in mobile learning are easy to remember and very helpful for the learners in their study. The learners are also very excited to take another mobile learning assisted course if the courses were relevant with their learning need. However, the bigger constraint using mobile learning is not easy for learners to access course content. This finding is consistent with our system. This is because our system does not provide the system which can allow the learners to assess. The learners also said that SMS is less exciting. It is because currently, related with our research, we only send facts to our learners. In our future plan, we plan put more fun elements like MMS,

GRPS and many more. The SMS features in the future will have more desirable functions as it can perform better. The more learners are recommended to use SMS, it shows that more learners believe in the effectiveness of SMS. SMS are identified as are an easy mechanism in manipulation and navigation, ubiquity and instantaneous response. Cooperation between mobile phone service providers has given a great impact on the SMS ease of use. Perhaps, SMS in Universiti Sains Malaysia is in the early stage of adoption. The benefits such as usefulness and can help learners in their study would be the most important drive of mobile learning and should not be unheeded in the development of new functions and enhancement of service features. It has some desirable functions that it can perform the more they would use SMS in the future.

ACKNOWLEDGMENT

The authors would like to thank Universiti Sains Malaysia for the support under RU grant and USM Fellowship scheme.

REFERENCES

[1] Kukulska-Hulme, A. (2005). Mobile Usability and User Experience. InA. Kukulska-Hulme & J. Traxler, J. (Eds.) *Mobile Learning: A handbook for educators and trainers* (pp. 45-56). London: Routledge.

[2] Mitchell, A., Heppel, S., & Kadirire, J. (2002). *Technology Watch Research Report*. Anglia: UltraLab.

[3] Naismith, L., Lonsdale, P., Vavoula, G. and Sharples, M. (2004) Literature Review in Mobile Technologies and Learning. FutureLab Report 11. http://www.futurelab.org.uk/resources/documents/lit_reviews/Mobile_Review.pdf.

[4] Kukulska-Hulme, A. and Traxler, J. (2007) Designing for mobile and wireless learning. In: Beetham, H. and Sharpe, R. (eds.), Rethinking Pedagogy for a Digital Age: Designing and Delivering E-Learning. London: Routledge, 180–192.

[5] Sharples, M. (ed.) (2006) Big issues in mobile learning. Report of a workshop by the Kaleidoscope Network of Excellence Mobile Learning Initiative, University of Nottingham, UK.

[6] MCMC (2009). *Fact & Figures (Statistics & Record)*. Retrieved September 11, 2009 form http://www.skmm.gov.my/facts_figures/stats/index.asp.

[7] Harris, P., Rettie, R. and Cheung, C.C. (2005). "Adoption and usage of m-commerce: A cross-cultural comparison of Hong Kong and the United Kingdom", *Journal of Electronic Commerce Research*, 6(3), 210-224.

[8] Gilligan, R. and Heinzmann, P. (2004). "Exploring how cultural factors could potentially influence ICT use: An Analysis of European SMS and MMS use", Cultural Difference Workgroup COST 269.

[9] Briggs, L. (2006). The age of the 'smart' cell phone. Campus Technology, 19(5), 24–57.

[10] CTIA (2007). *Wireless quick facts*. Retrieved August 20, 2009, from http://www.ctia.org/media/industry_info/index.cfm/AID/10323.

[11] Brown, H.T. (2005), "Towards a model for MLearning", *International Journal on E-Learning*, 4 (3), 299-315.

[12] Moses, O.O. (2008). Improving mobile learning with enhanced Shih's model of mobile learning. *US-China Education Review*, 5(11), 1-7.

[13] Rashmi Sinha. (2005). *Collaborative filtering strikes back (this time with tags)*. Retrieved August 24, 2009, from http://www.rashmisinha.com/archives/05_10/tags-collaborative-filtering.html.

[14] Cheong, L.S. (2004) Transfer of Learning among Malaysian Learners. *Masalah Pendidikan*, 27 , 99-108.

[15] Detterman, D. K. (1993). The case for prosecution: Transfer as an epiphenomenon. In D. K. Detterman & R. J. Sternberg (Eds.),

Transfer on Trial: Intelligence, Cognition, and Instruction (pp. 39–67) Stamford, CT: Ablex Publishing Corp.

[16] Koper, R. (2006). Current Research in Learning Design. *Educational Technology & Society*, 9 (1), 13-22.

[17] Nyiri, K. (2003). Mobile Communication: Essays on Cognition and Community. Vienna: Passagen Verlag.

[18] Liu, T.-C., Wang, H.-Y., Liang, J.-K., Chan, T.-W., & Yang, J.-C. (2002). Applying Wireless Technologies to a Build Highly Interactive Learning Environment. Paper presented at the IEEE International Workshop on Wireless and Mobile Technologies in Education, Växjö, Sweden.

[19] Anderson, T. (2002). *An Updated and Theoretical Rationale for Interaction*. IT Forum. Retrieved August 24, 2009, from http://it.coe.uga.edu/itforum/paper63/paper63.htm.

[20] Muirhead, B., & Juwah, C. (2003). *Interactivity in Computer-Mediated College and University Education: A Recent Review of the Literature*. International Forum of Educational Technology & Society. Retrieved August 14, 2009, from http://ifets.ieee.org/discussions/discuss_november2003.htm.l

[21] Muhlhauser, M., & Trompler, C. (2002). *Learning in the Digital Age: Paving a Smooth Path with Digital Lecture Halls.* Paper presented at the IEEE 35th Hawaii International Conference on System Sciences, Hawaii.

[22] Kukulska-Hulme, A. (2009). Will mobile learning change language learning?. *European Association for Computer Assisted Language Learning*, 21(2), 157–165.

[23] Bradley, K.D., Cunningham, J., Haines, R.T., Harris, W.E., Jr., Mueller, C.E., Royal, K.D., Sampson, S.O., Singletary, G. & Weber, J.A. (2006). Constructing and Evaluating Measures: Applications of the Rasch Measurement Model (pp. 1-54). University of Kentucky, Department of Educational Policy and Evaluation Studies. 131 Taylor Education Building, Lexington, KY 40506-0001.

[24] Linacre, J. M. (2004). optimizing rating scale category effectiveness. In E. V. Smith, JR. and R. M. Smith Introduction to Rasch Measurement (pp.258-278) Maple Grove, MN:JAM Press.

[25] Sekaran, U. (2000). Research Method for Business: A skill Building Approach. John Wiley & Sons. Inc.Singapore.

[26] Safie, N. (2004). *The use of Short Messaging System (SMS) as a supplementary learning tool in Open University Malaysia (OUM.).* A full paper presented for the 18th Annual Conference Association of Asian Open Universities (AAOU) Shanghai, China.

[27] Hassan, W.Z. & Sulaiman, W.A. (2007). Adoption of Short Messaging Service (SMS) in Malaysia. *In: International Conference on Global Research in Business and Economics, 27-29 Disember 2009, Bangkok, Thailand.*

[28] Woukeu, A., Millard, E.D., Tao, F. and Davis, C.H. (2005), "Challenges for semantic grid based mobile learning". Retrived August 18, 2009, from www.ubourgogne.fr/SITIS/05/download/Proceedings/Files/f135.pdf.

[29] Goh, T. and Kinshuk, D. (2006), "Getting ready for mobile learning – adaptation perspective", *Journal of Educational Multimedia and Hypermedia*, 15(2), 175-198.

AUTHORS

Issham Ismail is with the School of Distance Education, Universiti Sains Malaysia, Minden, Pulau Pinang, 11800 Malaysia (e-mail: issham@usm.my)

Rozhan M. Idrus is with the School of Distance Education, Universiti Sains Malaysia, Minden, Pulau Pinang, 11800 Malaysia. He is specialized in Open and Distance Learning Interactive Technologies and e-Learning (e-mail: rozhan@usm.my).

Azidah Abu Ziden is with the School of Education Studies, Universiti Sains Malaysia, Minden, Pulau Pinang, 11800 Malaysia (e-mail: azidah@usm.my)

Munirah Rosli is a student in the School of Distance Education, Universiti Sains Malaysia, Minden, Pulau Pinang, 11800 Malaysia and currently doing her master in Educational Technology in Universiti Sains Malaysia. (e-mail: munirahrosli@yahoo.com.my).

31

Mobile Learning Via SMS Among Distance Learners: Does Learning Transfer Occur?

Aznarahayu Ramli, Issham Ismail, and Rozhan Md. Idrus

Universiti Sains Malaysia, Penang, Malaysia

Abstract—The purpose of this study is to determine whether learners are willing to transfer learning in this mobile learning environment via SMS. The reason for this is to measure the effectiveness of the new method used in learning and education especially in distance education field. For this reason, students' responses are gathered which looked at three factors namely learner characteristic, learning design and learning environment. The data are gathered through a survey research design with questionnaires using five-point likert scale. The questionnaires was administered for 105 distance education students from four courses including Bachelor of Science, Bachelor of Arts, Bachelor of Social Science and Bachelor of Management. The Rasch Model Analysis was used to measure these dimensions. Rasch Model is a one-parameter logistic model within item response theory (IRT) whereby the amount of a given latent trait in a person and the amount of that same latent trait reflected in various items can be estimated independently yet still compared explicitly to one another. The result of the study showed that learning transfer occurred and being influenced most by learner's characteristics especially in term of their motivation as well as their perceive utility/value of the SMS learning to their job and academic performances.

Index Terms—Mobile Learning, SMS, Distance Education, Learning Transfer.

I. INTRODUCTION

The learning process slowly makes it away outside the classroom. The technology advancement has witnessed the emerging trend in learning and education both in real life and virtually. With the quick advance in mobile technologies and devices, comes the birth of new term in learning and education – *mobile learning*. In a study of this trend Muyinda [1], mobile learning is well positioned to champion these innovations.

Mobile learning becomes a new learning experience in the open and distance learning environments [2]. Quinn [3] defines mobile learning as learning through mobile computational devices while Shepherd [4] stated that *'M-Learning is not just electronic, it's mobile.'* Another definition by Trifonova [5] is defined mobile learning as a field which synchronizes two capable fields – mobile computing and E-learning.

The capability of mobile technologies in delivering and enhancing learning are quite difference between developed countries and developing countries. Traxler and Dearden [6] in their study found that in developed countries, mobile technologies are able to assist learning and education compared to developing countries due to the factors of their mains electricity, computer hardware and internet connectivity are stable, reliable, cheap and abundant. For example, in developed countries such as US and Europe, the Windows and Palm-base personal digital assistants (PDA) are widely used to provide media-rich educational content [7]. However, in developing Asian countries such as Malaysia, the high cost of ownership and connection of the devices makes it impossible to apply those devices in learning and educational context. In this case, the more accessible and popular mobile or cell phones can be used instead.

A. Background of the Study

The penetration rate of mobile phones in Malaysia has exceeded 100.0%. This occurs because of multiple subscriptions by the citizens. This phenomenon shows a significant growth of mobile phone adoption in Malaysia and the Malaysians are keen in owning and using mobile phone in which the data provided by Malaysian Communications and Multimedia Commission [8] showed that 90.3 % of Malaysians own mobile phones. The data also showed that 50.4 % of the mobile phones users sent more than 5 SMS per day.

The advantages of mobile devices such as limitless mobility and small size can bring new dimensions to the learning processes of the students especially for distance learning students who are always on the go and on the move [9]. Jacob and Issac [10] in their study, resounding the facts that, the goal of mobile learning is to facilitate learning and hoping that in future, learning can be ubiquitously with the aid of wireless convergence. They also addressed the issue regarding the challenges of mobile learning includes the instructional strategies, learning content and design that promotes to continuous learning activity.

Learners should experience mobile learning in order to know the learning process through that method. The purpose of the learning experience is for the learners to gain skills and knowledge through learning process and / or for there to be a change in attitudes and beliefs [11]. Huczynski and Lewis [12] also reported that the learning experience that learners gained will promote to the motivation to transfer learning. Transfers of learning refers to a situation whereby a student learns behaviors, skills, and knowledge in one context and applies them in another context. In which it means that, transfer of learning occurs when learners are able to apply what they learn through mobile learning in their study and job.

B. Purpose of the Study

The purpose of this study is to determine whether learners are willing to transfer learning in this mobile learning environment via SMS. The reason for this is to measure the effectiveness of the new method used in learning and education especially in distance education field. For this reason, students' responses are gathered which looked at three factors namely learner characteristic, learning design and learning environment. These three main factors have been proposed by Baldwin & Ford [13], the former researchers in learning transfer study who suggested that transfer factors would be represented by three domains: trainee characteristics, training design and work environment.

II. RESEARCH METHODOLOGY

A. Instrument

Questionnaire is used as the data collection instrument for this study. It is divided into two sections which are demographics questions and likert-type questions. A cover letter explaining the purpose of the questionnaire is also attached to the questionnaire. The demographic questions capture information of the respondents regarding gender, age, marital status, income level, and mobile device ownership. The likert-type questions are divided into three sections which are Learners' Characteristics, Learning Design and Learning Environment. Other than the questions on demographic information, all other questions used the five-point Likert Scale ranging from 1 to 5, in which 1 is for 'strongly disagree', 2 for 'disagree', 3 for 'neutral', 4 for 'agree' and 5 for 'strongly agree'.

The items used in this study are adopted from the study conducted by Holton et al., [14]. Their instrument offers several key strengths as it was based on a very large and extremely diverse sample and also builds on the results of several previous research efforts as well as followed generally accepted instrument development process. Thus, it provides a high level of confidence that the items being adopted will work well in this new learning environment (Mobile Learning via SMS). Holton et al., in this study also proposed that future research should combine the items with other instruments to assess transfer result more completely. Thus, in this study, the researchers also added some others items that are more focus on mobile learning environment which have been developed by University's experts.

B. Data Collection Procedure

A total of 119 questionnaires were distributed to students from four difference courses in School of Distance Education, Universiti Sains Malaysia (USM) which are Bachelor of Science, Bachelor of Arts, Bachelor of Social Science and Bachelor of Management. From the total, 105 questionnaires were returned hence making a response rate of 88 per cent. The high response rate could be attributed to the fact that the questionnaires are being sent via email hence making it more convenient, faster and time savvy. The responses are 43.8 percent from second year students, 53.3 percent from third year students and 2.9 percent from fourth year students.

This mobile learning assisted course was introduced in the second semester of the 2008/2009 academic session and covered two subjects for Management students which are Financial Principle for second year and International Business for third year. Besides that, for Science students specifically Physics students, the course that covered in this mobile learning assisted course are Mechanics and Optics. Both are for second year students. Two courses that covered for Economic students are Money and Banking for second year and Quantitative Economy for third year. During the duration of this program which was conducted for three months (February 2009 to April 2009), the registered students received text message once a day regarding their subject matter. This project is a preliminary project conducted to gather students' acceptance and readiness towards learning via mobile phone. Thus, at this stage it involves a one-way-communication whereby the students only can receive an SMS per day with no permission response back to the sender. The information being sent to the students are including short notes, terms, problem solving, equation, etc. related to their subject matter.

C. Data Analysis and Findings

Data for the study were collected using a survey research design. Questionnaires were sent out to 119 participants and 105 were returned hence making a response rate of 88 per cent. Based on the 105 usable responses, data for demographic were analyzed using Statistical Package for Social Science (SPSS) Version 12.0, while for the likert type questions, the RASCH Measurement Model were used to come out with the certain interpretation. RASCH Model is a one-parameter logistic model within item response theory (IRT) in which the amount of a given latent trait in a person and the amount of that same latent trait reflected in various items can be estimated independently yet still compared explicitly to one another.

III. RESULTS AND DISCUSSION

A. Frequency Analysis of Respondents

For the demographic, we interpreted the data based on frequency analysis. Frequency Analysis is the convenience and understandable way of looking at different value of variables. The items measured for this study includes Gender, Age, Marital Status, Income Level and Mobile Device Ownership. The results of frequency analysis is shown in Table I.

TABLE I.
FREQUENCY ANALYSIS OF RESPONDENTS

Items	Frequencies	Percentage
Gender		
Male	31	29.5
Female	74	70.5
Age		
20-29 years	44	41.9
30-39 years	46	43.8
40-49 years	12	11.4
50 years and above	3	2.9
Marital Status		
Single	49	46.7
Married	53	50.5
Single Parent	3	2.9
Income Level		
Below RM1500	22	21.0
RM1500-RM2000	28	26.7
RM2001-RM2500	24	22.9
RM2501-RM3000	15	14.3
RM3001-RM3500 and above	16	15.2
Mobile Device Ownership		
Mobile Phone	96	91.4
PDA/Pocket PC/Palmtop	3	2.9
Both	6	5.7

B. Item/Person Misfit Order Table

The item and person misfit value is also helpful in this study. The results represented the validation of the instrument whether suitable or not becoming as a tool in data collection [15]. In the other words, its aim is to determine whether the instrument used is a valid tool to utilize for data collection. This can be achieved by checking the unidimensionity of the items. Outfit is the outside factors that influence the items or variables while infit represent the "off-variable noise" that could have an impact on the items or variables. According to Fox [16], any items fall within the infit and outfit limit of 0.6 to 1.5 is considered acceptable. However, for Bond and Fox [17] if the sample is less than 500 participants, the acceptable value of outfit range are from 0 to 1.3. For this study, Table II measures the items of the variables fit the expectations of the RASCH Model.

Any items that fall outside the acceptable range are considered to be invalid for the reason that the respondents viewed the items differently than that intended by the researcher [18]. Out of 13 items being analyzed, 6 items have been rejected for the reasons they are fall outside the acceptable range. All the 6 items are presented in Table III.

TABLE II.
ITEM/PERSON MISFIT TABLE

Item	Description	Infit MNSQ	Outfit MNSQ
L4	My job performance improves when I apply new things that I have learned	1.34	1.33
L5	I am confident in my ability to use newly learned skills on the job	1.26	1.30
L7	I would take another mobile learning assisted course if relevant to my learning needs	0.78	0.75
L9	Mobile learning has helped me pace my studies in my distance education course	0.72	0.70
D10	The sequential presentation of lesson via SMS assisted in my studies	0.70	0.69
D11	Course learning objectives can be met by mobile learning	1.10	1.08
D15	The daily SMS messages assisted in my studies greatly	0.71	0.69
D17	Using mobile learning, it is easy for me to access course content	0.77	0.77
D18	I can easily remember the term that I received on my mobile phone	0.84	0.84
E21	My colleagues encourage me to use the skills and knowledge I have learned in my job	1.30	1.35
E22	My superior sets goals for me that encourage me to apply my learning on the job	1.21	1.16
E24	My workload allows me time to try the new things I have learned	1.09	1.08
E25	The resources needed for me to be able to apply what I've learned will be available to me after learning session	0.99	1.08
E26	I get feedback from people about how well I am applying what I've learned in my study	1.21	1.22

TABLE III.
ITEM FALL OUTSIDE THE ACCEPTABLE RANGE

Item	Description
L1	Before I further my study, I had a good understanding of how learning and education would fit my job-related development.
L3	It is waste for me if I do not utilize my learning.
D16	Receiving the SMS while at work (or anywhere) is not an inconvenience to me.
D19	I can easily remember the term that I received on my mobile phone
E23	My superior opposes the use of the techniques I've learned in my study.
E30	Does the system offer more than face-to-face meeting?

C. Variable Map

From the result in item and person misfit analysis, the data then being analyzed and interpret in Variable Map. This is to reveal what are factors among the acceptable items that are mostly influence students' willingness to transfer learning in a mobile learning environment. Variable map is another visual guide to information regarding relative scales [15]. It is used to present the items by ranking them according to the level of difficulty to endorse as well as presenting the respondents by ranking them according to their willingness to endorse the items. In this study, items which are easiest or favorable to endorse represented the main factors influencing students' willingness to transfer learning. The results are showed in Figure 1.

The result in red oval in Figure 1 shows items are ranked from least factors influencing learning transfer (at the top) to the crucial factors influencing learning transfer (at the bottom). The result on the other side of the figure showed the ID of the respondents who responded to each items. Figure 1 provided certain interpretations for this study.

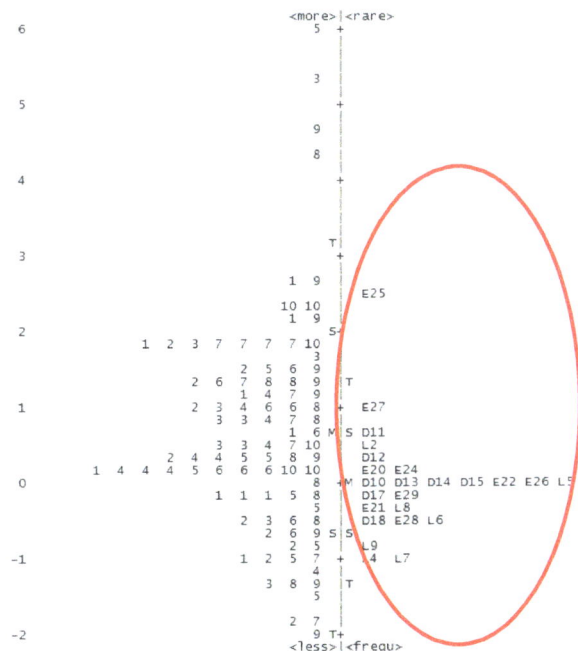

Figure 1.   Variable Map

Result in Figure 1 showed that item L4, L7 and L9 are the crucial factors influencing students' willingness to transfer learning. These three items represented Learners' Characteristics. Thus, it showed us that learner's characteristics are the most influential factors in learning transfer.

I. Learner's Characteristics

The result from this study showed that the factor that mostly influent learning transfer via SMS is the learner's characteristics and it is consistent with previous literature on psychology which stated that individual's ability and motivation which are referred to learner's characteristics are the primary factor that influences learning transfer [19].

Item L7 which is *"I would take another mobile learning assisted course if relevant to my learning needs"* measured learner's motivation to transfer learning via SMS. Learning motivation as defined by Tannenbaum & Yukl [20] is the learner's intensity and persistence of efforts that they apply in learning-oriented improvement activities, before, during and after learning. There are two motivation-relevant constructs that have been examined in previous research which are pre-training motivation [21] and motivation to learn [22] [23]. The study conducted by Facteau et al. [24] found the correlation between pre-training motivation and training transfer while Quinones [23] found the motivation to learn as the key principle in linking pre-training characteristics and training outcomes. Noe [20] on the studies in military settings found that motivation to learn have an impact on training effectiveness.

In this study, the result showed that learners are willing to take another mobile learning assisted courses if the courses are relevant to their learning needs. This indicates that most of the learners are motivated to continue this course via SMS and allowing the learning transfer to take place. However, their motivation level depends on the courses whether relevant or not to their learning needs. This means, for other courses via this method, they should be tailoring to learner's need and preferences in order to enhance their learning transfer.

Item L4 and L9 which are *"My job performance improves when I apply new things that I have learned"* and *"Mobile learning has helped me pace my studies in my distance education course"* separately, measured learner's perceiveness of utility / value of learning via SMS. Previous research found that learner's perceived utility or value of the learning program can influence learning transfer. This has been proved by the study conducted by Baumgartel et al. [25] which showed that managers who believe in the utility or value of the training are more likely to apply the skills learned in training. Axtell et al. [26] on the other hand found that learners who perceived training as relevant had higher levels of immediate skill transfer. Thus, Baldwin & Ford [13]; Clark et al. [27] have come to the conclusion that, in order for the learners to highly transfer the learning, they should perceive the new knowledge and skills will improve their performance in future.

The results in figure 1 showed that most of the learners believed in the mobile learning potential in improving their performance in their job as well as their study. The result also showed that the learners are interested and happy in using mobile learning as the system assists them in learning process by improving their academic perform-

ance and effectiveness. Because of that, it indicated that, when the learners perceived the skills learned via SMS as valuable, they are more willing to transfer the learning. In this study, most learners perceived mobile learning course as the method of learning that can improve their performance in their study as well as their jobs. These influence them to continue mobile learning in the future and affect the mobile learning field positively.

II. Learning Design

Learning design has become another group of factors that could influence learning transfer whether directly or indirectly. Figure 1 showed that item D18 is the most acceptable item for learning design. Item D18 which is *"I can easily remember the term that I received on my mobile phone"* is referred to the content relevant of the SMS learning. These questions revealed learner's perception regarding the content of the SMS learning. Bates [28] stated that for transfer to occur, the goals and materials for learning should be content valid. The last decade has proven that the content relevance is correlating with transfer outcomes [14] [29] [30]. Consistent with previous study done by Axtell et al. [26], found that content validity was highly correlated with transfer. On the other hand, Yamnill & McLean [31] in the study of Thai managers found that the primary factor in predicting trainee perception of successful transfer is the content relevance.

In this study, the terms being used to delivered message and learning contents to the learners via SMS are content valid. This is because, most of the learners agreed that they can easily remember the terms and it also showed that the terms are understandable and relevant. The result also indicated that, learning design also becoming one of the factors that should be considered in influencing learning transfer via SMS. The more attention being taken in the learning design especially the contents, the more learning transfer could occur.

III. Learning Enviroment

The third factor that was indentified in this study is learning environment. Learning environment was said to be a factor that influenced learning transfer by supervisory support and opportunity to perform [13]. As the result showed in Figure 1, items E28, E21 and E22 are the acceptable items for learning environment. Several studies explain the relationship between learning environment and learning transfer [32] [21]. These studies which supported previous both empirical and qualitative studies [33] [34] [35] [36] revealed the role of supervisory support in a positive aspect of influencing learning transfer.

Items E21 and E22 which are *"My colleagues encourage me to use the skills and knowledge I have learned in my job"* and *"My superior sets goals for me that encourage me to apply my learning on the job"* measured the learner's perception of support they received from their supervisors and also their colleagues. Support from peers and colleagues have proven to be the most influential factor in learning transfer than supervisory support [24]. In the study conducted by Chiaburu & Marinova [21], among individual and organization support for transfer, peer support having a highly significant relationship with learning transfer. It is consistent with this study on SMS learning whereby result from Figure 1 showed that item for peer support (E21) is situated in the more acceptable rank than item for supervisory support (E22). On the other hand, item E28 which is *"I would like to see the SMS learning to be used in next semester as well"* supported previous items by showing that with the support received from supervisors and peers, there open a good environment for learning via SMS. Thus, most of the learners are excited to continue the learning via SMS in future.

IV. CONCLUSION

The study has confirmed that learning transfer is yet to occur in mobile learning via SMS environment. Learner's characteristics especially in term of learner's motivation and learner's perceived utility/value in improving performance are proved to be the most influential factors in promoting learning transfer. Thus, learners must be motivated to learn and apply what they learned so that they could believe in SMS learning process and in improving their performance in their study and job. Besides that, there are also certain criteria of learning design that should be looking at in order for learning transfer to occur. They are regarding the content of the SMS learning being delivered to the learners which means, the learning contents via SMS should be designed properly especially concerning the terms, language, and the sequential presentation of the contents. Most importantly, the contents should be understandable and easy to be remembered by the learners. Last but not least, the support from supervisors and peers act as the unbeatable factors in enhancing learning transfer in this mobile learning environment. Responses and feedback from people around them could encourage learners to apply what they learned into their studies and jobs.

ACKNOWLEDGMENT

This work was supported by Universiti Sains Malaysia under RU grant and USM Fellowship scheme. A grateful appreciation goes to Universiti Sains Malaysia for the supports.

REFERENCES

[1] Muyinda, P. B. (2007). MLearning: Pedagogical, technical and organization hypes and realities. *Campus-Wide Information Systems*, 24 (2), 97-104. doi:10.1108/10650740710742709

[2] Safie, N. (2004). The use of Short Messaging System (SMS) as a supplementary learning tool in Open University Malaysia (OUM). *18th Annual Conference Association of Asian Open Universities (AAOU), 27 November – 30 November, 2004*, Shanghai, China, pp. 1-11

[3] Quinn, C. (2000). MLearning: Mobile, Wireless and In-Your-Pocket Learning. *Line Zine Magazine*, http://www.linezine.com/2.1/features/cqmmwiyp.htm

[4] Shepherd, C. (2001). M is for Maybe. *Tactix: Training and Communication Technology in Context*, http://www.fastrak-consulting.co.uk/tactix/features/mlearning.htm

[5] Trifonova, A. (2003). Mobile Learning-Review of Literature. Technical Report for Department of Information and Communication Technology, University of Trento.

[6] Traxler, J. & Dearden, P. (2005). The Potential for Using SMS to Support Learning and Organisation in Sub-Saharan Africa. *Proceedings of Development Studies Association*, http://www.asp2.wlv.ac.uk

[7] Ramos, A. J. & Trinona, J. (2008). Production of SMS Materials in Distance Education. *PANdora Distance Education Guidebook*, (eds)

[8] Malaysian Communications and Multimedia Commission (2008). *Hand phone users survey 2007*. Retrieved August 25, 2009, from http://www.skmm.gov.my

[9] Kinshuk, Suhonen, J., Sutinen, E. & Goh, T. (2003). Mobile Technologies in Support of Distance Learning. *The Asian Society of Open and Distance Education*, 1 (1), 60-68

[10] Jacob, M. S. & Issac, B. (2008). The Mobile Devices and its Mobile Learning Usage Analysis. *Proceedings of the International MultiConference of Engineers and Computer Scientists.*

[11] Seyler, D. L., Holton, E. F., Bates, R. A., Burnett, M. F. & Carvalho, M. A. (1998). Factors affecting motivation to transfer training. *International Journal of Training and Development*, 2 (1). doi:10.1111/1468-2419.00031

[12] Huczynskij, A. A. & Lewis, W. (2007). An Empirical Study into the Learning Transfer Process in Management Training. *Journal of Management Study*, 17 (2), 227-240. doi:10.1111/j.1467-6486.1980.tb00086.x

[13] Baldwin, T. T., & Ford, J. K. (1988). Transfer of training: A review and directions for future research. *Personnel Psychology*, 41, 63–105. doi:10.1111/j.1744-6570.1988.tb00632.x

[14] Holton, E. F., Bates, R., & Ruona,W. E. A. (2000). Development of a generalized learning transfer system inventory. *Human Resource Development Quarterly*, 11(4), 333–360. doi:10.1002/1532-1096(200024)11:4<333::AID-HRDQ2>3.0.CO;2-P

[15] Ren, W., Bradley, K.D., & Lumpp J.K. (2008). Applying the Rasch Model to evaluate an Implementation of the Kentucky Electronics Education Project. *J Sci Educ Technol*, 17, 618-625. doi:10.1007/s10956-008-9132-4

[16] Fox, C. (1999). An introduction to the partial credit model for developing nursing assessments. *Journal of Nursing Education*, 34 (8), 340

[17] Bond, T. & Fox, C. (2001). Applying Rasch model: fundamental measurement in the human sciences. *Mahwah, NJ: Lawrence Erlbaum Associates*

[18] Bradley, K. D., Cunningham, J. D., Haines, R. T., Harris, Jr. W. E., Mueller, C. E., Royal K. D., Sampson, S. O., Singletary, G. & Weber, J. (2006). Constructing and evaluating measures: Applications of the Rasch measurement model. Symposium presented at the Mid-Western Educational Research Association Annual Meeting, Columbus, OH, pp. 1-54

[19] Sackett, P. R., Gruys, M. L., & Ellingson, J. E. (1998). Ability personality interactions when predicting job performance. *Journal of Applied Psychology*, 83(4), 545–556. doi:10.1037/0021-9010.83.4.545

[20] Tannenbaum, S. I., & Yukl, G. (1992). Training and development in work organizations. *Annual Review of Psychology*, 43, 399–441. doi:10.1146/annurev.ps.43.020192.002151

[21] Chiaburu, D. S., & Marinova, S. V. (2005). What predicts skill transfer? An exploratory study of goal orientation, training self-efficacy and organizational supports. *International Journal of Training and Development*, 9, 110–123. doi:10.1111/j.1468-2419.2005.00225.x

[22] Noe, R. A. (1986). Trainee attributes and attitudes: Neglected influences on training effectiveness. *Academy of Management Review*, 11, 736–749. doi:10.2307/258393

[23] Quinones, M. A. (1995). Pretraining context effects: Training assignment as feedback. *Journal of Applied Psychology*, 80, 226–238. doi:10.1037/0021-9010.80.2.226

[24] Facteau, J. D., Dobbins, G. H., Russell, J. E. A., Ladd, R. T., & Kudisch, J. D. (1995). The influence of general perceptions of the training environment on pre-training motivation and perceived training transfer. *Journal of Management*, 21, 1–25. doi:10.1016/0149-2063(95)90031-4

[25] Baumgartel, H. J., Reynolds, M. J. I., & Pathan, R. Z. (1984). How personality and organizational climate variables moderate the effectiveness of management development programmes: A review and some recent research findings. *Management and Labour Studies*, 9(1), 1–16.

[26] Axtell, C. M., Maitlis, S., & Yearta, S. K. (1997). Predicting immediate and longer term transfer of training. *Personnel Review*, 26(3), 201–213. doi:10.1108/00483489710161413

[27] Clark, S. C., Dobbins, G. H., & Ladd, R. T. (1993). Exploratory field study of training motivation: Influence of involvement, credibility, and transfer climate. *Group & Organization Management*, 18, 292–307. doi:10.1177/1059601193183003

[28] Bates, R. A. (2003). *Managers as transfer agents: Improving learning transfer in organizations.* San Francisco, CA: Jossey Bass.

[29] Lim, D. H., & Morris, M. L. (2006). Influence of trainee characteristics, instructional satisfaction, and organizational climate on perceived learning and training transfer. *Human Resource Development Quarterly, 17*(1), 85–115. doi:10.1002/hrdq.1162

[30] Rodriguez, C. M., & Gregory, S. (2005). Qualitative study of transfer of training of student employees in a service industry. *Journal of Hospitality & Tourism Research, 29*, 42–66. doi:10.1177/1096348004270753

[31] Yamnill, S., & McLean, G. N. (2005). Factors affecting transfer of training in Thailand. *Human Resource Development Quarterly, 16*(3), 323–344. doi:10.1002/hrdq.1142

[32] Awoniyi, E. A., Griego, O. V., & Morgan, G. A. (2002). Person—environment fit and transfer of training. *International Journal of Training and Development, 6*(1), 25–35. doi:10.1111/1468-2419.00147

[33] Brinkerhoff, R. O., & Montesino, M. U. (1995). Partnerships for training transfer: Lessons from a corporate study. *Human Resource Development Quarterly, 6*(3), 263–274. doi:10.1002/hrdq.3920060305

[34] Broad, M. L., & Newstrom, J. W. (1992). *Transfer of training: Action packed strategies to ensure high payoff from training investments.* Reading, MA: Addison-Wesley.

[35] Burke, L. A., & Baldwin, T. T. (1999). Workforce training transfer: A study of the effect of relapse prevention training and transfer. *Human Resource Management, 38*(3), 227–243. doi:10.1002/(SICI)1099-050X(199923)38:3<227::AID-HRM5>3.0.CO;2-M

[36] Clarke, N. (2002). Job/work environment factors influencing training effectiveness within a human service agency: Some indicative support for Baldwin and Fords' transfer climate construct. *International Journal of Training and Development, 6*(3), 146–162. doi:10.1111/1468-2419.00156

AUTHORS

Issham Ismail is with the School of Distance Education, Universiti Sains Malaysia, Minden, Pulau Pinang, 11800 Malaysia (e-mail: issham@usm.my).

Aznarahayu Ramli is a student under the School of Distance Education, Universiti Sains Malaysia, Minden, Pulau Pinang, 11800 Malaysia and currently furthering her Masters Degree (e-mail: aznarahayu@gmail.com)

Rozhan M. Idrus is with the School of Distance Education, Universiti Sains Malaysia, Minden, Pulau Pinang, 11800 Malaysia. He is specialized in Open and Distance Learning Interactive Technologies and e-Learning (e-mail: rozhan@usm.my).

32

An Adaptation of E-learning Standards to M-learning

Daoudi Najima[1],[2] and Ajhoun Rachida [2]

[1]Ecole des sciences de l'information, ESI, Rabat, Morocco
[2] University Mohamed V, ENSIAS, Rabat, Morocco

Abstract—The exploitation of technological advances in learning has result in an exponential progress in this field through e-learning applications in the last decade, and currently through the emergence of a new concept called m-learning. M-learning is defined as the use of mobile technologies for learning; m-learning must benefit from e-learning technological advances in order to avoid reinventing the wheel. Nevertheless, m-learning, which is characterized by the use of mobile devices, permits, for example, the learners' mobility during their learning, and, as opposed to e-learning, allows a continuous change of the context. Moreover, m-learning faces some constraints caused by the use of its mobile technologies such as the limited screen size, reduced energy, resolution capacity and location change during an activity. Yet, there is an agreement among most research laboratories interested in e- and m- learning on the parallel use of these two learning environments. Therefore, it would be more sensible to allow communication and exchanges, to facilitate the sharing of learning subject matters and data between the two environments, and thereby to avoid the reproduction of contents that already exist. In other words, an educational heritage which is exploitable independently of the environment of its development must be created. The utilization of standards can offer pedagogical contents some structures which facilitate the interchangeability between e- and m- learning. In order to ensure the interoperability between e- and –m learning platforms and to take into account the specificities of m-learning, we have adopted the already existing standard LOM and the specification IMS LD.

Index Terms—e-learning, interoperability, m-learning,, LOM, IMS LD.

1. INTRODUCTION

This decade has witnessed a spectacular evolution in E-learning both technically and pedagogically, resulting in a significant increase in E-learning services. Shifting their concern from teaching resources development to course management, these services have played a key role not only in the diffusion and access to electronic resources, but also in managing the interactions among all the participants involved in an E-learning environment. The research tasks of standardization in learning covers several aspects such as the learner's profile and the structuring courses. Also, with the adoption of standard integrating pedagogical aspects, e-learning has reached an undeniable state of maturity.

In parallel, the extraordinary technological progress made in wireless networks and mobile data processing technology have allowed an effective integration of mobile devices in several applications, including those relating to learning. These developments have given birth to a new concept paralleling E-learning: M-learning, a new version of E-learning upgraded towards mobile technologies use.

Several research laboratories are interested in various aspects of mobile learning. The majority of them begin with research on its relationship to e-learning. Despite the diversity of their visions, there exists a consensus on the coexistence of these two learning environments

Consequently, the need seems tightly pressing to ensure the exploitation of the pre-existing assets of e-learning and to avoid any unnecessary reproduction. Moreover, it is important to ensure the communication, the exchanges, the sharing of teaching resources and the data between the two environments. Thus, we need to mask the heterogeneity of the devices and in particular the constraints imposed by mobile devices and then allow communication and data exchanges of contents developed on these environments. Also, it is equally important to exploit the existing contents independently of their environments of development and thus create an educational inheritance. In other words, it is necessary to ensure interoperability between these two environments of learning.

Our research orientation is articulated around the question of interoperability between e-learning and m-learning. In this article, we propose standardization as a solution since the use of a standardized structure will facilitate the exchanges and will allow resources sharing. It has been widely suggested that adopting a standard-based approach to M-learning could be a promising solution. However, in the absence of any standards peculiar to M-learning, the question as to what extent e-learning standards can be adapted to the needs imposed by mobile technologies use in M-learning seems urgently pressing. This is basically what the present paper attempts to explore.

2. M-LEARNING VS. E-LEARNING

The birth of m-learning beside e-learning induces a fundamental question concerning the relation between these two learning environments. Is M-learning a particular case of e-learning or vice versa? Are they two disjoined environments or do they converge on some common points? These questions reflect the different conceptions of M-learning which juxtapose it to E-learning and mobile technologies.

In order to answer these questions, we will try to compare E-learning and M-learning. As a preliminary step, these two concepts will be defined to trace the points on which they either converge and/or diverge.

Definitions of these two concepts abound in the literature on learning in virtual environments; however, only those which seem most exhaustive will be presented below.

E-learning is a learning environment based on the use of information and communication technologies to provide learning activities and services related to online training.

It also manages the interactions between the learners, the tutors, the author and the administrator during an online training course. As noted earlier, the exploitation of mobile technologies in the field of online training was behind the appearance of M-learning, a match between advances in E-learning and mobile technologies.

2.1 The common points

While m-learning and e-learning diverge on their "M" and "E", they obviously have similar characteristics as they are both concerned with online learning. For example, the participants in an M-learning as in an E-learning environment are the learners, the author, the administrator and the tutor. As it is the case with E-learning, M-learning provides teaching contents for training. Similarly, they can both be in real or remote time; thus, making use of the same transmission modes. In also both environments, a virtual tool of learning is required in order to allow a close follow-up of the training and management of the interactions between the various participants involved.

2.2 M-learning Specificities

M-learning is distinguished from E-learning by the use of mobile technologies. Consequently, the concept of mobility appears to have overcome more than ever space constraints. Thus, the learner can keep track of his learning activities from any location even while moving from one place to another on the condition that a wireless network service is available. This has multiplied possibilities for life-long learning in a more formal and informal setting regardless of space and time constraints.

Moreover, mobility has a considerable effect on the nature of activity offered because learners in m-learning can reach and move easily in geographical areas to practise trainings centred on the practical aspects. Indeed, in addition to the traditional ones such as courses and multiple choice exercises, m-learning provides a suitable environment for the training containers of the practical aspects. For example: assistance need, practical work, project realization since the learner can follow these activities in an authentic context. M-learning seems to cater for certain specialties more than others such as: agronomy, geology, archaeology, etc.

If the use of mobile technologies is behind the widening of activities type in m-learning, these technologies impose many constraints. Indeed, on the one hand, mobile devices are characterized by their small size and limited battery that impose the use of more voice, graphs and animation. On the other hand, the major problem encountered with

wireless networks which connect mobile devices to the internet is the period of disconnection generally due to the high cost of connection or to the lack of the necessary infrastructure.

For this reason, m-learning platforms must envisage services which take account of this constraint by supporting the periods of disconnection.

Although there are several points in common between e-learning and m-learning, the latter is characterized by specificities as discussed above. The pedagogical contents developed in such an environment are likely to be incompatible with the other. So one cannot reach the contents of an e-learning course automatically, nor carry out bidirectional exchanges between e-learning and m-learning in a transparent way. However, we must exploit the existing contents independently of their environments of production and hence create an educational inheritance. It is, thus, a problem of interoperability between e-learning and m-learning. To solve this problem, the existence of standards is essential in order to facilitate the exchanges and to allow the division of resources. The standardization guarantees the use of the same structure and, consequently, will facilitate the exchange of the contents between the two environments. For this reasons, we will study e-learning standardization field to see to what extent interoperability can be ensured.

In what follows we propose a study of these standards in order to satisfy the needs of mobile learning environment and ensure the portability of the digitized teaching equipment.

3.   ADAPTATION OF E-LEARNING STANDARDS TO M-LEARNING NEEDS

E-learning and m-learning finality is using advanced technologies in training. Currently there is a plethora of numerical resources of training which should not remain encapsulated in its environment of development in order to be able to be exploited in m-learning. In other words, one must ensure the profitability and the perenniality of these already produced matters of training and avoids reproducing contents which exist elsewhere. So as to create an interoperable environment of training allowing the teaching exchanges of contents and data between the two environments of m&e learning, we chose the use of the standards for courses structuring as a solution since the latter will make it possible to offer to e-learning and m-learning contents the same structure which will facilitate the exchange of these contents between the two environments. Standardization represents so a reliable way to satisfy the need for interoperability. We could raise, starting from the first section, that e-learning and m-learning have several common points but also some divergent points. Thus, the structures used for the courses should not be very different. Our approach consists of studying the structures suggested by the existing e-learning standards and improving them according to m-learning specificities.

3.1 The standards role

The standards play a very important part to make the access easier to the teaching contents and their diffusion and to enrich exchanges and communication between the platforms. They also allow the publication of these contents on heterogeneous environments. Moreover, "in a planetary world of circulation of the resources, only the tools and the standardized resources for teaching will have the possibility of resisting". Thus, in order to create an interoperable m&e learning environment, recourse to the study of the standards is of great importance. Our field of study is based on works of standardizations which have appeared since the advent of e-learning and the mostly used on an international scale in particular: LOM, IMS LD for the structuring of the teaching contents.

3.2 Adaptation of course structuring standards for mobile devices

In this part we will mainly study LOM standard and the specification IMS LD. This choice is warranted because LOM is a standard applied worldwide. As for IMS LD, it is a specification based on an approach directed process and focuses on the structuring of the teaching activities. The choice of this specification is dictated by the fact that it is the single specification which covers all the pedagogical approaches and it is based on LOM.

3.2.1 LOM

LOM 1484.12.1-2002 (Learning Object Metadata) is a standard of the IEEE approved in December 2002. It is the result of the work undertaken by the LTSC working group (Learning Technology Standards Committee) while being based on specifications produced by standardization organizations such as IMS, ADL, ARIADNE, DCMI. It offers the most detailed diagram of metadata. LOM includes nearly 80 hierarchical elements in 9 categories.

LOM structuring model presents a structuring model with units (curriculum vitae, course, and lesson) and 4 resources levels of various granularities. [Pern2006]

LOM was reinforced by the integration of 15 fields which constitute the Dublin Core model which is a definite model of generic metadata to be applied to any type of numerical document.

We propose here to study how LOM can be improved through the addition of some fields allowing the use of this standard for M-learning, and the widening of the significance of fields already existing in order to meet the need of m-learning.

a. General

The information contained in this category is used to describe and to identify the teaching object. Among these data we find : the identifier of the object, its title, its description, the list of the languages used, a list of key words, the extent of the resource (time, geography, culture...), the type of structure (collection, linear, hierarchical...), its level of granularity (from 1 to 4, 1 indicating a whole course).

The branch "**1.6 coverage**" may contain information relative to time and geography. In the case of m-learning, when the learner changes his localization, it's important to provide him with contents suitable with the new context by taking into account the information contained in this branch.

b. Technical

This part shows the design features necessary for the execution of the teaching object on an information processing system. Information on this category is: the browser (type, version), the operating system, data type or format (allowing to identify the software necessary to read them), numerical object size (in bytes), its physical localization (URL: Uniform Resource Locator or URI: UR Identifier), information to install the teaching object and the time it requires (in particular for audio files, animation or video).

Among the constraints imposed by mobile technologies there is the weak resolution, the small size of the screen, the limited memory size. These technical constraints can be easily integrated in this category by using all the branches. More particularly, one can use branch 4.4 requirement to identify the suitable device for each content. Thus, we propose the addition of a branch **4.4.1.5 device**. Moreover, we propose the use of **4.6 otherPlatformRequirements** to add all other requirements which can be drawn from research on mobile devices such as: resolution, graphic quality, battery and the screen size.

c. Educational

This category concerns the pedagogical description of the learning object. Information given here is related to the conditions of use of the standard resource: kind and level of interactivity, type of the resource (exercise, figure, index...), public Target (learner, teacher, author...), context of use (school, university, in-service training...). The age of learners to which the resource is addressed, the difficulty, the time of training, the user's language and suggestions for use. We have advanced in the previous section that m-learning offers a better opportunity for formal and informal training since the learner, using mobile devices, can move freely to follow an activity by having the possibility to realize the practical part of training in its real context. This category is very important because it will make it possible to take account of the technical constraints of mobile devices and more particularly of the battery and the periods of disconnection. Indeed, according to information contained in the branch 5.6 context, the most adapted resource in his context will be proposed wherever the learner moves. For example: A learner who is pursuing an informal training and has a mobile device with a weak battery can follow contents not requiring much energy. Branch 5.10 can contain proposals for uses of the resource in a particular environment, for example: the realization of a TP in a well defined context.

d. Relation

The relational aspect relates to the physical relations between the teaching objects. Is the type of relation mentioned as "is necessary for", "is a part of", "is version of", "is format of", "is referred to" etc.

As we have indicated before, the use of mobile devices requires the use of more than voice, video and animation. If the format text cannot be replaced, it must be adapted to the small screens available to the mobile devices. Consequently, we think that it's important to make possible the coexistence of several formats of the same contents each one appropriate with a device. Thus, we propose that this branch can be exploited to express the coexistence of several versions of a resource each one adapted with a device.

In spite of the broad use of LOM, this standard is not without gaps. Indeed, it was tender for comment to ISO SC36 WG4 and several gaps were highlighted.

We retain those which seem to us most relevant:

- No distinction made between resources, activities and units of training were among these gaps which go against good descriptions.

- Concentration on contents without taking into consideration the teaching approach to apply. In fact, the implicit choice of the transmissive relation restricted considerably the field of the possibilities: the cognitive step resting on induction (in the case of simulations for example) is not taken into account.[Arn2004]

Obviously, these remarks remain valid for m-learning. As a conclusion, in spite of the richness of its metadata being used for the description of the teaching object, LOM method is appropriate to some kinds of teaching and not to others. This led us to wonder about the relevance of other specifications especially IMS LD.

3.2.2 IMS LD

The beginning of the last decade was marked by the emergence of the pedagogical current in the e-learning environment. KOPER proposes a point of view which is radically different from the documentalist approach by affirming that in fact the objects of knowledge don't constitute the key of success of an environment of learning, but the activities which are associated with it.[Eca2005].

IMS LD proposes a conceptual meta-model describing the learning situation by defining the relations between (1) the objectives in terms of knowledge or skills, (2) the actors of the learning, (3) the activities carried out and (4) the environment and the contents necessary to the installation of a learning situation. [Per,Lej2004]

IMS LD was inspired from the Educational Modeling Language (EML). The latter had as objectives to describe a situation of training with the following elements (and their relations)

1) Objectives: knowledge or skills to acquire
2) Roles: actors of learning
3) Activities carried out
4) Environment of training
5) Contents. [Eca2005]

The aim of IMS LD is to allow the application of the teaching approaches according to the need and to guarantee the exchange and the teaching interoperating of the learning contents. It defines the structure of a learning

unit as a theatrical part gathering a whole of acts made up of partitions where activities are in relation with roles. An activity is located in an environment including (chat, forum, transport...) as well as resources of contents described using the LOM.

The strong point of IMS LD lies in its proposal of three levels of implementation:

Level A: Contains the core of the teaching design of IMS (roles, the elementary activities and resources) and their coordination thanks to the elements: method, play, and act. The activities of training are simply ordered in time, to be carried out by learning, by using the objects and/or the services of training. [Dan2006]

Level B: It adds to level **A** properties, conditions, tutorial services, and elements acting together "It provides specific means to create complex structures and experiments of learning. The properties can be used as variables, local or total, storing or withdrawing information for a user alone, an implied group, or even all concerned persons. Through these mechanisms, the process of learning can change during the execution time of the unit. Decisions can be made taking into account dynamic aspects. [Dan2005]

In m-learning, learners can use several devices during the follow-up of a learning scenario. However, the mobile devices used in m-learning generally present a potential source of constraints relative to their physical characteristics, for example: reduced screen size, restricted methods of entry, limited memory and battery. Moreover, wireless networks present sometimes problems of disconnection caused by the weak cover or the price of connection. Thus, we can consider that among the conditions which make it possible to decide in favour of the evolution of a teaching scenario at a given time, there is the type of networks and the device used.

Level C: The level C adds notifications to the level B which can start another activity making it possible to have dynamic scenarios.

Like for the level B, we propose to consider the context during an m-learning activity as being an event which makes it possible to start a new activity more suitable with the new context. For example: The learner, with his mobile devices, can follow activities anywhere. The localization can impose a change in the scenario of learning. We illustrated this idea in **Figure n° 1**.

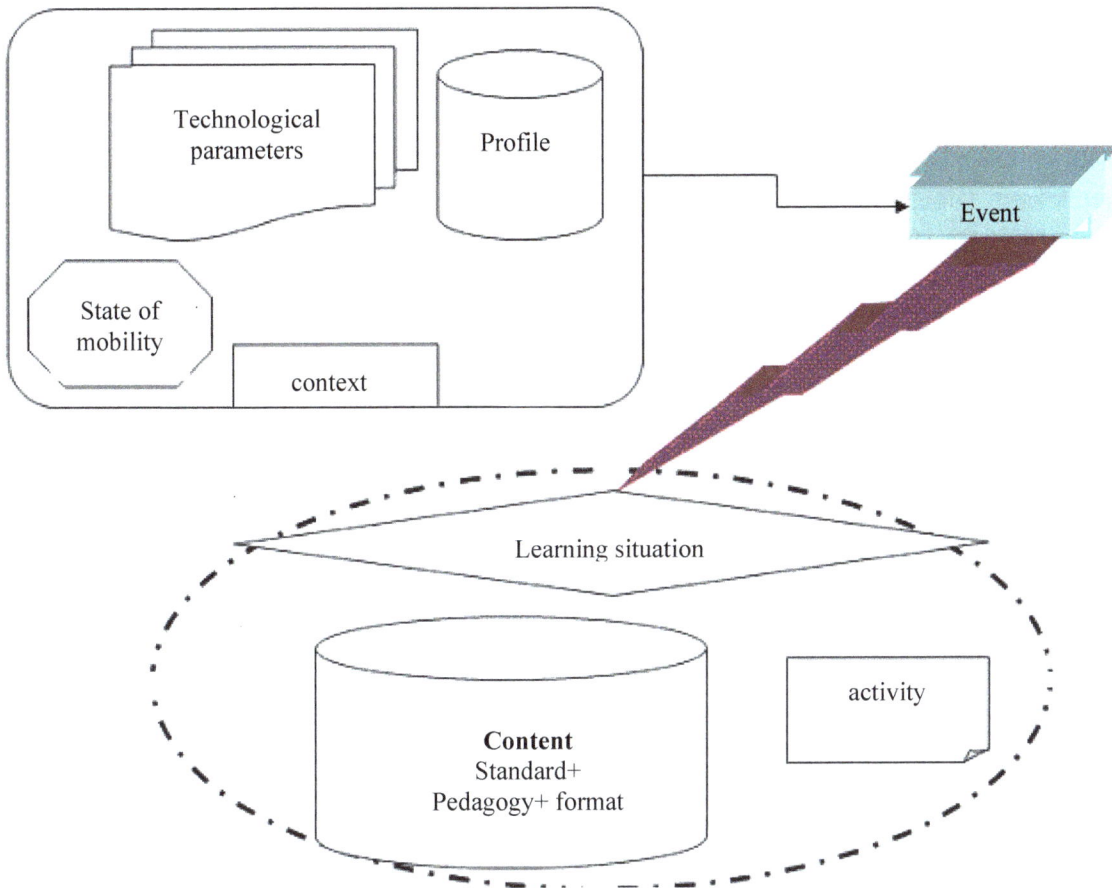

Figure 1.   Model of taking into account of m-learning specificities in IMS LD (B and C).

We think that the context of learning is being aggregated from the technological parameters (mobile device and the wireless network used), state of mobility (which is the level of mobility during learning) and learner profile (its own parameters). The change of these parameters will influence the learning scenario. Consequently, a new activity, more adapted to the new context, must propose to the learner contents under an appropriate format taking into consideration of the technical constraints of mobile devices. A specific teaching approach adequate to the state of mobility will be used. And finally, the contents must be designed according to a standard of course structuring facilitating its exchange when the learner changes his device.

IMS LD Specification presents an undeniable asset for the traditional distant or mobile learning since it proposes a modeling in three levels which remains rather broad and where we can act to take into account m-learning specificities.

To summarize our ideas about IMS LD, we consider that IMS LD is the most appropriate specification that answers m-learning needs thanks to the two levels B and C which are not detailed in the specification. Figure 1 proposed a model which represents the taking into account of specificities of the m-learning in the form of event of IMS LD.

4.   CONCLUSION

In a context marked by the development of communication technologies used in training, we witness the emergence of m-learning, in addition to e-learning which existed before.

The coexistence of these two environments imposes itself, as e-learning and m-learning both aim at fostering training, hence the need to take advantage of contents already produced by e-learning. Thus communication exchanges as well as the sharing of learning subject matters and data between the two environments must be performed. In other words, it is necessary to create an educational heritage exploitable independently of the environment of the teaching matters development. To meet the interoperability need, we think it is fundamental to underlie the important role of the standardisation of the structure of teaching matters, which will facilitate the exchanges between the two environments.

In this article, we have tried to take into account the specificities of m-learning in order to propose a structuring of pedagogical contents according to the LOM standard and the specification IMS LD. The results of our study constitute an important stage before the definition of an interoperable e&m learning architecture. In our project, we are going to use IMS LD specification, due to its richness in taking into account pedagogical approaches.

REFERENCES

[Per2006] Pernin Jean-Philippe. LOM, SCORM et IMS-Learning Design : ressources, activités et scénarios. *Babel - edit -,* L'indexation des ressources pédagogiques numériques. ENSSIB - janvier 2006.

[Arn2004] Michel Arnaud, La gestion des ressources avec les métadonnées, Journée "Normes et standards éducatifs », Mars 2004, Lyon, France.

[Der2006] Alain Derycke, Tutoriel UBIMOB'06 Du E-Learning au. Pervasive-Learning.septembre 2006

[Eca2005] Ecaterina Pacurar Giacomini, une plate-forme basée sur IMS LD, pour la conception de cours en ligne dans le cadre du projet CEPIAH : (Conception et Evaluation des Polycopiés Interactifs pour l'Apprentissage Humain). Thèse de doctorat de l'Université de Technologie de Compiègne : netUniversité, novembre 2005.

[Per2004] Jean-Philippe PERNIN, Anne LEJEUNE, « Dispositifs d'apprentissage instrumentés par les technologies : vers une ingénierie centrée sur les scénarios »,

[Dan2006] Daniel K. Schneider, Version: 0.8 (modifié le 13/12/06) La norme learning design.

[Dan2005] Daniel Burgos, Michel Arnaud, Patrick Neuhauser et Rob Koper IMS Learning Design : la flexibilité pédagogique au service des besoins de l'e-formation.

AUTHORS

Daoudi Najima is with Ecole des sciences de l'information, ESI, Rabat, Morocco

Ajhoun Rachida is with the University Mohamed V, ENSIAS, Rabat, Morocco

Permissions

All chapters in this book were first published in IJIM, by International Association of Online Engineering; hereby published with permission under the Creative Commons Attribution License or equivalent. Every chapter published in this book has been scrutinized by our experts. Their significance has been extensively debated. The topics covered herein carry significant findings which will fuel the growth of the discipline. They may even be implemented as practical applications or may be referred to as a beginning point for another development.

The contributors of this book come from diverse backgrounds, making this book a truly international effort. This book will bring forth new frontiers with its revolutionizing research information and detailed analysis of the nascent developments around the world.

We would like to thank all the contributing authors for lending their expertise to make the book truly unique. They have played a crucial role in the development of this book. Without their invaluable contributions this book wouldn't have been possible. They have made vital efforts to compile up to date information on the varied aspects of this subject to make this book a valuable addition to the collection of many professionals and students.

This book was conceptualized with the vision of imparting up-to-date information and advanced data in this field. To ensure the same, a matchless editorial board was set up. Every individual on the board went through rigorous rounds of assessment to prove their worth. After which they invested a large part of their time researching and compiling the most relevant data for our readers.

The editorial board has been involved in producing this book since its inception. They have spent rigorous hours researching and exploring the diverse topics which have resulted in the successful publishing of this book. They have passed on their knowledge of decades through this book. To expedite this challenging task, the publisher supported the team at every step. A small team of assistant editors was also appointed to further simplify the editing procedure and attain best results for the readers.

Apart from the editorial board, the designing team has also invested a significant amount of their time in understanding the subject and creating the most relevant covers. They scrutinized every image to scout for the most suitable representation of the subject and create an appropriate cover for the book.

The publishing team has been an ardent support to the editorial, designing and production team. Their endless efforts to recruit the best for this project, has resulted in the accomplishment of this book. They are a veteran in the field of academics and their pool of knowledge is as vast as their experience in printing. Their expertise and guidance has proved useful at every step. Their uncompromising quality standards have made this book an exceptional effort. Their encouragement from time to time has been an inspiration for everyone.

The publisher and the editorial board hope that this book will prove to be a valuable piece of knowledge or researchers, students, practitioners and scholars across the globe.

List of Contributors

O. Zawacki-Richter
University of Oldenburg, Oldenburg, Germany

T. Brown
Midrand Graduate Institute, Midrand, South Africa

R. Delport
University of Pretoria, Pretoria, South Africa

S. Martín, E. Sancristobal, R. Gil, M. Castro and J. Peire
UNED (Spanish University for Distance Education, Madrid, Spain

David Guralnick and Christine Levy
Kaleidoscope Learning, New York, New York, USA

J.Y-K. Yau and M.S. Joy
University of Warwick, Coventry, UK

N. Baya'a and W. Daher
Al-Qasemi Academic College of Education, Baqa El-Garbiah, Israel

Zarina Dzolkhifli, Hamidah Ibrahim and Lilly Suriani Affendey
Universiti Putra Malaysia, Serdang, Selangor, Malaysia

Praveen Madiraju
Marquette University, Milwaukee, WI, USA

N. Eteokleous
Frederick University Cyprus, School of Education, Nicosia, Cyprus

D. Ktoridou
University of Nicosia,School of Business, Nicosia, Cyprus

F. Fotouhi-Ghazvini
Qom University, Qom, Iran
University of Bradford, Bradford, UK

R. Earnshaw and D. Robison
University of Bradford, Bradford, UK

P. Excell
Glyndŵr University, Wrexham, UK

C. A. Scolari, H. Navarro Güere, I. García and H. Pardo Kuklinski
Universitat de Vic, Vic, Spain

J. Soriano
Universitat Autónoma de Barcelona, Barcelona, Spain

J. Al-Sadi and B. Abu-Shawar
Arab Open University, Amman, Jordan

I. Nataatmadja and L. E. Dyson
University of Technology Sydney, Australia

Issham Ismail, Hanysah Baharum and Rozhan M. Idrus
Universiti Sains Malaysia, Penang, Malaysia

M. Rosselle and D. Leclet
MIS-UPJV Laboratory, Amiens, France

B. Talon
Université Lille Nord de France, Calais, France

K. Oudidi, A. Habbani and M. Elkoutbi
University Mohammed V- Souissi, Rabat, Morocco

Issham Ismail, Siti Sarah Mohd Johari and Rozhan Md. Idrus
Universiti Sains Malaysia, Penang, Malaysia

B. T. David, R. Chalon, O. Champalle, G. Masserey and C. Yin
Laboratory LIESP/Ecole Centrale de Lyon, Lyon, France

V. Glavinic
University of Zagreb, Zagreb, Croatia

S. Ljubic
University of Rijeka, Rijeka, Croatia

M. Kukec
College of Applied Sciences, Varazdin, Croatia

A. Thatcher and G. Mooney
University of the Witwatersrand, Johannesburg, South Africa

Frank Allan Hansen and Niels Olof Bouvin
Aarhus University, Århus, Denmark

P.L. Kubben
Maastricht University, The Netherlands

Dr. Sahar Idwan
The Hashemite University, Zarqa, Jordan

Yahya S. H. Khraisat and Anwar Al-Mofleh
Al-Balqa Applied University/Al-Huson University College, Amman, Jordan

E. Alepis and M. Virvou
University of Piraeus, Piraeus, Greece

K. Kabassi
TEI of the Ionian Islands, Zakynthos, Greece

Hans L. Cycon
FHTW Berlin, Berlin, Germany

Thomas C. Schmidt and Gabriel Hege
HAW Hamburg, Hamburg, Germany

Matthias Wählisch
HAW Hamburg, Hamburg, Germany
link-lab, Berlin, Germany

Mark Palkow
daViKo GmbH, Berlin, Germany

Muzhir Al-Ani
Amman Arab University, Amman, Jordan

Wael Al-Sawalmeh
Philadelphia University, Amman, Jordan

Teija Vainio
Tampere University of Technology, Tampere, Finland

Mona Laroussi
Riadi Laboratory /ENSI , Tunis, Tunisia
LIFL Laboratory - Lille 1 University Lille France

Alain Derycke
LIFL Laboratory - Lille 1 University Lille France

Cruz-Flores, René and López-Morteo and Gabriel
Engineering Institute, Autonomous University of Baja California, Baja California, México
Marília Mendes and Ana Lisse Carvalho
University of Fortaleza (UNIFOR, MIA), Fortaleza, Brazil

Elizabeth Furtado and Placido Rogerio Pinheiro
University of Fortaleza (UNIFOR, MIA), Fortaleza, Brazil
University of State of Ceará (UECE), Fortaleza, Brazil

Issham Ismail, Rozhan M. Idrus, Azidah Abu Ziden and Munirah Rosli
Universiti Sains Malaysia, Penang, Malaysia

Aznarahayu Ramli, Issham Ismail and Rozhan Md. Idrus
Universiti Sains Malaysia, Penang, Malaysia

Daoudi Najima
Ecole des sciences de l'information, ESI, Rabat, Morocco
University Mohamed V, ENSIAS, Rabat, Morocco

Ajhoun Rachida
University Mohamed V, ENSIAS, Rabat, Morocco

Index

A

Activity Theory, 111, 113-114, 119-121, 186

Ad Hoc Networks, 84, 90-92

Adaptive Systems, 105, 153

Adaptive User Interfaces, 105

Affective Computing, 153, 157

Amplitude Shift Keying, 145-147

Antenna, 145-147, 150, 167-169

Avc Software Codec, 159

B

Bluetooth, 4, 7, 11, 42, 52, 58, 84, 91, 103, 127, 132-133, 141-142, 144, 188-189, 191

C

Cmc, 62, 64

Collaboration, 13-15, 22, 27-29, 38, 42-44, 62, 102, 127, 135, 159-160, 177, 187-188, 201

Computer Augmented Environment, 100, 104

Conferences, 11, 13-15, 46, 64, 157-161, 171, 203

Context-aware Hypermedia, 122-123, 125, 128, 130, 134

Cooperative Activities, 83, 100

Coverage Area Measurements, 164

D

Data Caching, 31

Digital Cellular System, 164

Distance Education, 1, 3, 5-6, 11-12, 21, 46, 74-75, 77, 95-99, 204, 206, 209-211, 213-214

E

E-learning, 1, 13-15, 37, 39, 46, 62, 64-70, 74, 77-78, 83, 93-94, 98-101, 158, 181, 185, 204, 207-209, 214-219

Education Innovation, 1

Educational Games, 48-49, 51-52

Educational Systems, 37, 39, 45, 111, 153-154, 177, 185

F

Faculty Perspectives, 37, 40

H

Handheld, 16-17, 21, 23, 37, 39, 46, 52, 62-63, 67, 94, 103, 108, 130, 134-135, 138-139, 173-174, 176, 180, 184, 196, 205

Handover, 164-165, 167-168

Higher Education, 1, 37-41, 43-46, 52, 61, 64, 69, 73-74, 78, 98, 112, 120-121

I

Information Systems Embedded Application, 84

Integrity Constraints, 31-33, 36

Integrity Tests, 31-35

L

Language Learning, 48, 51, 208

Large Class Teaching, 111-112, 118

Learning By Doing, 100

Learning Java Programming, 16

Learning Objects, 16-17, 19

Learning Schedule, 16-21

Lms, 62-67, 185

Location-based Services, 6

M

M-learning, 6, 37-38, 41-42, 44-46, 49, 52, 62-63, 66-67, 69-70, 72-73, 76, 78-79, 83, 93, 95, 98, 100-101, 104-106, 109, 111-112, 114, 117, 120, 157, 163, 181, 183, 186-188, 195, 204-205, 207, 215-219

Mathematics Education, 22, 29

Mavbt, 141-142, 144

Mcommunication, 54-57, 60

Medicine 2.0, 138-139

Microcontroller, 145-146, 149-150

Middle School Students, 22-23

Mlearning, 1, 21, 73, 98, 100, 195, 207, 213, 215

Mobile Blog, 79

Mobile Conferencing, 159, 161

Mobile Database, 31, 36

Mobile Devices, 1, 3-9, 11, 14, 16-17, 19, 21, 31, 37-46, 54-56, 58-60, 62-63, 66-68, 70, 74-75, 93-94, 96, 98, 103, 105-106, 109-111, 120, 123, 126, 128, 130, 136, 138-139, 153-154, 157, 159-160, 164, 170-178, 180, 184, 187-198, 203, 205, 207, 209, 213, 215-219

Mobile Gaming, 54, 58

Mobile Learning, 1, 3-5, 11, 13, 16-17, 20-23, 29, 37-38, 40-46, 48, 52, 62, 66-68, 73-74, 76-79, 93-98, 100-102, 105, 114, 120-123, 126-127, 132, 162-163, 177-178, 180-182, 186-187, 195-196, 204, 206-213, 215-216, 219

Mobile Marketing, 54, 56, 144

Mobile Phone Messaging, 111

Mobile Phones, 3-4, 6, 8-9, 11, 22-29, 37, 42, 44-45, 48, 52, 54, 63, 65, 77, 95, 100, 112-113, 115-116, 118-120, 126, 130, 135, 141-142, 144, 159-160, 162-163, 168, 177, 195, 204, 209

Mobile Positioning, 164

Mobile Video-based Learning, 159

Mobility, 6, 23, 31, 37-38, 45, 67, 74, 79-80, 84-94, 100-101, 112, 163-166, 169, 177, 181-182, 185, 188, 195, 205, 215-216, 219
Mobility Quantification, 84, 86
Multipoint Relays, 84, 90

N
New Technologies, 6, 39, 41, 43-44, 46, 62
Node Mobility, 84, 86, 88, 90
Nomadic Learning, 122
Nomadism, 79-83

O
Olsr Protocol, 84, 90-91
Open Source, 62, 139, 160

P
Pda, 9, 31, 37, 42, 44, 46, 63, 75-76, 79, 102-103, 109, 111, 138-141, 180, 198, 205, 209-210
Pedagogical Aspects, 37, 40, 42, 44, 194, 215
Peer-to-peer Group Communication, 160
Perceptions Of Mathematics Learning, 22
Personalization, 6, 45, 66, 105, 107-109, 158, 184
Physical Linking, 122
Podcast, 46, 69-70, 72-73, 83
Pre-service Teachers, 22-23
Professional Networking, 13

R
Rasch Model, 74-75, 78, 93, 95-98, 205-206, 209-211, 213

Receiver, 7, 56, 74, 124, 127, 132-134, 145, 147-149, 151, 165, 167, 169
Rfid, 6-12, 83, 100-103, 123, 125, 130, 135

S
Self-regulated Learning, 16-17, 21
Sms, 4, 23, 42, 48, 52, 55, 57-59, 74-78, 93-98, 115-117, 120, 141, 204-213
Social Networking, 13-14

T
Tagging, 122-123, 125, 127, 135
Technological Aspects, 41, 187, 189
Text Message, 77, 95, 114-120, 210
Transmitter, 141, 145-150, 166-167

U
Ubiquitous Learning, 21, 69, 79, 111, 132, 135, 185
Universal Access, 105, 108-110
Universal Usability, 105, 108-109
Usb Key, 79-83

V
Vocabulary Learning, 48, 51-52

W
Wap, 42, 48-49, 58, 62-68, 93, 98, 144
Wearable Computer, 100, 102, 104
Wiki, 83, 138-139
Wireless Sensor Network, 84, 91
Wireless Technology, 46, 62, 66-68, 74, 78, 93, 111-112, 141-142

www.ingramcontent.com/pod-product-compliance
Lightning Source LLC
Chambersburg PA
CBHW080534200326
41458CB00012B/4435